I0001990

Tributes

Volume 33

Logic and Computation

Essays in Honour of Amílcar Sernadas

Volume 24
Modestly Radical or Radically Modest. Festschrift for Jean Paul Van Bendegem on the Occasion of his 60th Birthday
Patrick Allo and Bart Van Kerkhove, eds.

Volume 25
The Facts Matter. Essays on Logic and Cognition in Honour of Rineke Verbrugge
Sujata Ghosh and Jakub Szymanik, eds.

Volume 26
Learning and Inferring. Festschrift for Alejandro C. Frery on the Occasion of his 55th Birrthday
Bruno Lopes and Talita Perciano, eds.

Volume 27
Why is this a Proof? Festschrift for Luiz Carlos Pereira
Edward Hermann Haeusler, Wagner de Campos Sanz and Bruno Lopes, eds.

Volume 28
Conceptual Clarifications. Tributes to Patrick Suppes (1922-2014)
Jean-Yves Béziau, Décio Krause and Jonas R. Becker Arenhart, eds.

Volume 29
Computational Models of Rationality. Essays Dedicated to Gabriele Kern-Isberner on the Occasion of her 60th Birthday
Christoph Beierle, Gerhard Brewka and Matthias Thimm, eds.

Volume 30
Liber Amicorum Alberti. A Tribute to Albert Visser
Jan van Eijck, Rosalie Iemhoff and Joost J. Joosten, eds.

Volume 31
"Shut up," he explained. Essays in Honour of Peter K. Schotch
Gillman Payette, ed.

Volume 32
From Semantics to Dialectometry. Festschrift in Honour of John Nerbonne.
Martijn Wieling, Martin Kroon, Gertjan van Noord, and Gosse Bouma eds.

Volume 33
Logic and Computation. Essays in Honour of Amílcar Sernadas
Carlos Caleiro, Fransciso Dionísio, Paula Gouveia, Paulo Mateus and João Rasga, eds.

Tributes Series Editor
Dov Gabbay dov.gabbay@kcl.ac.uk

Logic and Computation

Essays in Honour of Amílcar Sernadas

edited by

Carlos Caleiro,

Francisco Dionísio,

Paula Gouveia,

Paulo Mateus

and

João Rasga

© Individual authors and College Publications 2017. All rights reserved.

ISBN 978-1-84890-248-0

College Publications
Scientific Director: Dov Gabbay
Managing Director: Jane Spurr

http://www.collegepublications.co.uk

Cover design by Laraine Welch

Printed by Lightning Source, Milton Keynes, UK

All rights reserved. No part of this publication may be reproduced, stored in a retrieval system or transmitted in any form, or by any means, electronic, mechanical, photocopying, recording or otherwise without prior permission, in writing, from the publisher.

Contents

Preface

This volume celebrates Amílcar Sernadas (1952–2017) highly significant and original scientific contributions, as well as his outstanding academic career. This compilation of articles by colleagues, former students and friends was prepared in the months that followed the Conference in Honour of Amílcar (**festschrift.math.tecnico.ulisboa.pt**) which was held at the Instituto Superior Técnico in April 2016, to celebrate his 64th birthday. Unfortunately Amílcar could not live to see the final result, which we believe would have pleased him.

Amílcar was an internationally leading researcher in logic and the foundations of computing. He could, in a remarkable way, bridge the gap between purely mathematical aspects of logic and their applications to various areas of computer science. His ideas influenced the growth and maturing of the area of abstract algebraic specifications and formal software development. Amílcar was also one of the true pioneers in the area of combination of logics. His later research interests included probabilistic and quantum logics.

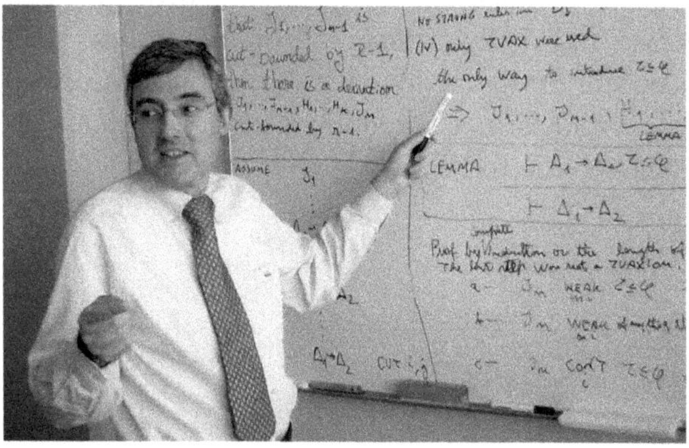

Amílcar's charismatic personality and passion for discovery was a driving force in his areas of interest, and he gained exceptional scientific reputation with his work. He was very productive and his work was published in the most prestigious international journals. Amílcar created and led research groups who have greatly benefitted from his supervision, high standards, and novel ideas. He always managed to ask the right uncomfortable questions to get to the heart of a problem. A biography of Amílcar Sernadas is included below, containing a summary of his scientific and academic careers. A 2017 snapshot of his scientific genealogical tree, which for sure will continue to grow is presented on page xii.

This volume contains 13 papers. The contributions are from researchers that admired Amílcar's work and were inspired by his ideas and many were presented in the conference in honour of Amílcar referred above. The topics covered range quite nicely the very wide spectrum of Amílcar's scientific interests. Their organization somehow reflects a chronology of his scientific activity, from the compositional specification of software systems, to more fundamental questions in mathematical logic, and ultimately to his more recent profound interest in probabilistic and quantum systems, and logics. Two articles are authored by Amílcar himself: the first paper "Local Specification of Distributed Families of Sequential Objects" with Hans-Dieter Ehrich represents one very fruitfull research direction, and the 10th paper "Epistemic nature of quantum reasoning" with Alfredo Barbosa Henriques testifies Amílcar's universal curiosity. All of us know that Amílcar for strong ethical reasons would never allow papers authored by him in his Festschrift. But we believe that now that he is no longer among us these papers should be included in this volume.

The editors would like the take the opportunity to thank all the authors that contributed to this volume and also Dov Gabbay and Jane Spurr from College Publications. We thank Pedro Adão for the scientific genealogical tree and José Carmo's help with Amílcar's biography. We also thank Cristina Sernadas for all the obvious reasons. On behalf of all his colleagues and former students, it is an enormous privilege for us to dedicate this volume to Amílcar's memory.

The editors

Biography

Amílcar Sernadas was born on April 20, 1952, in Angola. He graduated in Electrotechnical Engineering from Instituto Superior Técnico (IST), in 1975. He received his Ph.D. degree in Computer Science from the University of London, in 1980, and obtained his habilitation in Numerical

Analysis and Computation from the University of Lisbon, in 1982. He joined the Department of Mathematics of IST in 1986, became Associate Professor in 1987, and Full Professor in 1990.

He was awarded the title of "*Professor Distinto*" (Distinguished Professor) of IST in 2016 in the presence of the president of the Portuguese Republic.

His most important scientific contributions, that granted him international recognition, were in the areas of temporal system specification, algebraic object specification, combination of logics and probabilistic and quantum logics.

Research in the area of temporal system specification consisted in the definition of temporal logics for specification and verification of information systems. Research in the area of algebraic object specification contributed for the foundations of the object-oriented paradigm, with application to information systems development. In the late 1990s, using the categorial paradigm, he led the systematic and rigorous approach to combination of logics, a subject that emerged at that time as an independent area of research. In the first decade of the 21st century he explored the exogenous approach to probabilistic logics and proposed a similar approach to quantum logics. He also proposed automata and Turing machines integrating classic and quantum components.

Amílcar coauthored more than 140 publications with more that 3200 citations in Google Scholar. Five of those have more than 100 citations and fifteen of them more that 50. He published in the most prestigious journals of his research areas such as Information and Computation, Theoretical Computer Science, the Journal of Logic and Computation, Logica Universalis, the Logic Journal of the IGPL, Studia Logica and the Journal of Symbolic Logic.

Amílcar created and led research groups in the field of Logic and Computation (see below). He received the "*Prémio Descartes do Instituto de Informática*" (Descartes prize from the *Instituto de Informática*) in 1992 with Cristina Sernadas and José Félix Costa. He supervised or co-supervised twelve Ph.D. theses and six post-docs. Three of the Ph.D. theses were awarded IBM prizes and one of them the "Anastácio da Cunha" prize from the Portuguese Mathematical Society. His scientific genealogy has 50 descendants, including a great-great-grandson.

He was a member of the scientific committee of the Advanced Studies in Mathematics and Logic series and of the Formal Aspects of Computing journal. He was affiliated with 16 scientific organisations including AMS, SPM (Portuguese Mathematical Society), ACM, EATCS, and several IFIP working groups. From 1996 to 2004 he was member of the board of directors of CIM (International Center for Mathematics) and since 2005 member of CIM's scientific council.

He participated in more that 60 jurys for Full or Associate professor, more than 40 habilitation and more than 30 Ph.D. defense jurys, and more that 30 M.Sc. defense jurys.

Besides his intensive research activities Amílcar also participated in setting up several degrees of IST, such as the bachelor degrees LMAC (Applied Mathematics and Computation) and LEIC (Computer Science and Engineering), the master degree MMA (Mathematics and Applications) and the Mathematics Ph.D. program of IST. These degrees played a vital role in creating a new generation of mathematicians and computer scientists. He was responsible or co-responsible for the syllabus of dozens of courses, in the fields of Programming, Category Theory, Operations Research, Mathematical Logic, and Theory of Computation. He taught many of these courses and authored many unpublished course materials and two published textbooks: *Introdução à Programação em Mathematica* (Introduction to Programming in Mathematica, in Portuguese) and *Fundamentos de Lógica e Teoria da Computação* (Foundations of Logic and Theory of Computation, also translated to English). He and Cristina Sernadas edited the collection *Cadernos de Lógica e Computação* from College Publications, London.

Amílcar was very involved in academic life, always pursuing strategies for excellency and transparent solutions that would implement such strategies. He was a member of the scientific Council of IST since 2009 and a member of IST's *Assembleia de Representantes* (Representatives Assembly) in several terms. He was a member of the Coordinating Committee of the Scientific Council of IST in several terms. He was member of the Senate of *Universidade Técnica de Lisboa* since 1997 and member of the Senate of *Universidade de Lisboa* since 2013.

Research units

Amílcar Sernadas led - and created - several research units: he was coordinator of the area *Computation: Foundations and languages* of the Center for Statistics and Applications of the University of Lisbon (1980–1985), leader of the *Computation* project of INESC: Institute for Systems Engineering and Computers (1987–1994), coordinator of the *Theory of Computation Group* at ISR: Institute for Systems and Robotics (1994–1996), coordinator of the area *Logic and Computation* of the Center of Applied Mathematics at IST (1997–2001), founder of the Center of Logic and Computation at IST and its president from 1997–2001, founder and first coordinator of SQIG: Security and Quantum Information Group from IT: *Instituto de Telecomunicações* (2006–2015) and was member of

CMAF-CIO: *Centro de Matemática, Aplicações Fundamentais e Investigação Operacional* from the Faculty of Sciences of the University of Lisbon (2015–2017).

Research projects

He participated and led national and international research projects, including pioneer projects with the industry in interface areas between Mathematics and Engineering.

International projects

IS-CORE: Information Systems – COrrectness and REusability (1989–1991), OBLOG: OBject LOGic (1989–1994), **IS-CORE2** (1993–1996), **COMPASS**: Comprehensive Algebraic Approach to System Specification and Development (1993–1996), **FIREworks**: Feature Integration in Requirements Engineering (1997–2000), **ASPIRE**: Advanced modeling and SPecification of distributed InfoRmation systEms (1997–2000), **LANDAUER**: Operating ICT basic switches below the Landauer limit (2012–2015).

National projects

INFOLOG: INFOrmation LOGic (1983–1985), **INDOC**: INtelligent DOCumentation (1988–1989), **NORMLOG**: NORM LOGic (1988–1989), **FAC3**: Fundamentos Algébricos e Categoriais da Ciência da Computação (1990–1993), **SitCalc**: Situation Calculus–Modularization, reification and distribution (1996–1998), **ACL**: Algebraic Combination of Logics (1997–1999), **LOGCOMP**: LOGic and COMPutation (1997–2000), **Problog**: Probabilistic Methods in Logic of Reactive Systems (1999–2001), **FibLog**: Fibring Logics (2002–2004), **QuantLog**: Logic in Quantum Computation and Information (2005–2007), **KLog**: Kleistic Logic (2007–2010), **QSec**: Quantum Security (2007–2010).

Co-authors and co-editors

The following persons co-authored or co-edited publications with Amílcar Sernadas: Alberto Zanardo, Alfredo Barbosa Henriques, André Souto, Antoni Olivè, António Hespanha, António Pacheco, António Rodrigues, Carlos Caleiro, Cláudia Nunes, Cristina Sernadas, Daowen Qiu, Filipe Santos, Francisco Miguel Dionísio, Georg Reichwein, Graça Gaspar, Gritt Denker, Gunter Saake, Guy Lohman, Hans-Dieter Ehrich, Helder Coelho, Hugo Lourenço, Jaime Ramos, Janis Bubenko, Javier Pinto, João Gouveia, João Pedro Sousa, João Rasga, José Carmo, José Félix

Costa, José Granado, José Luiz Fiadeiro, José Valença, Joseph Goguen, Klemens Böhm, Luca Viganò, Luís Alcácer, Luís Alho Andrade, Luis Cruz Filipe, Lvzhou Li, Manuel Cabral Morais, Marcelo Coniglio, Mark Ryan, Martin Gogolla, Paula Gouveia, Paulo Blauth Menezes, Paulo Mateus, Paulo Mendes, Pedro Baltazar, Pedro Resende, Pierre-Yves Schobbens, Ralf Jungclaus, Rafael Camps, Renwei Li, Robert Meersman, Rogério Carapuça, Rohit Chada, Sérgio Mascarenhas, Till Mossakowsky, Tom Maibaum, Walter Carnielli, Yasser Omar.

Scientific descendants (see also page xii)

- José Carmo (1988)
  - Renwei Li (1993)
  - Filipe Santos (1998)
  - Olga Pacheco (2002)
- Rogério Carapuça (1988)
- José Fiadeiro (1988)
  - Isabel Nunes (1998)
  - Antónia Lopes (1999)
    - André Leal Santos (2009)
    - Liliana Rosa (2012)
  - Pedro Ramos (1999)
  - Michel Wermelinger (1999)
    - Angela Rodriguez (2009)
    - Liliana Montrieux (2013)
  - Ahmed M. Al-Ghamdi (2008)
  - João Abril de Abreu (2009)
  - Osama El-Hassam (2009)
  - Georgios Koutsoukos (2011)
  - Ionut Tutu (2015)
- José Félix Costa (1992)
  - Manuel Campagnolo (2001)
    - Daniel Graça (2007)
      - Amaury Pouly (2015)

  - João Neto (2002)
  - Hélia Guerra (2004)
  - Luís Gomes (2007)

- Diana Santos (1996)
  - Rachel Xavier Aires (2005)
  - Marcirio Chaves (2009)

- Paulo Menezes (1997)
  - Carlos Campany (2005)

- Pedro Resende (1998)
  - Misha Protin (2008)
  - Elias Rodrigues (2009)

- Sofia Guerra (1999)

- Jaime Ramos (2000)

- António Ravara (2000)
  - Maxime Gamboni (2010)

- Carlos Caleiro (2000)
  - João Marcos (2005)
  - Ricardo Gonçalves (2008)
  - Bruno Conchinha (2014)
  - Andreia Mordido (2017)

- Paulo Mateus (2001)
  - Pedro Adão (2006)
  - Pedro Baltazar (2010)
  - Manuel Martins (2016)
  - David Henriques (2016)
  - Ricardo Loura (2017)
  - João Rodrigues (2017)

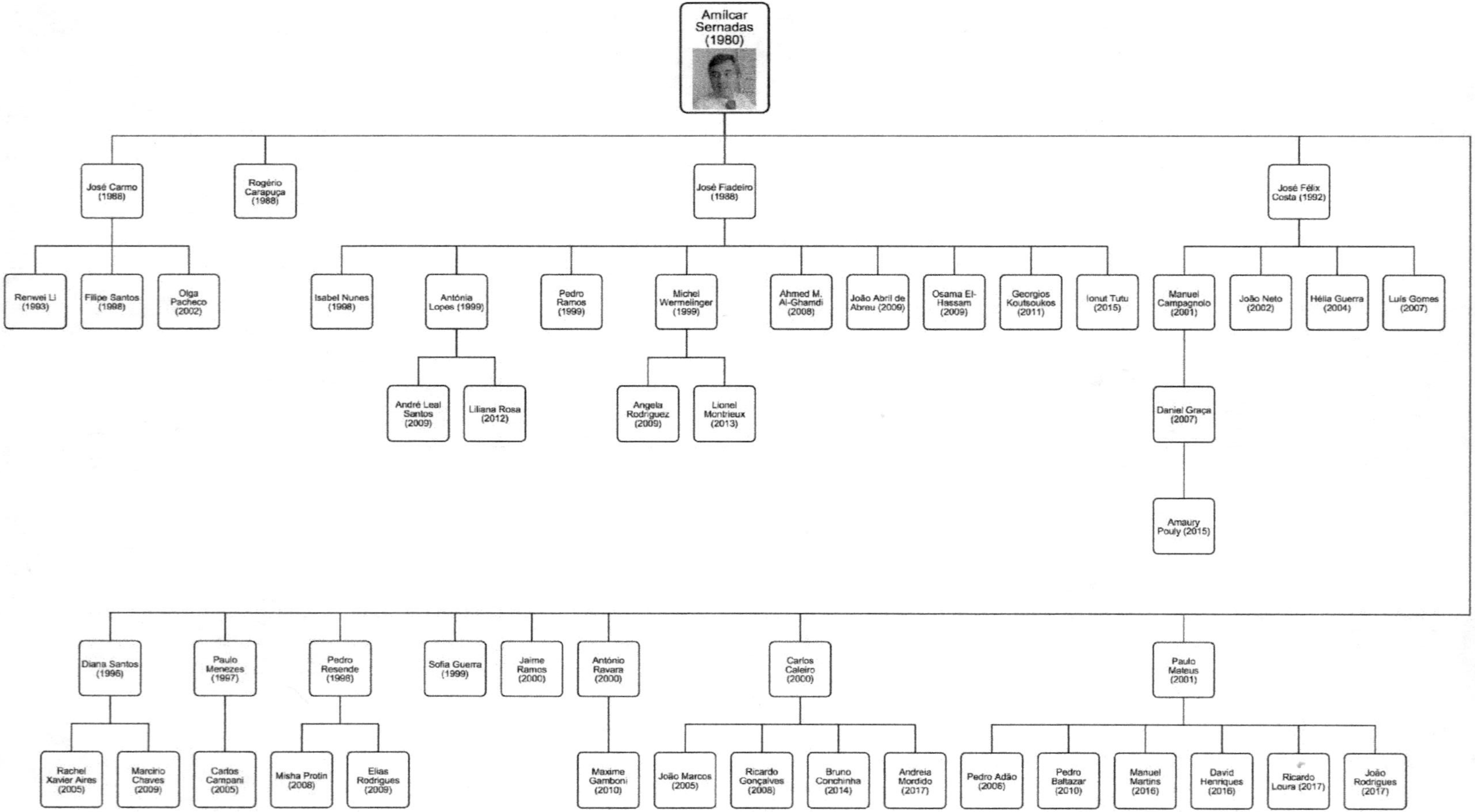

Amílcar Sernadas (1980)
José Carmo (1988)
Rogério Carapuça (1988)
José Fiadeiro (1988)
José Félix Costa (1992)
Renwei Li (1993)
Filipe Santos (1998)
Olga Pacheco (2002)
Isabel Nunes (1998)
Antónia Lopes (1999)
Pedro Ramos (1999)
Michel Wermelinger (1999)
Ahmed M. Al-Ghamdi (2008)
João Abril de Abreu (2009)
Osama El-Hassam (2009)
Georgios Koutsoukos (2011)
Ionut Tutu (2015)
Manuel Campagnolo (2001)
João Neto (2002)
Hélia Guerra (2004)
Luís Gomes (2007)
André Leal Santos (2009)
Liliana Rosa (2012)
Angela Rodriguez (2009)
Lionel Montrieux (2013)
Daniel Graça (2007)
Amaury Pouly (2015)
Diana Santos (1996)
Paulo Menezes (1997)
Pedro Resende (1998)
Sofia Guerra (1999)
Jaime Ramos (2000)
António Ravara (2000)
Carlos Caleiro (2000)
Paulo Mateus (2001)
Rachel Xavier Aires (2005)
Marcirio Chaves (2009)
Carlos Campani (2005)
Misha Protin (2008)
Elias Rodrigues (2009)
Maxime Gamboni (2010)
João Marcos (2005)
Ricardo Gonçalves (2008)
Bruno Conchinha (2014)
Andreia Mordido (2017)
Pedro Adão (2006)
Pedro Baltazar (2010)
Manuel Martins (2016)
David Henriques (2016)
Ricardo Loura (2017)
João Rodrigues (2017)

1

Local Specification of Distributed Families of Sequential Objects†

Hans-Dieter Ehrich[1] and Amílcar Sernadas[2]

[1]Abteilung Datenbanken, Technische Universität, Braunschweig, Germany
HD.Ehrich@tu-bs.de

[2]Dep. de Matemática, Instituto Superior Técnico, Universidade de Lisboa, Portugal
amilcar.sernadas@tecnico.ulisboa.pt

Abstract

Fully concurrent models of distributed object systems are specified using linear temporal logic that does not per se cope with concurrency. This is achieved by employing the principle of local sequentiality: we specify from local viewpoints assuming that there is no intra-object concurrency but full inter-object concurrency. Local formulae are labelled by identity terms. For interaction, objects may refer to actions of other objects, e.g., calling them to happen synchronously. A locality predicate allows for making local statements about other objects. The interpretation structures are global webs of local life cycles, glued together at shared communication events. These interpretation structures are embedded in an interpretation frame that is a labelled locally sequential event structure. Two initiality results are presented: the category of labelled locally sequential event structures has initial elements, and so has the full subcategory of those satisfying given temporal axioms. As in abstract data type theory, these initial elements are obvious candidates for assigning standard semantics to signatures and specifications.

---

†**Remembering and honoring Amílcar:** *I gratefully acknowledge his significant influence in everything we did together. The present paper was our last joint paper. It appeared in the Proceedings of the 10th Workshop on Abstract Data Types 1994, LNCS 906, and is reprinted here by friendly permission of Springer Verlag under licence number 4071401417196. I am very grateful that Amílcar and I could cooperate for more than a decade. Thanks are also due to Cristina and our groups at IST and TUBS. Our work was partly supported by the EU in several ESPRIT working groups. This set the surrounding conditions for a very effective and enjoyable exchange of ideas and results with many colleagues in Europe and elsewhere. HDE*

1.1 Introduction

In abstract data type theory, higher-order model classes like isomorphism classes of many-sorted algebras are specified with (conditional) equational logic that does not per se allow for specifying such classes. The trick is to employ some general higher-order principle to specifiable classes. A popular principle of this kind is initiality, i.e., restriction to initial models.

This paper suggests an analogous trick for distributed systems specification. The higher-order principle is *local sequentiality*. Employing local sequentiality, we can specify fully concurrent models of distributed computation using a logic that does not per se cope with concurrency.

Local sequentiality means that we distinguish between local objects and global families of objects, making the general assumption that there is no intra-object concurrency but full inter-object concurrency.

Objects are locally specified in a sequential process logic, linear temporal logic in our case. Local formulae are labelled by identity terms. Interaction is specified by locally referring to actions of other objects, e.g., calling them to happen synchronously. For making statements about other objects, a locality predicate is introduced.

The interpretation structures for linear temporal logic are life cycles, i.e., linear causality chains of events. Accordingly, our interpretation structures for local specification are global webs of local life cycles, glued together at shared communication events. These interpretation structures are embedded in an interpretation frame that is a labelled locally sequential event structure. Interpretation frames are models of processes.

This extends the simple models we used in our previous work where interpretations were (sets of) life cycles [SEC90, EGS91] or menues [ES91]. The life cycle model has been used as a semantic basis for object specification languages OBLOG [SSE87, SSG$^+$91], TROLL [JSHS91, SJE92] [Ju93, HSJHK94], TROLL *light* [CGH92, GCH93], and GNOME [SR94].

Our work has been greatly influenced by related work on specification languages and theoretical foundations. In a sense, we integrate work on algebraic specification of data types [EGL89, EM85, EM90] and databases [Eh86, EDG88], process specification [Ho85, Mi89], the specification of reactive systems [Se80] [MP89, Sa91a], conceptual modeling [Ch76, Bo85, EGH$^+$92, SF91, SJH93] and knowledge representation [ST89] [MB89].

Approaches to logic and algebraic foundations of object-orientation and concurrency have given essential input to the work reported here. The results in [FSMS91, FM91, FM92, SSC92] have been influential.

FOOPS [GM87, GW90] has provided insights in the algebraic nature of objects. Algebraic approaches to concurrency are given in [AR92] [MM93, Br93].

The local specification logic and interpretation structures put forward in this paper are influenced by the n-agent logic in [LMRT91], but we deviate from that logic in essential respects: we have more elementary temporal operators, and our interpretations are quite different.

In a companion paper [ESSS95], the foundations outlined here are applied to the semantic description of an abstract object specification language.

1.2 Object signatures and interpretations

1.2.1 Data signatures

We assume that the reader is familiar with data signatures and their algebraic interpretations, but we briefly introduce our notation and terminology.

A *data signature* is a pair $\Sigma_D = (S_D, \boldsymbol{\Omega}_D)$ where S_D is a set of *data sorts*, and $\boldsymbol{\Omega}_D = \{\Omega_{x,s}\}_{x \in S_D^*, s \in S_D}$ is an $S_D^* \times S_D$-indexed family of sets of *data operation symbols.* Given an S_D-indexed set $\boldsymbol{X} = \{X_s\}_{s \in S_D}$ of variable symbols, the Σ_D*-terms over* $\boldsymbol{X}$ are denoted by $\boldsymbol{T}_{\Sigma_D}(\boldsymbol{X})$.

If $x = s_1 \ldots s_n$, we write $\omega : s_1 \times \ldots \times s_n \to s$ or $\omega : x \to s$ $[\in \boldsymbol{\Omega}]$ for $\omega \in \Omega_{s_1 \ldots s_n, s}$.

An *interpretation* of Σ_D is a Σ_D-algebra $\boldsymbol{U}$ with carrier sets s_U for each $s \in S_D$, and operations $\omega_U : x_U \to s_U$ for each operator $\omega : x \to s \in \boldsymbol{\Omega}$. If $x = s_1 \ldots s_n \in S_D^*$, then $x_U = s_{1U} \times \ldots \times s_{nU}$ is the cartesian product. The interpretation of a term $t \in \boldsymbol{T}_{\Sigma_D}(\boldsymbol{X})$ in $\boldsymbol{U}$ with a given variable assignment θ is denoted by t_U^θ.

The class of all Σ_D-algebras is structured by Σ_D*-algebra morphisms.* A Σ_D-algebra morphism $\boldsymbol{h} : \boldsymbol{U} \to \boldsymbol{V}$ is a family of maps $\boldsymbol{h} = \{h_s : s_U \to s_V\}_{s \in S_D}$ such that, for each operator $\omega : x \to s \in \boldsymbol{\Omega}$ and each element $\boldsymbol{a} \in x_U$, we have $h_s(\omega_U(\boldsymbol{a})) = \omega_V(\boldsymbol{h}_x(\boldsymbol{a}))$.

The nice properties of the category Σ_D-**alg** of Σ_D-algebras and Σ_D-algebra morphisms are well known and have been utilized for elegant semantic constructions for abstract data type specifications [EGL89, EM85] [EM90].

1.2.2 Class and instance signatures

Classes in the sense of what follows are *object* classes, not classes in the sense of set theory.

Definition 1.1 *Let Σ_D be a data signature. A class signature over Σ_D is a triple $\Sigma_C = (S_O, I, A)$ where S_O is a set of object sorts, $I = \{I_{x,b}\}_{x \in S^*_{DO}, b \in S_O}$ is an $S^*_{DO} \times S_O$-indexed set family of instance operators, and $A = \{A_{x,b}\}_{x \in S^*_{DO}, b \in S_O}$ is an $S^*_{DO} \times S_O$-indexed set family of action operators. Here, $S_{DO} = S_D \cup S_O$.*

Like for data operators, we use the notation $i : x \to b$ for $i \in I_{x,b}$ and $a : x \to b$ for $a \in A_{x,b}$.

Example 1.2 *We give a signature for a class of flip-flops using an intuitive ad-hoc notation that easily translates to our formalism. The specification says that flip-flops can be created, set, reset and destroyed. Moreover, we have an infinite set of flip-flop identities:* `F1`, `F2`, `R(`n`)` *for all natural numbers n, and a "next" flip-flop* `N(`f`)` *for every flip-flop f. Each flip-flop identity is associated with a flip-flop instance. This does not necessarily mean that the system has infinitely many flip-flops, an instance may have several "alias" names. This aspect is not persued in the present paper.*

```
class  flip-flop;
   object sort  FF;
   data sort  nat;
   actions  create, set, reset, destroy;
   instances  F1, F2, R : nat, N : FF;
end  flip-flop.
```

The notation is intended to mean the following class signature. The underlying data signature is assumed to contain the sort `nat` *of natural numbers.*

$S_O = \{\mathtt{FF}\}$
$I = \{\mathtt{F1} : \ \to \mathtt{FF},\ \mathtt{F2} : \ \to \mathtt{FF},\ \mathtt{R} : \mathtt{nat} \to \mathtt{FF},\ \mathtt{N} : \mathtt{FF} \to \mathtt{FF}\}$
$A = \{\mathtt{create} : \ \to \mathtt{FF},\ \mathtt{destroy} : \ \to \mathtt{FF},\ \mathtt{set} : \ \to \mathtt{FF},\ \mathtt{reset} : \ \to \mathtt{FF}\}$

There are infinitely many flip-flop identities, e.g., `F1`, `F2`, `R(0)`, `R(1)`, ... , `N(F1)`, `N(F2)`, `N(R(0))`, `N(N(R(0)))`, *etc. flip-flops can be created, set, reset, and destroyed. The "value" of a flip-flop is represented by the fact that* `set` *or* `reset`, *respectively, is enabled (cf. section 3).* □

The interpretation of a class signature Σ_C is indirectly given by (1) extending the underlying data signature Σ_D to cover the identities and actions specified in the class signature, and (2) deriving an instance signature Σ_I for the identities and individual action alphabets of all object instances.

The data signature extension is defined as follows. Each object sort b goes with the two data sorts of object identities b^i and object actions b^a, respectively. Thus, the object sorts S_O give rise to two sets of data sorts called S_O^i and S_O^a. Let $S^i = S_D \cup S_O^i$ and $S^a = S_D \cup S_O^a$. For $x = s_1 \ldots s_n \in S_{DO}^*$, we define $x^i = s_1^i \ldots s_n^i \in S^{i*}$ where, for $j \in \{1, \ldots, n\}$, $s_j^i = s_j$ if $s_j \in S_D$. The notation x^a is defined correspondingly.

Definition 1.3 *Given a data signature* $\Sigma_D = (S_D, \boldsymbol{\Omega}_D)$ *and a class signature* $\Sigma_C = (S_O, I, A)$ *over* Σ_D, *the extended data signature* $\Sigma = (S, \boldsymbol{\Omega})$ *is given by*

$$S = S_D \cup S_O^a \cup S_O^i$$

where $S_O^a = \{b^a \mid b \in S_O\} \cup \{ac\}$ *and* $S_O^i = \{b^i \mid b \in S_O\} \cup \{id\}$, *and*

$$\boldsymbol{\Omega} = \boldsymbol{\Omega}_D \cup \boldsymbol{\Omega}_O^a \cup \boldsymbol{\Omega}_O^i \cup \boldsymbol{\Omega}_O^{ai}$$

where, for every object sort $b \in S_O$ *and every* $x \in S_{DO}^*$, *we have* $\Omega^a_{O;b^i x^a, b^a} = A_{x,b}$, $\Omega^i_{O;x^i,b^i} = I_{x,b}$, $\Omega^{ai}_{O;b^a p^i, bool} = \{ai_{bp}\}$. *The other sets in the families are empty. Moreover, for every object sort* $b \in S_O$, *we assume* $b^i \leq id$ *and* $b^a \leq ac$.

Before we explain the transformation, we illustrate it by the `flip-flop` example.

Example 1.4 *The* `flip-flop` *class signature in example 1.2 transforms to the following data signature extension.*

sorts $\mathtt{FF}^i \leq$`id`, $\mathtt{FF}^a \leq$`ac`
ops `F1, F2 :` $\to \mathtt{FF}^i$
`R : nat` $\to \mathtt{FF}^i$
`N :` $\mathtt{FF}^i \to \mathtt{FF}^i$
`create, set, reset, destroy :` $\mathtt{FF}^i \to \mathtt{FF}^a$
$\mathtt{ai}_{\mathtt{FF},\mathtt{FF}}$ `:` $\mathtt{FF}^a \times \mathtt{FF}^i \to$ `bool` □

The sort *id* is to be interpreted by all identities in the system. Because of the standard unique naming assumption in object-oriented systems, this should be the disjoint union of the interpretations of the sorts b^i for $b \in S_O$. Using order sorting, we have $b^i \leq id$ for every object sort $b \in S_O$.

The sort *ac* is to be interpreted by all actions in the system, i.e., by the union of the interpretations of the sorts b^a for $b \in S_O$. We do not require disjointness here, we will allow for overlap expressing that actions are shared among objects. But still, using order sorting, we have $b^a \leq id$ for every object sort $b \in S_O$.

The data identity operators $i : x^i \to b^i \in \boldsymbol{\Omega}_O^i$ are derived from the object identity operators $i : x \to b \in I$. They may be parameterized over data, including identities but not actions. So we can cope with objects identified by other objects.

The data action operators $a : b^i \times x^a \to b^a \in \boldsymbol{\Omega}_O^a$ are derived from the object action operators $a : x \to b \in A$. They may have actions ("methods") as parameters. Implicitly, this allows for identity parameters as well since each action carries the identity of its object with it. Data action operators are associated with object instances, so we have identities of sort b as additional parameter. If $a(i, t_1, \ldots, t_n) \in T_\Sigma(\boldsymbol{X})_{b^a}$, we call i the *identity of action* $a(i, t_1, \ldots, t_n)$.

For any object sorts $b, p \in S_O$, the id-of-action operators $ai_{bp} : b^a \times p^i \to bool$ are to be interpreted by relations associating with each individual action the identities of its objects. Of course, the equation $ai_{bb}(a(u, y), u) = true$ should hold for every $a : b^i \times x^a \to b^a \in \boldsymbol{\Omega}_O^a$. Moreover, if the occurrence of action a' of sort $i' : b'$ implies the occurrence of action a'' of sort $i'' : b''$ (e.g., via synchronous action calling), then a' should also belong to the actions of i'', i.e., $ai_{b'b''}(a', i'') = true$. The idea is that $ai_{bp}(a, i) = true$ should hold whenever action a affects object i.

Definition 1.5 *An instance signature is a pair $\Sigma_I = (Id, \mathbf{Ac})$ where Id is a set of identities, and $\mathbf{Ac} = \{Ac_i\}_{i \in Id}$ is an Id-indexed family of action alphabets.*

An instance signature is a distributed action alphabet, providing a local action alphabet for each object. Communication is established by actions shared among two or more objects.

Definition 1.6 *The partners of an action α are $P(\alpha) = \{i \in Id \mid \alpha \in Ac_i\}$.*

Let $\Sigma_C = (S_O, I, A)$ be a class signature over the data signature $\Sigma_D = (S_D, \boldsymbol{\Omega}_D)$. Let $\boldsymbol{U}$ be an interpretation of the extended data signature Σ.

Definition 1.7 *The instance signature determined by Σ and $\boldsymbol{U}$ is $\Sigma_I = (Id, \mathbf{Ac})$ where $Id = id_U$ is the global set of identities, and $\mathbf{Ac} = \{Ac_i\}_{i \in Id}$ is the Id-indexed family of action alphabets $Ac_i = \{\alpha \in ac_U \mid ai_{qb}(\alpha, i) = true$ for some sort $q \in S_O$ and the sort b of $i\}$.*

1.2.3 An event-based interpretation of instance signatures

The interpretations of instance signatures are concurrent processes synchronized via shared events, i.e., action occurrences.

Suitable interpretation structures can be based on any distributed process model, for instance Petri nets or distributed transition systems. But the interpretation structures should fit to the specification logic. Choosing the latter suggests the appropriate process model.

As an example, for simplicity, and because of its proven practicality [MP89] [Se80, Sa91a, Sa91b], we use linear temporal logic for specifying the local properties of individual objects. The suitable interpretation structures for linear temporal logic are linear traces of events, also called life cycles [SEC90, EGS91].

There is an obvious distributed model based on individual life cycles and event sharing, namely global webs of life cycles glued together at shared communication events. This is the model we adopt here. A frame for this model is an event structure that is locally sequential.

Event structures were introduced by Winskel [Wi80]. A recent survey of models for concurrency including event structures is [WN93]. We briefly give the definition and introduce our notation. Our main deviation from standard notation is that we use $\to^*$ for causality instead of the usual $\leq$ because we use the latter for order sorting, and we want to use event structures where causality is the reflexive and transitive closure of a base relation $\to$ of "step" causality. The following definition is taken from [WN93].

Definition 1.8 *A (discrete prime) event structure is a triple*

$$E = (Ev, \to^*, \#)$$

where Ev is a set of events and $\to^*, \# \subseteq Ev \times Ev$ are binary relations called causality and conflict, respectively. Causality $\to^*$ is a partial ordering, and conflict $\#$ is symmetric and irreflexive. For each event $e \in E$, its local configuration $\downarrow e = \{e' \mid e' \to^* e\}$ is finite. Conflict propagates over causality, i.e., $e\#e' \to^* e'' \Rightarrow e\#e''$ for all $e, e', e'' \in Ev$. Two events $e, e' \in Ev$ are concurrent, $e \, co \, e'$ iff $\neg(e \to^* e' \vee e' \to^* e \vee e\#e')$.

In the sequel, the order-theoretic notions refer to causality.

Definition 1.9 *Let E be an event structure. A life cycle in E is a maximal totally ordered sub-event structure of E.*

Definition 1.10 *A sequential event structure E is an event structure in which (1) there is a unique minimal element $\varepsilon \in Ev$, (2) every local* *configuration* $\downarrow e$ *is totally ordered, and (3) causally independent events are always in conflict, i.e., for all events* $e, f \in Ev$, *we have* $e \# f$ *iff neither* $e \rightarrow^* f$ *nor* $f \rightarrow^* e$ *holds. The events* $Ev_+ = Ev - \{\varepsilon\}$ *are called the proper events.*

Intuitively, the causally minimal event represents an imaginary "initial event" where no action occurred so far. This represents the "prenatal" state of an object where nothing happened yet (not even a "birth" action).

With respect to causality, a sequential event structure E is a rooted tree. It represents a set of life cycles grouped together by equal prefixes. A single life cycle is also a sequential event structure.

Sequential event structures are our model of objects. There is no intra-object concurrency. Thus, conflict is a derived concept. We omit it from notation. Assuming that causality $\rightarrow^*$ is the reflexive and transitive closure of a base relation $\rightarrow$ of "step" causality, we arrive at the notation

$$E = (Ev, \rightarrow)$$

for sequential event structures. We will extend this notation to any event structure where conflict is determined from causality by a general assumption. Actually, this holds for all event structures we consider here.

The interpretation structures we envisage for a given instance signature Σ_I are labelled locally sequential Σ_I-event structures. This is our model for fully concurrent families of sequential objects. For modelling communication, the local event sets Ev_i may overlap: an event $e \in Ev_i \cap Ev_j \cap \ldots$ is shared by objects $i, j, \ldots \in Id$.

We make these notions precise, but first we need some notation. Given a set I and an I-indexed family $\boldsymbol{M} = \{M_i\}_{i \in I}$ of sets, we denote their union by $\bigcup \boldsymbol{M} = \bigcup_{i \in I} M_i$. If $\boldsymbol{E} = \{E_i\}_{i \in I}$ is a family of event structures, we denote their union by $\bigcup \boldsymbol{E} = (\bigcup_{i \in I} Ev_i, \bigcup_{i \in I} \rightarrow_i)$.

The notation is justified by the assumption that global conflicts are only those inherited by local conflicts: for all events $e, f \in E$, we have $e \# f$ iff, for some object i and some events $e', f' \in Ev_i$, we have a local conflict $e' \#_i f'$ while $e' \rightarrow^* e$ and $f' \rightarrow^* f$. That is, we globally assume $e \, co \, f$ as a default.

Definition 1.11 *Let* $\Sigma_I = (Id, \boldsymbol{Ac})$ *be an instance signature. A* Σ_I-*event structure* E *is the union* $\bigcup \boldsymbol{E}$ *of an* Id*-indexed family* $\boldsymbol{E} = \{E_i\}_{i \in I}$ *of sequential event structures where* $E_i = (Ev_i, \rightarrow_i)$. *The set family of proper events is denoted by* $\boldsymbol{Ev}_+ = \{Ev_{i+}\}_{i \in Id}$, *and* $Ev_+ = \bigcup \boldsymbol{Ev}_+$.

That is, Σ_I-event structures are *locally sequential.* This model is similar to the n-agent model described in [LMRT91].

Definition 1.12 *Given a Σ_I-event structure E, a distributed life cycle $L = (Lc, \rightarrow)$ in E is an event structure $L \subseteq E$ that is the union of a family $\boldsymbol{L} = \{L_i\}_{i \in Id} \subseteq \boldsymbol{E}$ of life cycles in E, i.e., $L_i \subseteq E_i$ for every $i \in Id$.*

A distributed life cycle is a system of life cycles for the individual objects, glued together at shared "communication" events.

Our interpretation structures are *labelled* distributed life cycles within labelled Σ_I-event structures, so we have to introduce labelling.

Definition 1.13 *Let $E = (Ev, \rightarrow)$ be a Σ_I-event structure. A labelling for E is a map $\bar{\alpha} : Ev_+ \rightarrow \bigcup \boldsymbol{Ac}$ that is the union $\bar{\alpha} = \bigcup \alpha$ of a family of maps $\alpha = \{\alpha_i : Ev_{i+} \rightarrow Ac_i\}_{i \in Id}$ satisfying the following condition: for all events $e', e'' \in Ev$, we have $\bar{\alpha}(e') \neq \bar{\alpha}(e'')$ whenever there is an event $e \in Ev$ such that $e \rightarrow e'$ and $e \rightarrow e''$.*

Intuitively, each proper event is the occurrence of an action, and the label is supposed to be this action. The imaginary start events ε_i do not have labels. The labels of the immediate successors of an event e are the actions *enabled* at e (cf. definition 1.24). These must be distinct for different successor events.

Definition 1.14 *An interpretation frame for a given instance signature Σ_I is a labelled Σ_I-event structure $\bar{E} = (E, \bar{\alpha})$. An interpretation structure within $\bar{E}$ is a labelled distributed life cycle $\bar{L} = (L, \bar{\alpha} \mid_L)$ where $L \subseteq E$.*

Now we define one particular interpretation frame determined by a given instance signature Σ_I. The idea is obvious: the events are defined to be all possible occurrences of actions.

However, not every combination of local configurations is a meaningful context for an action to occur. For instance, if I invite you to meet, then both of us must "remember" that invitation, otherwise the meeting cannot take place. More precisely, when we meet, the two of us must not be in local configurations where you are sure I invited you, and I am sure I never did.

The relevant concept is that of a consistent configuration for an action in which it can possibly occur. In such a configuration, any two objects must agree on their past communication. Generalizing the notation for local configurations $\downarrow e$ to sets, we define $\downarrow C = \{e' \mid \exists e \in C : e' \rightarrow^* e\}$ for any subset $C \subseteq Ev$ of events.

Definition 1.15 *Given a Σ_I-event structure $E = (Ev, \rightarrow)$, a configuration in E is a set of events $Cf \subseteq Ev$ with the properties (1) $Cf = \downarrow Cf$ and (2) Cf is conflict-free, i.e., $(Cf \times Cf) \cap \# = \emptyset$. If $I \subseteq Id$ is a set of object identities, then a configuration for I in E is a configuration that is the range $\gamma(I)$ of a map $\gamma : I \rightarrow Ev$ such that $\gamma(i) \in Ev_i$ for every $i \in I$.*

A configuration Cf for I in E contains one event $e_i = \gamma(i) \in Ev_i$ for every $i \in I$. These events e_i need not be distinct. Two different objects $i, j \in I$ may share an event in Cf, i.e., we may have $e_i \in Ev_j$ as well. In this case, e_i and e_j must be causally related because otherwise they would be in conflict.

Definition 1.16 *Given an instance signature $\Sigma_I = (Id, \boldsymbol{Ac})$, the interpretation frame $\bar{E}(\Sigma_I) = (E(\Sigma_I), \bar{\alpha})$ is inductively defined as follows. $E(\Sigma_I) = \bigcup \boldsymbol{E}(\Sigma_I)$ where $\boldsymbol{E}(\Sigma_I) = \{E_i\}_{i \in Id}$ where $E_i = (Ev_i, \rightarrow_i)$, for each object $i \in Id$.*

(1) $\varepsilon_i \in Ev_i$ for every object $i \in Id$.

(2) if α is an action and Cf is a configuration for α's partners $P(\alpha)$, then $Cf\,\alpha \in Ev_i$ for every partner $i \in P(\alpha)$; as for labelling and causality: we have $\bar{\alpha}(Cf\,\alpha) = \alpha$, and $e \rightarrow_i Cf\,\alpha$ for every $e \in Cf \cap Ev_i$ and every $i \in P(\alpha)$.

The basis of this inductive definition is given by birth events that happen in configurations consisting of one or more "initial events" ε_i. The event $Cf\,\alpha$ represents α occurring in the configuration Cf. This event is shared by the partners of α.

We prove that the construction is sound.

Theorem 1.17 *For each instance signature Σ_I, $\bar{E}(\Sigma_I)$ is an interpretation frame for Σ_I.*

Proof: We have to show that every local event structure $E_i = (Ev_i, \rightarrow_i)$, $i \in Id$, is sequential, and that the labels are ok, i.e., immediate successors have different labels.

We prove the first by induction over the structure of $\bar{E}(\Sigma_I)$.

Let $i \in Id$ be an identity, and let $e \in E_i$. If $e = \varepsilon_i$, then $\downarrow_i e = \{\varepsilon_i\}$ is trivially totally ordered, where $\downarrow_i e = \downarrow e \cap E_i$. Otherwise, we have $e \in Cf\,\alpha$ for some action $\alpha \in Ac_i$ and some configuration Cf for $P(\alpha)$. Assume that every life cycle prefix so far, i.e., every local configuration $\downarrow_j f$ for $f \in Cf$ and $j \in P(\alpha)$, is totally ordered. Then we have to show that $\downarrow_j Cf\,\alpha$ is totally ordered as well for every $j \in P(\alpha)$.

Assume that, for some partner $j \in P(\alpha)$, $g, g' \in \downarrow_j Cf$. Since $\downarrow_j Cf$ is a linear trace, g and g' are causally related in E_j, say $g \rightarrow_j^* g'$. Since both are causal for $Cf\,\alpha$, $\downarrow_j Cf\,\alpha$ is totally ordered as well.

As for labelling, we have to show that labels of events with a common immediate predecessor are different. But this is obvious from construction since the events are *defined* by the actions applied to the configurations representing the current states of the action's partners. □

Labelled Σ_I-event structures are related by *morphisms*. Adapting general event structure morphisms from [WN93] to our structures, we arrive at the following definition. Let $\Sigma_I = (Id, \boldsymbol{Ac})$ be an instance signature.

Definition 1.18 *Let* $E_1 = (Ev_1, \rightarrow_1)$ *and* $E_2 = (Ev_2, \rightarrow_2)$ *be two* Σ_I*-event structures. A* Σ_I*-event structure morphism* $h : E_1 \rightarrow E_2$ *is the union* $\bigcup \boldsymbol{h}$ *of an* Id*-indexed family of partial surjective maps* $\boldsymbol{h} = \{h_i : Ev_{1i} \cdots\!\twoheadrightarrow Ev_{2i}\}_{i \in Id}$ *such that* $h : Ev_1 \cdots\!\twoheadrightarrow Ev_2$ *is a partial surjective map, and for all* $e_1, e_2 \in Ev_1$*, if* $h(e_1)$ *and* $h(e_2)$ *are both defined, then* $h(e_1) \rightarrow_2 h(e_2)$ *iff* $e_1 \rightarrow_1 e_2$.

These morphisms are easily extended to *labelled* Σ_I-event structures: if $h(e)$ is defined, then $\bar{\alpha}_2(h(e)) = \bar{\alpha}_1(e)$.

Labelled Σ_I-event structures and their morphisms form a category Σ_I**-evt**. We cannot go into a detailed analysis of this category, we just mention that it has initial elements*. In fact, we can prove that the labelled Σ_I-event structure constructed in definition 1.16 is an initial element.

Theorem 1.19 $\bar{E}(\Sigma_I)$ *is initial in* Σ_I**-evt**.

Proof: The initial morphism $h : \bar{E}(\Sigma_I) \rightarrow \bar{F}$ to some labelled Σ_I-event structure $\bar{F}$ is defined by sending the minimal element ε_i to the corresponding minimal element in F_i, for every $i \in Id$. The other events are mapped as follows. If $h(Cf)$ is defined for (every event in) a configuration Cf, then we define, for every action α, $h(Cf\,\alpha) = h(Cf)\alpha$ if the latter exists, otherwise $h(Cf\,\alpha)$ is undefined. By $h(Cf)\alpha$ we mean the immediate successor of all events in $h(Cf)$ labelled α. It is not hard to see that this defines a morphism and that it is unique. □

*For obvious reasons, we avoid the usual term "object" for an element of a category.

1.3 Object specification

Our logic for object class specification is a linear temporal logic with locality. We illustrate the idea by means of the `flip-flop` example.

It is beyond the scope of this paper to develop an abstract specification language for class signatures like the one in the `flip-flop` example 1.2 (cf. [ESSS95]). We give the "target" logic directly into which such a language is to be translated.

That is, we give a specification logic for the extended data signature $\Sigma = (S, \boldsymbol{\Omega})$.

The S-indexed family $\boldsymbol{T}_\Sigma(\boldsymbol{X})$ of sets of *terms* is defined as usual, employing an S-indexed family $\boldsymbol{X}$ of sets of *variables*.

Definition 1.20 *The set $L_\Sigma(\boldsymbol{X})$ of formulae is inductively defined as follows:*

- $(t_1 =_{ss} t_2) \in L_\Sigma(\boldsymbol{X})$ *provided that* $t_1, t_2 \in T_\Sigma(\boldsymbol{X})_s$*;*
- $\tau(i) \in L_\Sigma(\boldsymbol{X})$ *provided that* $i \in T_\Sigma(\boldsymbol{X})_{id}$*;*
- $\alpha(i),\ \rhd\alpha(i) \in L_\Sigma(\boldsymbol{X})$ *provided that* $\alpha(i) \in T_\Sigma(\boldsymbol{X})_{ac}$ *where* $i \in T_\Sigma(\boldsymbol{X})_{id}$ *is the identity of* α*;*
- $(\neg\,\varphi) \in L_\Sigma(\boldsymbol{X})$ *provided that* $\varphi \in L_\Sigma(\boldsymbol{X})$*;*
- $(\varphi \Rightarrow \varphi') \in L_\Sigma(\boldsymbol{X})$ *provided that* $\varphi, \varphi' \in L_\Sigma(\boldsymbol{X})$*;*
- $(\exists x\ \varphi) \in L_\Sigma(\boldsymbol{X})$ *provided that* $x \in X_s$ *for some* $s \in S$ *and* $\varphi \in L_\Sigma(\boldsymbol{X})$*;*
- $(\mathsf{X}_i\,\varphi), (\mathsf{F}_i\,\varphi), (\mathsf{Y}_i\,\varphi), (\mathsf{P}_i\,\varphi) \in L_\Sigma(\boldsymbol{X})$ *provided that* $i \in T_\Sigma(\boldsymbol{X})_{id}$ *and* $\varphi \in L_\Sigma(\boldsymbol{X})$.

Definition 1.21 *The set $L^{\mathcal{T}}_\Sigma(\boldsymbol{X})$ of local formulae is the set* $\{i : \varphi \mid i \in T_\Sigma(\boldsymbol{X})_{id},\ \varphi \in L_\Sigma(\boldsymbol{X})\}$.

The locality predicate $\tau(i)$ says that the formula is local to i, i.e., the "owner" of the local formula is communicating with i. The predicate $\alpha(i)$ means that action α has just occurred in object i. The predicate $\rhd\alpha(i)$ means that action α is enabled in object i, i.e., it may happen next. Of course, actions must be enabled when they occur (i.e., just before they have just occurred). The symbols for the local temporal operators have the following meaning: X_i means next in i, F_i means sometime in the future of i, P_i means sometime in the past of i, and Y_i means previous (*yesterday*) in i. X_i and Y_i are meant to be the strong versions, i.e., the next and previous states, respectively, have to exist.

We apply the usual rules for omitting brackets, and we introduce further connectives through abbreviations, e.g., $(\varphi \vee \varphi')$ for $((\neg\varphi) \Rightarrow \varphi')$. The same applies to temporal operators, e.g., $(\mathsf{X}_i^? \varphi)$ for $(\neg(\mathsf{X}_i(\neg\varphi)))$, $(\mathsf{G}_i \varphi)$ for $(\neg(\mathsf{F}_i(\neg\varphi)))$, $(\mathsf{Y}_i^? \varphi)$ for $(\neg(\mathsf{Y}_i(\neg\varphi)))$, and $(\mathsf{H}_i \varphi)$ for $(\neg(\mathsf{P}_i(\neg\varphi)))$.

Definition 1.22 *An object specification is a pair* $= (\Sigma, \Phi)$ *where* Σ *is an extended data signature, and* $\Phi \subseteq L_\Sigma^\tau(\boldsymbol{X})$ *is a set of local formulae as axioms.*

Example 1.23 *We specify a class of flip-flops, based on the signature given in example 1.4.*

sorts nat, $\mathrm{FF}^i \leq$id, $\mathrm{FF}^a \leq$ac

ops F1, F2 : $\rightarrow \mathrm{FF}^i$
R : nat $\rightarrow \mathrm{FF}^i$
N : $\mathrm{FF}^i \rightarrow \mathrm{FF}^i$
create, set, reset, destroy : $\mathrm{FF}^i \rightarrow \mathrm{FF}^a$
$\mathtt{ai}_{\mathtt{FF,FF}}$: $\mathrm{FF}^a \times \mathrm{FF}^i \rightarrow$ bool

axioms $\forall f, g$:FF, n:nat

f: $\mathtt{create}(g) \Rightarrow \tau(g)$
f: $\mathtt{set}(g) \Rightarrow \tau(g)$
f: $\mathtt{reset}(g) \Rightarrow \tau(g)$
f: $\mathtt{destroy}(g) \Rightarrow \tau(g)$

f: $\rhd\mathtt{create}(f) \Leftrightarrow \neg\rhd\mathtt{set}(f) \wedge \neg\rhd\mathtt{reset}(f)$
f: $\mathtt{create}(f) \Rightarrow \mathsf{Y}_f \rhd\mathtt{create}(f)$
f: $\mathtt{create}(f) \Rightarrow \rhd\mathtt{set}(f)$
f: $\mathtt{create}(f) \Rightarrow \mathsf{G}_f(\neg\mathsf{P}_f\,\mathtt{destroy}(f) \Rightarrow \neg(\rhd\mathtt{set}(f) \Leftrightarrow \rhd\mathtt{reset}(f)))$
f: $\mathtt{set}(f) \Rightarrow \mathsf{Y}_f \rhd\mathtt{set}(f)$
f: $\mathtt{set}(f) \Rightarrow \rhd\mathtt{reset}(f)$
f: $\mathtt{reset}(f) \Rightarrow \mathsf{Y}_f \rhd\mathtt{reset}(f)$
f: $\mathtt{reset}(f) \Rightarrow \rhd\mathtt{set}(f)$
f: $\mathtt{destroy}(f) \Rightarrow \mathsf{Y}_f \rhd\mathtt{destroy}(f)$
f: $\mathtt{destroy}(f) \Rightarrow \neg\rhd\mathtt{set}(f) \wedge \neg\rhd\mathtt{reset}(f)$

f: $\mathtt{set}(f) \Rightarrow \mathsf{X}_f\,\mathtt{set}(\mathtt{N}(f))$

F1: $\mathtt{set}(\mathtt{F1}) \Rightarrow \mathtt{set}(\mathtt{F2})$
F2: $\mathtt{destroy}(\mathtt{F2}) \Rightarrow \mathsf{P}_{\mathtt{F1}}\,\mathtt{set}(\mathtt{F1})$

R(n): $\mathtt{set}(\mathtt{R}(n)) \Rightarrow \mathsf{F}_{\mathtt{R}(n)}\,\mathtt{set}(\mathtt{R}(n+1))$

The first four axioms say that only shared actions may occur locally. Actually, these axioms need not be given explicitly because they are satisfied anyway in every interpretation (cf. definition 1.24 below).

Axiom five gives a necessary and sufficient condition for a flip-flop's creation to be enabled: it must neither be in a set nor in a reset state. Axiom six is the instance of a general rule: no action occurs unless it is enabled. Axioms seven and eight describe the state after creation: the flip-flop is set, and from now on it is always either set or reset as long as it is not destroyed. Axioms nine to twelve give the obvious preconditions and effects of set and reset. Axiom thirteen is another instance of the general rule mentioned. The fourteenth axiom puts a flip-flop after destruction in the same state as before creation: resurrection is possible!

The fifteenth axiom puts a clockwork of flip-flops into operation: flip-flops in the chain N *are set step by step, one after the other, once an initial one is set.*

The sixteenth axiom talks about two particular flip-flops, it says that whenever F1 *has been set, then* F2 *has been set at the same time. This is an instance of synchronous action calling. As a consequence, we have* $\mathtt{set(F1)} \Rightarrow \tau(\mathtt{F2})$ *and thus* $ai_{\mathtt{FF},\mathtt{FF}}(\mathtt{set(F1)},\mathtt{F2}) = \mathtt{true}$, *i.e.*, set(F1) *is also an action of* F2. *The seventeenth axiom says that* F2 *can only be destroyed if it has at least once been set by* F1 *via action calling.*

The eighteenth and last axiom says that whenever $\mathtt{R}(n)$ *has been set, then* $\mathtt{R}(n+1)$ *will eventually be set.* R *is a "lazy" version of clockwork* N. *Consider the following alternative:*

$$\mathtt{R}(n) : \mathtt{set}(\mathtt{R}(n)) \Rightarrow \mathsf{F}_{\mathtt{R}(n+1)}\, \mathtt{set}(\mathtt{R}(n+1))$$

This does not say the same: in the former case, a communication between $\mathtt{R}(n)$ *and* $\mathtt{R}(n+1)$ *is required at the moment when* $\mathtt{R}(n+1)$ *is set. In the latter case, the communication must take place right now and reassure* $\mathtt{R}(n)$ *that its successor* $\mathtt{R}(n+1)$ *will eventually be set.*

Here is a simple fact that is entailed by the axioms:

$$\mathtt{F1}: \ \mathtt{set(F1)} \Rightarrow \tau(\mathtt{F2}) \wedge \mathsf{P}_{\mathtt{F2}}\, \mathtt{create(F2)}.$$

□

The logic $L_\Sigma(\boldsymbol{X})$ over an extended data signature is interpreted over an instance signature $\Sigma_I = (Id, \boldsymbol{Ac})$ based on a data signature Σ_D and a data universe $\boldsymbol{U}$, and an interpretation structure $\bar{L} = (L, \bar{\alpha}\,|_L)$ within an interpretation frame $\bar{E} = (E, \bar{\alpha})$, as described in section 2. Please remember that $L = \bigcup\{L_i\}_{i \in Id}$ where $L_i = (Lc_i, \rightarrow_i)$, and $E = \bigcup\{E_i\}_{i \in Id}$ where $E_i = (Ev_i, \rightarrow_i)$.

In particular, the local formulas are interpreted in a local configuration $\downarrow e$ of L and a variable assignment θ. Of course, data terms are to be interpreted globally in $\boldsymbol{U}$.

Definition 1.24 *The satisfaction relation* $\models$ *is inductively defined by the following rules:*

- $\bar{L}, \downarrow e, \theta \models i : (t_1 =_{ss} t_2)$ *iff* $t_{1}{}^{\theta}_{U} = t_{2}{}^{\theta}_{U}$;
- $\bar{L}, \downarrow e, \theta \models i : \tau(j)$ *iff* $e \in Lc_{i^{\theta}_{U}} \cap Lc_{j^{\theta}_{U}}$;
- $\bar{L}, \downarrow e, \theta \models i : \alpha(j)$ *iff* $e \in Lc_{i^{\theta}_{U}}$ *and* $\bar{\alpha}(e) = \alpha(j)$;
- $\bar{L}, \downarrow e, \theta \models i : \rhd\alpha(j)$ *iff* $e \in Lc_{i^{\theta}_{U}}$ *and, for some* $e' \in Ev_{i^{\theta}_{U}}$, $e \to_{i^{\theta}_{U}} e'$ *and* $\bar{\alpha}(e') = \alpha(j)$;
- $\bar{L}, \downarrow e, \theta \models i : (\neg\varphi)$ *iff* $e \in Lc_{i^{\theta}_{U}}$ *and not* $\bar{L}, \downarrow e, \theta \models i : \varphi$;
- $\bar{L}, \downarrow e, \theta \models i : (\varphi \Rightarrow \varphi')$ *iff* $e \in Lc_{i^{\theta}_{U}}$ *and* $\bar{L}, \downarrow e, \theta \models i : \varphi'$ *or not* $\bar{L}, \downarrow e, \theta \models i : \varphi$;
- $\bar{L}, \downarrow e, \theta \models i : (\exists x\ \varphi)$ *iff* $e \in Lc_{i^{\theta}_{U}}$ *and* $\bar{L}, \downarrow e, \theta' \models i : \varphi$ *for some* x*-equivalent assignment* θ';
- $\bar{L}, \downarrow e, \theta \models i : (\mathsf{X}_j\ \varphi)$ *iff* $e \in Lc_{i^{\theta}_{U}}$ *and* $\bar{L}, \downarrow e', \theta \models j : \varphi$ *for some event* $e' \in Lc_{j^{\theta}_{U}}$ *such that* $e \to_{i^{\theta}_{U}} e'$ *or* $e \to_{j^{\theta}_{U}} e'$;
- $\bar{L}, \downarrow e, \theta \models i : (\mathsf{F}_j\ \varphi)$ *iff* $e \in Lc_{i^{\theta}_{U}}$ *and* $\bar{L}, \downarrow e', \theta \models j : \varphi$ *for some event* $e' \in Lc_{j^{\theta}_{U}}$ *such that* $e \to^{*} e'$;
- $\bar{L}, \downarrow e, \theta \models i : (\mathsf{Y}_j\ \varphi)$ *iff* $e \in Lc_{i^{\theta}_{U}}$ *and* $\bar{L}, \downarrow e', \theta \models j : \varphi$ *for some event* $e' \in Lc_{j^{\theta}_{U}}$ *such that* $e' \to_{i^{\theta}_{U}} e$ *or* $e' \to_{j^{\theta}_{U}} e$;
- $\bar{L}, \downarrow e, \theta \models i : (\mathsf{P}_j\ \varphi)$ *iff* $e \in Lc_{i^{\theta}_{U}}$ *and* $\bar{L}, \downarrow e', \theta \models j : \varphi$ *for some event* $e' \in Lc_{j^{\theta}_{U}}$ *such that* $e' \to^{*} e$.

This requires some explanation.

The first rule is straightforward, equations between data terms are interpreted globally. The second rule says that j is local in i iff i's current event is shared by j. The third rule says that, in any local configuration of a proper event, there is precisely one action that just occurred (possibly shared with some other object j), given by the label.

The forth rule is a little unusual in that it is not interpreted in an isolated life cycle but in a life cycle *in context*. Intuitively, $\alpha(j)$ is enabled if it may happen in some next step in the *frame*, not necessarily in the life cycle. A more classic way to capture this would use life cycles with one-step look-ahead ("barbed wires") as interpretation structures.

Rules five to seven are adapted from predicate calculus.

The eighth rule requires some thought: the events e and e' may belong to different life cycles! For i to know that φ holds for j tomorrow, i and j may communicate either today or tomorrow.

Also in the nineth rule, e and e' may belong to different objects. $i : \mathsf{F}_j \varphi$ holds in a local configuration $\downarrow e$ for i iff $j : \varphi$ holds in some future configuration $\downarrow e'$ for j where e' causally depends on e. This causal dependency may involve a chain of objects $i = i_1, \ldots, i_n = j$ where successive objects communicate via some shared event.

The last two rules are the past-directed analoga of future-directed rules eight and nine.

Definition 1.25 *A labelled distributed life cycle $\bar{L}$ satisfies a local formula $i : \varphi$, written $\bar{L} \models i : \varphi$, iff $\bar{L}, \downarrow e, \theta \models i : \varphi$ for every local configuration $\downarrow e$ and every variable assignment θ. A local formula $i : \varphi$ is valid in the interpretation frame $\bar{E}$, written $\bar{E} \models i : \varphi$, iff $\bar{L} \models i : \varphi$ for every distributed life cycle $\bar{L}$ in $\bar{E}$.*

Interpretation structures represent single runs of processes. Since we are interested in entire processes, we may ask which interpretation frames represent specified processes, i.e., have the property that all axioms are valid. In particular, it is interesting to look at these interpretation frames within the category Σ_I-**evt** of all interpretation frames, with morphisms as defined in definition 1.18. Since there is an initial element in Σ_I-**evt** (theorem 1.19), the obvious question to ask is whether the same holds for the subcategory of interpretation frames satisfying given axioms Φ. The answer is positive.

Let *Ospec* $= (\Sigma, \Phi)$ be an object specification (cf. definition 1.22), and let Σ_I be the instance signature determined by Σ and a given data universe $\boldsymbol{U}$ (cf. definition 1.7). Let *Ospec*-**evt** be the full subcategory of all elements in Σ_I-**evt** satisfying Φ.

Theorem 1.26 *Ospec*-**evt** *has initial elements.*

Proof idea: an initial element is given by the largest labelled sub-event structure of $\bar{E}(\Sigma_I)$, the initial element of Σ_I-**evt** (cf. theorem 1.19), containing all distributed life cycles satisfying all axioms in Φ. □

As in abstract data type theory, the initial elements of Σ_I-**evt** and *Ospec*-**evt** are obvious candidates for assigning standard semantics to signatures and specifications, respectively.

1.4 Concluding Remarks

The local specification logic and distributed semantics presented in this paper deserve further study. An obvious next step is to provide a proof system. [LMRT91] give a sound and complete proof system for their propositional n-agent logic, but that does not carry over to our case. It is not clear whether a complete proof system for our logic exists. But still, there are software tools for analysing and animating temporal logic specifications [Sa91b] that can be adapted to our approach.

Our theory will be extended towards *structured* specifications. That means that we have to introduce and study appropriate signature and specification morphisms and corresponding forgetful functors on interpretation frames and structures. This would provide the basis for studying composition as well as parameterization. For composition, there are two aspects: composing sequential objects from components [ESSS95], and composing concurrent families from subfamilies.

There is work in progress to incorporate reification issues in our framework [De94, De95]. It is well known that the temporal operators of our logic are not suitable for action refinement, the logic in [CE94] (which is similar to Hennessy-Milner logic [HM85]) may be better suited. This is an indication that we should work in a family of related logics, as put forward in [MM94].

While reification requires a more "operational" logic, specification expressiveness and comfort suggest to move in the opposite direction. An obvious extension of our logic is to introduce branching-time operators. Branching-time formulae cannot be interpreted in single life cycles. That means that our interpretation frames will play the role of interpretation structures.

Acknowledgments

The authors are grateful to their colleagues in the IS-CORE and COMPASS projects for inspiration about the nature of objects and how to reason about them. Special thanks are due to Cristina Sernadas and Gunter Saake for attention, hints, suggestions, objections, and understanding. The work of José Fiadeiro and Tom Maibaum has been influential. Also many thanks to Grit Denker, and to the Hildesheim process logic group of Ulla Goltz for surprising insights in the nature of process logics, concurrency, and locality. Finally, we are grateful to the anonymous referees for friendly comments and constructive criticism.

This work was partly supported by the EU under ESPRIT-III BRA WG 6071 IS-CORE (Information Systems - COrrectness and REusability) and BRA WG 6112 COMPASS (COMPrehensive Algebraic approach to System Specification and development) and by ESDI under research contract OBLOG (OBject LOGic). To appear in *Recent Trends in Data Type Specification*, E. Astesiano, G. Reggio and A. Tarlecki, eds. LNCS 906, Springer-Verlag, Berlin 1995.

Bibliography

[AR92] E. Astesiano and G. Reggio. Algebraic Specification of Concurrency. Recent Trends in Data Type Specification, LNCS 655, Springer-Verlag, Berlin 1992.

[Bo85] A. Borgida. Features of Languages for the Development of Information Systems at the Conceptual Level. *IEEE Software* 2 (1985), 63-73.

[Br93] M. Broy. Functional Specification of Time-Sensitive Communicating Systems. *ACM Transactions on Software Engineering and Methodology* 2 (1993), 1-46.

[CE94] S. Conrad and H.-D. Ehrich. An Elementary Logic for Object Specification and Verification. In U. Lipeck and G. Vossen, editors, *Workshop Formale Grundlagen für den Entwurf von Informationssystemen, Tutzing*, pages 197–206. Technical Report Univ. Hannover, No. 03/94, 1994.

[CGH92] S. Conrad, M. Gogolla, and R. Herzig. TROLL *light*: A Core Language for Specifying Objects. Informatik-Bericht 92–02, TU Braunschweig, 1992.

[Ch76] P. P Chen. The Entity-Relationship Model—Toward a Unified View of Data. *ACM Transactions on Database Systems*, Vol. 1, No. 1, 1976, 9–36.

[De94] G. Denker. Object Reification (Extended Abstract). Working Papers of the International Workshop on Information Systems – Correctness and Reusability, IS-CORE'94. R. Wieringa and R. Feenstra, eds. Technical Report IR-357, VU Amsterdam 1994.

[De95] G. Denker. Transactions in Object-Oriented Specifications. *Proceedings of the 10th Workshop on Abstract Data Types*, 1994, LNCS 906.

[EDG88] H.-D. Ehrich, K. Drosten, and M. Gogolla. Towards an Algebraic Semantics for Database Specification. In: R. Meersmann and A. Sernadas (eds.). *Proc. 2nd IFIP WG 2.6 Working Conf. on Database Semantics "Data and Knowledge" (DS-2)*, Albufeira (Portugal), 1988. North-Holland, Amsterdam, 119-135.

[EGH+92] G. Engels, M. Gogolla, U. Hohenstein, K. Hülsmann, P. Löhr-Richter, G. Saake, and H.-D. Ehrich. Conceptual modelling of database applications using an extended ER model. *Data & Knowledge Engineering, North-Holland*, Vol. 9, No. 2, 1992, 157-204.

[EGL89] H.-D. Ehrich, M. Gogolla, and U. Lipeck. Algebraische Spezifikation Abstrakter Datentypen. Teubner–Verlag, Stuttgart 1989.

[EGS91] H.-D. Ehrich,J. Goguen, and A. Sernadas. A Categorial Theory of Objects as Observed Processes. Proc. REX/FOOL School/Workshop, J. W. deBakker et. al. (eds.), LNCS 489, Springer-Verlag, Berlin 1991, 203-228.

[Eh86] H.-D. Ehrich. Key Extensions of Abstract Data Types, Final Algebras, and Database Semantics. In: D. Pitt et al. (eds.): *Proc. Workshop on Category Theory and Computer Programming.* Springer, Berlin, LNCS series, 1986, 412-433.

[EM85] H. Ehrig and B. Mahr. Fundamentals of Algebraic Specification 1. Springer-Verlag, Berlin 1985.

[EM90] H. Ehrig and B. Mahr. Fundamentals of Algebraic Specification 2. Springer-Verlag, Berlin 1985.

[ES91] H.-D. Ehrich and A. Sernadas. Fundamental Object Concepts and Constructions. Information Systems – Correctness and Reusability, Proc. ISCORE Workshop'91 (G. Saake and A. Sernadas, eds.), Informatik-Berichte 91-03, Techn. Univ. Braunschweig 1991, 1-24.

[ESSS95] H.-D. Ehrich, G. Saake, A. Sernadas, and C. Sernadas. Distributed Temporal Logic for Concurrent Object Families (Extended Abstract). Proc. ISCORE Workshop '94, R. Wieringa, ed. World Scientific Publishers. To appear 1995.

[FM91] J. Fiadeiro and T. Maibaum. Towards Object Calculi. Information Systems – Correctness and Reusability, Proc. ISCORE Workshop'91 (G. Saake and A. Sernadas, eds.), Informatik-Berichte 91-03, Techn. Univ. Braunschweig 1991, 129-178.

[FM92] J. Fiadeiro and T. Maibaum. Temporal Theories as Modularisation Units for Concurrent System Specification. *Formal Aspects of Computing* 4 (1992), 239-272.

[FSMS91] J. Fiadeiro, C. Sernadas, T. Maibaum, and G. Saake. Proof-Theoretic Semantics of Object-Oriented Specification Constructs. In: R. Meersman, W. Kent, and S. Khosla (eds.). *Object-Oriented Databases: Analysis, Design and Construction (Proc. 4th IFIP WG 2.6 Working Conference DS-4, Windermere (UK))*, Amsterdam, 1991. North-Holland, 243-284.

[GCH93] M. Gogolla, S. Conrad, and R. Herzig. Sketching Concepts and Computational Model of TROLL *light*. In A. Miola, editor, *Proc. 3rd Int. Conf. Design and Implementation of Symbolic Computation Systems (DISCO'93)*, pages 17-32. Springer, LNCS 722, 1993.

[GM87] J. A. Goguen and J. Meseguer. Unifying functional, object-oriented and relational programming with logical semantics. *Research Direction in Object-Oriented Programming*, B.Shriver,P.Wegner (eds.), MIT Press 1987, 417-477.

[GW90] J. A. Goguen and D. Wolfram. On Types and FOOPS. In: R. Meersman, W. Kent, and S. Khosla (eds.). *Object-Oriented Databases: Analysis, Design and Construction (Proc. 4th IFIP WG 2.6 Working Conference DS-4, Windermere (UK))*, Amsterdam, 1991. North-Holland.

[HM85] M. Hennessy and R. Milner. Algebraic Laws for Nondeterminism and Concurrency. Journal of the ACM 32 (1985), 137-161.

[Ho85] C. A. R. Hoare. Communicating Sequential Processes. Prentice-Hall, Englewood Cliffs, NJ, 1985.

[HSJHK94] T. Hartmann, G. Saake, R. Jungclaus, P. Hartel, and J. Kusch. Revised Version of the Modeling Language TROLL. Informatik-Bericht 94-03, TU Braunschweig 1994.

[JSHS91] R. Jungclaus, G. Saake, T. Hartmann, and C. Sernadas. Object-Oriented Specification of Information Systems: The TROLL Language. Informatik-Bericht 91-04, TU Braunschweig, 1991.

[Ju93] R. Jungclaus. Modeling of Dynamic Object Systems, a Logic-based Approach. Advanced Studies in Computer Science. Vieweg Verlag, Braunschweig/Wiesbaden, 1993.

[LMRT91] K. Lodaya, M. Mukund, R. Ramanujam, and P.S. Thiagarajan. Models and Logics for True Concurrency. in P.S. Thiagarajan (ed.): Some Models and Logics for Concurrency. Advanced School on the Algebraic, Logical and Categorical Foundations of Concurrency. Gargnano del Garda, 1991.

[MB89] J. Mylopoulos and M. Brodie, (eds.). Readings in Artificial Intelligence & Databases. Morgan Kaufmann Publ. San Mateo, 1989.

[Mi89] R. Milner. Communication and Concurrency. Prentice-Hall, Englewood Cliffs, 1989.

[MM93] N. Martí-Oliet and J. Meseguer. Rewriting Logic as a Logical and Semantic Framework. Report SRI-CSL-93-05, SRI International, Menlo Park 1993.

[MM94] N. Martí-Oliet and J. Meseguer. General Logics and Logical Frameworks. In: D. M. Gabbay (ed.). *What is a Logical System?*. Oxford University Press 1994. To appear

[MP89] Z. Manna and A. Pnueli. The Anchored Version of the Temporal Framework. In: J. deBakker, W. deRoever, and G. Rozenberg (eds.). *Linear Time, Branching Time and Partial Order in Logics and Models for Concurrency.* LNCS 354, Springer-Verlag, Berlin, 1989, 201-284.

[Sa91a] G. Saake. Conceptual Modeling of Database Applications. In: Karagiannis, D. (ed.): *Proc. 1st IS/KI Workshop, Ulm (Germany), 1990.* Springer, Berlin, LNCS 474, 1991, 213-232.

[Sa91b] G. Saake. Descriptive Specification of Database Object Behaviour. *Data & Knowledge Engineering* 6 (1991), 47-74.

[Se80] A. Sernadas. Temporal Aspects of Logical Procedure Definition. *Information Systems*, Vol. 5, 1980, 167–187.

[SEC90] A. Sernadas, H.-D. Ehrich, and J.-F. Costa. From processes to objects. *The INESC Journal of Research and Development 1:1*, pages 7-27, 1990.

[SF91] C. Sernadas and J. Fiadeiro. Towards Object-Oriented Conceptual Modelling. *Data & Knowledge Engineering* 6 (1991), 479-508.

[SJE92] G. Saake, R. Jungclaus, and H.-D. Ehrich. Object-Oriented Specification and Stepwise Refinement. In J. de Meer, V. Heymer, and R. Roth, editors, *Proc. Open Distributed Processing, Berlin (D), 8.-11. Okt. 1991 (IFIP Transactions C: Communication Systems, Vol. 1)*, pages 99-121. North-Holland, 1992.

[SJH93] G. Saake, R. Jungclaus, and T. Hartmann. Application Modelling in Heterogenous Environments Using an Object Specification Language. *International Journal of Intelligent and Cooperative Information Systems* 2 (1993), 425-449.

[SR94] A. Sernadas and J. Ramos. The GNOME Language: Syntax, Semantics and Calculus. Tech. Report, Instituto Superior Técnico, Lisboa 1994.

[SSC92] A. Sernadas, C. Sernadas, and J.F. Costa. Object Specification Logic. Internal report, INESC, University of Lisbon, 1992. (to appear in Journal of Logic and Computation).

[SSE87] A. Sernadas, C. Sernadas, and H.-D. Ehrich. Object-Oriented Specification of Databases: An Algebraic Approach. In P.M. Stoecker and W. Kent, editors, *Proc. 13th Int. Conf. on Very Large Databases VLDB'87*, pages 107-116. VLDB Endowment Press, Saratoga (CA), 1987.

[SSG+91] A. Sernadas, C. Sernadas, P. Gouveia, P. Resende, and J. Gouveia. OBLOG – Object-Oriented Logic: An Informal Introduction. Technical report, INESC, Lisbon, 1991.

[ST89] J. W. Schmidt and C. Thanos (eds.). Foundations of Knowledge Base Management. Springer-Verlag, Berlin, 1989.

[Wi80] G. Winskel: Events in Computation. PhD thesis, University of Edinburgh.

[WN93] G. Winskel and M. Nielsen. Models for Concurrency. Report DAIMI PB – 463, Computer Science Department, Aarhus University 1993.

2

REPRESENTATION OF LEGAL KNOWLEDGE AND EXPERT SYSTEMS IN LAW

Paulo de Sousa Mendes
Faculdade de Direito, Universidade de Lisboa, Portugal
paulosousamendes@fd.ulisboa.pt

Abstract

Can a judge be substituted by a computer, or to put it more dramatic, by a robot, and if this is possible, to which degree? There has not yet been any fundamental or applied research project aiming the substitution of a human decision-maker by a machine in the area of law. But there has been great enthusiasm regarding the creation of expert systems in order to support the court ruling. In the legal field, since the decade of the 1990s the expert systems were not longer considered to be part of the buzzwords. Did they simply get out of fashion and vanished without any trace? In this paper, I try to list some of the reasons that caused the gradual lack of interest of the academics and legal practitioners for these informational applications and I will also try to defend the urgent need for new strategies for the connection between artificial intelligence and law.

2.1 Introduction

May a judge come to be replaced by a computer or – if we wish to make the question even more dramatic – be replaced by a robot?

Is the machine-judge an impending reality, perhaps? That is, a machine-judge that clearly does more than pronounce the words of the Law. There is no doubt that the abomination which was the surprising object of Montesquieu's praise may, today, literally become a reality without too much difficulty; in other words, a robot capable of recognising the words of the Law and pronouncing them in an audible voice, with human mannerisms. But this is not what matters.

There are machines, which are capable of winning the most difficult game in the world, beating the best chess player of all time[1]. If that is so, shouldn't judges be on their guard regarding the possible appearance of machines, which are capable of producing better legal decisions that they themselves do?

The question is purely rhetoric: there has never been any basic or applied research, which sought to replace a human decision-maker by a machine in the field of Law. Yet, there has been great enthusiasm regarding the creation of Expert Systems (also called Knowledge-Based Systems) to support judicial decisions. In the legal field, expert systems ceased to be among the buzz expressions from the 1990s on (Tiscornia, 2014: 441). Have they simply gone out of fashion and disappeared without a trace? In this paper I will try to list the reasons, which have led to jurists' gradual lack of interest in these Information Technology (IT) applications and I will attempt to defend the urgent need for new strategies to make the interconnection between Artificial Intelligence (AI) and Law.

[1]Deep Blue is a supercomputer (with the respective software) created by IBM especially to play chess, made up of 256 co-processors capable of analysing around 200 million positions per second. In May 1997, Deep Blue beat Kasparov in a match of 6 games, racking up 2 victories, 3 ties and 1 defeat (final score: 3.5 to 2.5), thereby making it the first computer to beat a world champion in a clash which complied fully with the Rules of the World Chess Federation, including the maximum duration of the games.

2.2 Expert Systems

It is worth clarifying first that expert systems are computer programs which are intended to solve complex problems in a manner similar to that of experts (heuristic programs). The idea is that they will be capable of maintaining relatively open dialogue with the user, thereby distinguishing them from common IT applications which require us to fill in all the fields reserved for answers and which become blocked every time we miss the response required by the system (versatile programs). Also, expert systems must be capable of clarifying the "why" of the questions they pose to the user when that user needs additional information in order to give his answers, and they must justify their solutions and the route taken to reach them, otherwise the user would not be able to scrutinise the suitability of the answers obtained by the program (transparent programs) (Susskind, 1989: 9).

Expert systems include a "knowledge base", where relevant data is formalised, and an "inference engine", which contains the algorithms needed to solve the particular problems of the field of knowledge in question.

2.2.1 The early days of expert systems

Dendral was the first expert system written, that appeared in 1969. It was used to determine the molecular structure of unknown molecules (Susskind, 1989: 10).

In the 1970s and 1980s expert systems were the source of much enthusiasm and they mobilised the efforts of many interdisciplinary teams of researchers across the world. IT applications thus arose in such varied areas as writing and voice recognition, location of mineral deposits, diagnosis of illnesses and weather forecasting. The following systems can be regarded as successful expert systems, some of them were also successful in commercial terms: Prospector, Mycin, Caduceus and Casnet (Susskind, 1989: 10-11).

2.2.2 Expert systems in Law

Jurists also accompanied this wave of interest in the topic and sought to associate themselves with the development of IT applications in the field of Law. Around 25 programs in the legal field appeared in the 1970s and 80s: Taxman I/II, Judith, Tax Advisor, CCLIPS, Legol/Norma, Sara, Polytext/Arbit, etc. (Susskind, 1989: 16-18; Tiscornia, 2014: 439-411, 448; Susskind & Susskind, 2015: 183).

I myself was a member of a multidisciplinary group of knowledge engineers, logicians, mathematicians and jurists which was active between

1988 and 1995, dedicated to representing and formalising legal knowledge with a view to developing models which could become operational. The initial formation of the group was based on the interest of the team, led by Amílcar Sernadas, now honoured, and Cristina Sernadas at the Institute for Systems and Computer Engineering (INESC), in developing a language model to represent object-oriented knowledge, called Oblog. Amílcar Sernadas contacted António Hespanha in order to put together a team of jurists who might be interested in the project, assuming that Law would be a prime area to test the model in question (Hespanha, 2000: 140). A basis for understanding between people from different knowledge areas had to be created and there was also training in deontic logic, but the fundamental point was to really make advances in representing and formalising legal knowledge.

2.3 Representing and Formalising Legal Knowledge

We must now relate the above developments with the representation and formalisation of legal knowledge.

2.3.1 Well-structured and ill-structured legal problems

Some attempts to represent and formalise legal knowledge suffered from over-confidence in the possibility of dealing with Law as if it were a system of rules already formalised by itself in the natural language of the sources of regulation, which would supposedly facilitate the task of modelling and resolving problems (Hespanha: 2000, 148). Thus, there was a belief that there were certain areas of Law where problems could be considered to be "well-structured", insofar as they could be fully described in quantitative terms, they allowed for the possibility of describing the object to be achieved via an "objective" function and there were algorithms capable of obtaining and describing the solutions quantitatively (Bento & Côrte-Real, 2000: 21). In fact, there are some very simple legal questions which allow this type of knowledge modelling and there are even specific examples in which modelling has been developed to the point where operational and effective logic programs have been created, as is the case of the representation of the British Nationality Act of 1981 produced by the team led by Marek Sergot at Imperial College, London (Hespanha, 2000: 147). Indeed, some problems that jurists have considered hard to crack have proved, after all, to be trivial in the eyes of knowledge engineers, logicians and mathematicians. This is so much so that the team that I myself was a part of at INESC created, thanks to Susana Brito, on the jurists' side, and José Carmo, on the logicians' side, a program which worked particularly well regarding the issue of renvoi (return) in private international law. Hespanha recalls "[the] epic session where we showed the results to Professor Isabel Magalhães Collaço; epic because of the efficacy of the robot-jurist, but also because of the reaction of our dear Professor, who was not particularly ready to accept that a machine might resolve a conundrum of legal intelligence with relative ease" (Hespanha, 2000: 142). The program was never published since, from the IT point of view, it was not very interesting, having been built in a basic programming language, PASCAL (Hespanha, 2000: 142). What remained was only a record of the experiment in the short article by Susana Brito, entitled "Uma experiência do Projeto Normlog – O sistema do reenvio concebido para apoio ao ensino" ["An experiment of the Normlog Project – The system of renvoi designed as a teaching support"] (Brito, 1991: 529-542). In particular, the program allowed the standard user, a university student in Law, to: (1) test whether he had correctly solved a given case of renvoi considered in abstract; (2) to request a solution from the program; (3) to observe an illustrative scheme of the solution found by the program, or, lastly, (4) to follow the various steps in solving the problem (Brito, 1991: 540-541).

Legal problems, as a rule, cannot be considered "well-structured", but, on the contrary, "ill-structured". The knowledge engineers, logicians and mathematicians had to discover at their own cost that the supposed formalisation of knowledge in the natural language of the legislative texts was, after all, just an illusion, which is, in fact, typical of the first contact laypeople have with the Law. Jurists have always been aware of the "ill-structured" nature of the vast majority of legal problems. In the field of philosophy of Law, Kelsen has already explained that "[the] scientific interpretation [of the sources of rules] is pure cognitive determination of the meaning of legal rules" (Kelsen, 1984: 472). So, the scientific interpretation describes the meanings which fit within the text of the sources of rules, but is relatively inconclusive, since "[...] the result of a legal interpretation can only be ascertaining the framework of the law to be interpreted, and thereby the cognition of several possibilities within that framework" (Kelsen, 1984: 467). One might add that, if it were called on to establish connections between the legal rules, the scientific interpretation would not be able, in the light of positive law which is technically imperfect and full of loopholes, to produce either logical or complete results, which prevents the scientific interpretation from presenting solutions for specific cases (Bjarup, 2013: 193).

In the application of the Law by a court, the scientific interpretation (obtained by a knowledge operation) of the Law to be applied combines with an act of will in which the applying body makes a choice between possibilities revealed via that same cognitive interpretation or, ultimately, covers the gaps in the Law when the legislator did not foresee the case submitted for a judicial decision (Kelsen, 1984: 470).

The application of the Law to specific situations is an activity of creating meaning and referring to values, which necessarily transcend the scientific interpretation of the sources of the rules. Merely theoretical-objective approaches to sources of rules cannot, therefore, resolve most legal problems, which is why expert systems will not be able to obtain promising results if they are limited to these types of approaches.

2.3.2 Deontic logic

The challenge of representing and formalising legal knowledge gave rise, in the meantime, to more sophisticated approaches. The importance of deontic logic and computational linguistics is frequently recognised for defining legal information expert systems in rule-based approaches (Sernadas et al., 1989: 781-782; Tiscornia, 2014: 440).

When rule-based approaches are adopted, then statutory laws become a collection of rules. Such rules are relevant for providing the conditions that must be fulfilled for the application of a given statutory law. The problem with this kind of approach is that thousands of rules are expected even in trivial cases (Sernadas et al., 1989: 781).

These difficulties explain, to a large extent, why rule-based representation of legal knowledge was abandoned, and deontic logic itself eventually became mere "ontology", which is the term normally used in the field of AI – perhaps with some philosophical incorrectness – to refer to the explanation of abstract concepts (Araújo, 1999: 32; Tiscornia, 2014: 445-446; John, Caro & Boella, 2016: 3; Schweighofer, 2016: 4).

It is on this merely theoretical ground that the studies, which persist in deontic logic are currently written (Rotolo & Sartor, 2012: 15-16).

2.3.3 Representation and formalisation via objects

Other possible strategies are by means of representing and formalising object-oriented legal knowledge. This was the strategy followed by the INESC group mentioned previously.

Through an object-oriented approach (Oblog) we can represent both the statutory norms and the legal dogmatics[2], adopting the object, in-

[2]Legal doctrine is often called "legal dogmatics" (Rechtsdogmatik). The term has an established meaning well known among continental law theorists. In Anglo-

stead of the formula, as the basic structuring unit. Since the complete integration of both static and dynamic aspects is inherent to the object concept, we can also represent the evolution of the entities (Sernadas et al., 1989: 782). For illustration purposes, we choose the crime of omission of help in the Portuguese Criminal Code, where the balance between statutory norms and doctrinal rules is typical of the importance of the legal dogmatics (Sernadas et al., 1989: 782).

The Oblog view of the Universe of Discourse (UoD) stresses clearly the different levels on which knowledge is organised. On a paramount level, a subset of the meta-theory of legal dogmatics, directing the organisation of knowledge concerning any particular infringement, is represented. Items like the theory of infringement (with their summit qualifications or their components), including the categories of justification reasons and exculpating aspects are represented at this level, as well as the references to the statutory sources of the penal norms that support this theory. Relations of logical inclusion between penal norms are also represented at this level. For example, for each justification reason there is a self-contained module (an object) containing the relevant conceptual entities (Sernadas et al., 1989: 785).

The main structuring units in the Oblog approach are the object and the object type. Objects can be used to define entities, values, processes etc. An object has a set of attributes, providing the respective definition at a particular moment, and a set of events, indicating the main things that happen in the life of an object. Moreover, the description of an object also includes the characterisation of the possible evolutions of the object, i.e. the sequences of events that are admissible in its life, as well as rules for stating how the possible events affect the values of the attributes. Objects can be grouped in the so-called object types. An object type description includes a surrogate mechanism for stating how the instances of the object type are constructed and identified and a template describing the different instances as objects (Sernadas et al., 1989: 786).

This work was directed by Cristina Sernadas, on the logicians' side, and by myself, on the jurists' side, thanks to my training in criminal law.

The conclusion drawn by two of the participants in the Oblog Project is somewhat similar in recognising that the representation and formalisation of knowledge on the crime of omission of help punished by the Portuguese Criminal Code was relatively unsuccessful, notwithstanding

---

American legal theory the term "legal dogmatics" is not so well known, however. It also produces misunderstandings among legal researchers who dislike the word "dogmatic" because it calls up the idea of "narrow-mindedness", or something like it (Pattaro, 2005: 2).

the progress made in terms of the modelling which we intended. Thus, Susana Brito recalls that "[the crime of omission of help] obtained satisfactory representation in Oblog (object-based language), although, due to technical and human difficulties, an expert system never came to be made" (Brito, 1992: 536-537). António Hespanha also states that, [d]espite significant progress in the modelling, [...], taking a step back, I believe that we failed in the ambition of our purpose. Besides which, we continued not to have a shell [i.e. an executable IT applicable which would serve as a support to introduce specific knowledge] which would create executable programs" (Hespanha, 2000:141).

Nowadays I am convinced that the strategy we followed, despite being innovative in seeking to include formalisation of the elements making up a given infringement in the more general context of the theory of criminal infringement, ended up leading to the well-known phenomenon of "combinatorial explosion of possibilities". This was inevitable as the representation of the theory of infringement grew to cover in detail the elements of the concepts of "agency", "complicity", "unlawfulness", "culpability" and "liability" (Sernadas et al., 1989: 790-791) or as the representation of the elements making up omission of help covered the concepts of "act and omission" in criminal law and "concurrency of crimes", not to mention the problems of the applicable sanction and the determination of the specific punishment.

This being so, the main merit of the approach followed seems to me to lie in the fact of exposing the profound conceptual structure offered by legal dogmatics itself via the theory of criminal infringement (of German origin) which is behind the interpretation of the provisions of the Criminal Code (in this case, the Portuguese Code) and which is essential for an understanding of the method of reasoning of the applier of the Law, especially in the criminal arena. The representation of that knowledge had, moreover, the merit of revealing the dynamic aspects of criminal responsibility, incorporating some concepts of criminal procedure which could not but be present in the model, insofar as the perpetrator of the crime has to be selected as a suspect and classified as a defendant before he can be found innocent or guilty (Sernadas et al., 1989: 808). However, these merits once again end up transforming the analysis into a mere ontology. The diagrams themselves are essential for representing and formalising knowledge by objects, which is not a small feat, above all if we recall the importance that is currently given in certain cutting edge sectors of legal knowledge to the representation of abstract thinking through images, the so-called "legal visualisation" (Schefbeck, 2014: XVII-XXII; Broekman, 2014: 72-74; Čyras, 2014: 175-183; Schweighofer, 2016: 6). But we are very far from the original aim of creating an expert system to support judicial decisions.

2.3.4 Neural networks

An interesting alternative is characterised by the simulation of neural networks (artificial neurons)[3].

"Artificial neural networks are generally presented as systems of interconnected 'neurons', which send messages to each other. The connections have numeric weights that can be tuned based on experience, making neural nets adaptive to inputs and capable of learning". "Historically, the use of neural networks models marked a paradigm shift in the late eighties from high-level (symbolic) AI, characterised by expert systems with knowledge embodied in if-then rules, to low-level (sub-symbolic) machine learning, characterised by knowledge embodied in the parameters of a dynamical system".[4]

Indeed, neural networks have the advantage of dispensing with the reproduction of the complexities of legal reasoning. A neural network is not transparent, since the knowledge that it contains does not exist explicitly, other than as a set of numbers (a set of values associated to connections between units). Neural networks do not, therefore, provide any explanation regarding how they process information and knowledge. They have a layer of inputs and a layer of outputs, interconnected by a hidden layer(s) of intermediate units.

In Ludwig-Maximilians-Universität München, Lothar Philipps was an enthusiastic supporter of neural networks, having published numerous promising theoretical works (Philipps, 1999: 820-827)[5]. However, these were always and only theoretical works.

2.4 New "Back to the Future" Strategies

The interest in developing expert systems to support judicial decisions decreased, in fact, from the end of the 1990s, which was undoubtedly due to the scarce results obtained from the different approaches tested (Susskind & Susskind, 2015: 183-184, 187). The lack of interest, moreover, appears to continue to the present date, although I believe it is inevitable that the scientific community will have to return to the matter. Indeed, there is much to do and that can be done. It all depends on the new "back to the future" strategies.

---

[3]In 1969, Arthur E. Bryson and Yu-Chi Ho discovered the back propagation algorithm of neural networks, which only gained relevance later, when the rediscovery of this same algorithm, in 1986, finally led to the development of neural networks.

[4]http://en.wikipedia.org/wiki/Artificial_neural_network.

[5]The INESC group was always very close to Lothar Philipps, who was invited to give presentations in Portugal and who frequently invited me to give presentations in Munich.

2.4.1 Tools for questions of fact

Firstly, it is important to note that the attempts to represent and formalise legal knowledge have to date grossly ignored the approach of production techniques for proving facts, which are decisive for the criminal investigation, amongst others, and for the actual judicial decision.

It is in this field that highly powerful IT applications have appeared, both in the commercial area (EnCase, Nuix) or in open source (Total Commander), which may be used as technologies for the search and gathering of digital forensic evidence and also for analysing it (E-Discovery, Data Mining).

It is also important to use cryptography techniques to guarantee the integrity of the digital documents (e.g. emails, files), ensuring at the time of the search, inspection or audit that these have not been altered, and also guaranteeing that they cannot be, and have not been, altered after they have been copied (certified copy).

We cannot forget that we live in a digital era, in which criminal investigation, alongside others, is mostly or almost exclusively conducted in this environment.

There is an urgent need for the academic community to take an interest in this area. It will not warrant the pompous name of expert systems, but it will certainly be of greater practical use. One might add that it will also contribute to an understanding of the theory of evidence law, which has received so little attention in the Roman-Germanic Law countries.

2.4.2 Tools to support decisions on questions of law

New approaches are also necessary in terms of representing and formalising questions of law.

Case-based reasoning has been recognised as an appropriate approach for dealing with knowledge which is not totally understood and adequate for open domains (Bento & Côrte-Real: 2000, 44).

Jurists with Roman-Germanic training often tend to think that case-based reasoning is typical of Common Law systems. But this is not necessarily so (Sousa Mendes, 2014: 143-152).

Some very interesting approaches are those that have been taken by the Research Laboratory for Law, Logic & Technology (LLT Lab) of Hofstra University, New York, led by Vern Walker. Based on studies of case law (for example, a sample of 35 judicial decisions from courts of first instance in cases regarding the civil liability of pharmaceutical companies for the side effects of vaccines sold), Vern Walker's team represented in a flow chart and formalised with symbols the criteria used in sequence by the North American courts, in the cases of the relevant sample, to determine the causality of the offences to the physical integrity of the patients as a result of the administration of the vaccines (Walker et al., 2013: 196). It should be said that the representation is not executable via a computer program, and is not yet, therefore, similar to an expert system to support legal decisions, and neither was it developed with this aim. Besides this, the sample used is very incomplete and also lacked mechanisms which automatically enabled linguistic analysis of the set of legal decisions in this area, in order to discover if the decision-making criteria of the causal issue are uniform or not. Nevertheless, the approach is very promising, besides allowing us to predict, with a certain amount of certainty, the probable sense of future judicial decisions, something which almost certainly will not happen with rule-based expert systems or even object-based ones.

2.5 Conclusion

The difficulties of representation of legal knowledge and the extreme complexity of connections between the uncountable elements of Law (Tiscornia, 2014: 442) cannot determine our lack of interest in this line of research, but, on the contrary, must conduct us to the search for more and better AI tools applicable in the field of Law.

However, it is not advisable for an expert system to propose single solutions for each difficult legal problem. Law and its performers are not being asked to operate as "machines", with no room for more than one valid solution (Palma, 2014: 682-683).

Bibliography

Adebayo Kolawole John, Luigi di Caro, Guido Boella, "Annotating Legal Documents with Ontology Concepts", in: *Jusletter IT*, No. 25 (Feb. 2016), pp. 1-12 (online: www.weblaw.ch).

Antonino Rotolo, Giovanni Sartor, *Deontic Logic – Introductory materials*, 2012 (draft).

António Manuel Hespanha, "Os juristas que se cuidem... Dez anos de inteligência artificial e Direito", in: *Themis*, Year I, No. 1 (2000), pp. 139-169.

Cristina Sernadas, Paulo de Sousa Mendes, António M. Hespanha, Filipe Santos, Sérgio Mascarenhas, Amílcar Sernadas,

"An Object-oriented Representation of the Dogmatics of Omission to Help in Criminal Law", in: Vv.Aa., *Pre-proceedings of the III International Conference on Logica, Informatica, Diritto*, Florence: Antonio A. Martino, 1989, pp. 781-809.

DANIELA TISCORNIA,"About 'Good Law'?", in: Vv.Aa., *Zeichen und Zauber des Rechts – Festschrift für Friedrich Lachmayer* (ed.: E. Schweighofer et al.), Bern: Weblaw, 2014, pp. 437-449.

ENRICO PATTARO, *A Treatise of Legal Philosophy and General Jurisprudence*, Vol. I (The Law and The Right), Berlin: Springer, 2012.

ERICH SCHWEIGHOFER, "The Role of AI & Law in Legal Data Science", in: *Jusletter IT 25* (Feb. 2016), pp. 1-8 (online: www.weblaw.ch).

FERNANDO ARAÚJO, "Lógica jurídica e informática jurídica: Da axiomatização deôntica às estruturas não-monotónicas do raciocínio rebatível", in: Vv.Aa., *Direito da Sociedade de Informação*, Vol. I, Coimbra: Coimbra Editora, 1999, pp. 7-71.

GÜNTHER SCHEFBECK, "Von Kreisen, Brücken und Bildern: Assoziationen zu Friedrich Lachmayer", in: Vv.Aa., *Zeichen und Zauber des Rechts – Festschrift für Friedrich Lachmayer* (ed.: E. Schweighofer et al.), Bern: Weblaw, 2014, pp. XVII-XXII.

HANS KELSEN, *Teoria Pura do Direito*, (translation by João Baptista Machado from 2nd Ed. of the original *Reine Rechtslehre*, 1960), 6th Ed., Coimbra: Arménio Amado, 1984.

JAN M. BROEKMAN, "Law, Cubism, Semiotics", in: Vv.Aa., *Zeichen und Zauber des Rechts – Festschrift für Friedrich Lachmayer* (ed.: E. Schweighofer et al.), Bern: Weblaw, 2014, pp. 55-75.

JES BJARUP, "Kelsen and Hägerström: Clearing Up Misunderstandings and Mapping Out the Common Ground", in: Vv.Aa., *Kelsen Revisited – New Essays on the Pure Theory of Law* (ed.: Luís Duarte d'Almeida, John Gardner, Leslie Green), Oxford and Portland, Oregon: Hart Publishing, 2013, pp. 163-194.

JOÃO BENTO, JORGE CÔRTE-REAL, *Tópicos de Inteligência Artificial: Notas de apoio às aulas*, Lisbon: IST, 2000.

JOHNATHAN JENKINS, "What Can Information Technology Do for Law?", in: *Harvard Journal for Law and Technology*, No. 2 (2008), pp. 589-607.

LOTHAR PHILIPPS, "Naheliegende Anwendungen neuronaler Netze in der Rechtswissenschaft", in: *Jur* PC, No. 1/90 (Jan. 1990), pp. 820-827.

MARIA FERNANDA PALMA, "O Direito como máquina jurídica e o controlo de si mesmo como fundamento da responsabilidade", in: *Revista Portuguesa de Filosofia*, Vol. 70, No. 4 (2014), pp. 681-694.

PAULO DE SOUSA MENDES, "Em defesa do particularismo moral e do pluralismo liberal: em especial no domínio do Direito Penal", in: Vv.Aa., *Multiculturalismo e Direito Penal* (ed.: Teresa Pizarro Beleza, Pedro Caeiro, Frederico de Lacerda da Costa Pinto), Coimbra: Almedina, 2014, pp. 143-152.

RICHARD SUSSKIND, *Expert Systems in Law*, 1rst Ed. Pbck., Oxford: Clarendon Press, 1989 (1rst Ed., 1987).

RICHARD SUSSKIND, DANIEL SUSSKIND, *The Future of the Professions – How Technology Will Transform the Work of Human Experts*, Oxford: Oxford University Press, 2015.

SUSANA DE BRITO, "Uma experiência do Projeto Normlog – O sistema do reenvio concebido para apoio ao ensino", in: Vv.Aa., *Colóquio Informática e Tribunais, Bases de Dados Administrativas e Jurídicas* (ed.: Gabinete do Director da Informatização Judiciária), Lisboa: Ministério da Justiça, 1991, pp. 529-542.

VERN WALKER ET AL., "A Process Approach to Inferences of Causation: Empirical Research from Vaccine Cases in the United States", in: *Law, Probability and Risk*, Vol. 12, Nos. 3-4 (2013), pp. 189-205.

VYTAUTAS ČYRAS, "Friedrich Lachmayer's Landscape of Legal Informatics", in: Vv.Aa. *Zeichen und Zauber des Rechts – Festschrift für Friedrich Lachmayer* (ed.: E. Schweighofer et al.), Bern: Weblaw, 2014, pp. 175-183.

3

A TIME-MODEL INVARIANT FRAGMENT OF METRIC TEMPORAL LOGIC

Antónia Lopes[1] and José Luiz Fiadeiro[2]
[1]Faculdade de Ciências, Universidade de Lisboa, Portugal
malopes@ciencias.ulisboa.pt

[2]Dep. of Computer Science, Royal Holloway, University of London, UK
Jose.Fiadeiro@rhul.ac.uk

Abstract

Hybrid systems, such as cyber-physical systems, typically contain components that are best described using a continuous time model, and others that are more naturally described using a discrete one, even several discrete ones if different components operate over different time granularities. Consequently, in order to specify and reason about the global behaviour of these systems using a logical formalism, one would like to put together formulas that are meant to be interpreted in different time models. In this paper, we present a fragment of Metric Temporal Logic that is time-model invariant, i.e., its formulas have the same meaning under a continuous and a discrete semantics. We show that this fragment is expressive enough to capture classes of properties that are traditionally used for hybrid systems and, in particular, that its formulas do not constrain the timing of environment observations, which is another important property for specifying systems that can be interconnected to form larger systems.

3.1 Introduction

Hybrid systems, such as those that are now operating in cyberspace, encompass physical devices and software components that often have real-time requirements. Whilst the former are best described under a continuous time model, the latter are more naturally described under a discrete one, even several discrete ones if different software components operate over platforms with different time granularities. This means that, in order to deduce global properties of a system from specifications of its components, one needs to consider properties expressed under different time models. In this paper, we characterise a fragment of Metric Temporal Logic that is time-model invariant, i.e., its formulas have the same meaning under a continuous and a discrete semantics, which makes it easier to specify and reason about such global properties of hybrid systems.

Metric Temporal Logic (MTL)[Koy90, AH91] has been considered to be particularly suited for the specification of real-time properties of hybrid systems. MTL has a discrete point-based semantics defined over traces, and also a continuous semantics defined over signals, which is why we chose it as a formalism for reasoning about networks of systems that may work under different time granularities, i.e., about heterogeneous timed systems [FL14]: the continuous semantics offers a time domain that is closed for time refinement, i.e., the interpretation of any formula does not change if the time granularity of a component is refined; on the other hand, the discrete semantics is more amenable to automated analysis (MTL under the continous semantics being undecidable [AFH96]), and therefore more convenient for reasoning about the behaviour of systems, but is sensitive to the chosen time granularity, which is problematic for physical devices and heterogenous timed software components.

Given this, and in order to get the best of both worlds, we investigated a non-trivial sub-language of MTL that we could use to specify real-time properties of systems without having to be concerned with the time model in which they are meant to be interpreted. Whilst it is easy to check that a specification asserting bounded response properties of the form "every request p is followed by a response q within t time units" has the same meaning under the two time models (which in a sense justified our endeavour), the challenge was to find a way to systematically identify a sub-language of MTL with the ability to express many other properties that have the same interpretation under both models.

In this paper, we present such a non-trivial time-model invariant fragment of MTL, which we designate by HMTL, and give an account of how this fragment was arrived at. We introduce two functions that connect the continuous and the pointwise time domains (instants and points), and we investigate conditions under which the traditional continuous semantics of MTL over signals and that over timed traces [SP07] are equivalent (the latter being closer to the intended discrete domain). Then we show that the timed traces that satisfy the formulas of HMTL under the continuous and pointwise semantics are exactly the same.

Finally, we present a time-related notion of closure for timed traces through which we can characterise the class of properties that do not impose constraints on the timing of the observations that the environment can make, in the same way that the notion of closure for traces defined by the Cantor topology helps to characterise safety properties [AS87]. The formulas of HMTL are shown to have this property, making HMTL suitable for the specification of components that can be integrated in a system with heterogeneous time granularities.

The structure of the paper is as follows. Section 3.2 presents the models of timed behaviour and notation used through the paper. Section 3.3 introduces a time-related refinement relation and the corresponding closure operation over timed traces. The main observations and results that led to the identification of a time-model invariant fragment of MTL are presented in Section 3.4. The paper finishes with an overview and discussion of related work in Section 3.5, and with concluding remarks in Section 3.6. Proofs are collected in an appendix.

3.2 Preliminaries

This section sets out the terminology and notation that is used throughout the paper. We start by recalling a few standard concepts related to models of timed behaviours.

Hybrid systems, such as cyber-physical systems, typically have physical components that are best described considering a continuous time model. Behaviours under such model are generally expressed in terms of boolean signals over the time domain $\mathbb{R}_{\geq 0}$. Signals are functions that determine the truth value of propositions at any non-negative real number. As usual, we consider only signals in which every proposition changes its truth value only a finite number of times in any bounded interval of time [HR04].

Definition 3.1 (Signal) *Let A be a finite set. A* signal *for A is a function $f : \mathbb{R}_{\geq 0} \to 2^A$ with finite variability, that is, with only finitely many discontinuities in any bounded interval. We denote the set of signals over A by $F(A)$.*

In contrast, a continuous observation model is not required for software components and, hence, their behaviour is usually captured using a discrete time model. Behaviours in this model can be expressed in terms of timed traces over a set A of propositions, i.e., infinite sequences of pairs of an instant of time and the set of propositions that are true at that time. Usually, the elements of A represent actions that a component or a system can execute. In this case, a timed trace comprises the executed actions timestamped with non-negative real numbers.

Definition 3.2 (Timed traces) *Let A be a finite set.*

- *A* trace λ *over A is an element of A^ω, i.e., an infinite sequence of elements of A.*
- *A* time sequence τ *is a trace over $\mathbb{R}_{\geq 0}$ such that: (1) $\tau(0) = 0$; (2) $\tau(i) < \tau(i+1)$ for every $i \in \mathbb{N}$; (3) the set $\{\tau(i) : i \in \mathbb{N}\}$ is unbounded, i.e., time progresses (the 'non-Zeno' condition).*
- *A* timed trace *over A is a pair $\langle \sigma, \tau \rangle$ where σ is a trace over 2^A and τ is a time sequence. We denote the set of timed traces over A by $\Lambda(A)$.*

Notice that, by allowing empty sets in the sequences σ, we can model finite behaviours through timed traces of the form $\langle \pi.\emptyset^\omega, \tau \rangle$ where $\pi \in (2^A)^*$ — the component stops executing actions after a certain point in time whilst still part of a system. The empty set also allows us to model observations that are triggered by actions performed by the environment in which the components execute, i.e., by other components of the wider system of which the component is a part.

Timed traces define signals in a straightforward way. Reciprocally, the process of sampling with a given schedule (for instance, a periodic rate) induces a map from signals to timed traces.

Definition 3.3 (Signals vs timed traces) *Let A be a finite set.*

1. *A timed trace $\lambda = \langle \sigma, \tau \rangle$ over A defines the signal $\boldsymbol{f}_\lambda$ over A where, for every i, $\boldsymbol{f}_\lambda(\tau(i)) = \sigma(i)$ and $\boldsymbol{f}_\lambda(t) = \emptyset$ everywhere else.*
2. *A signal f over A and a time sequence τ define a timed trace $\boldsymbol{\lambda}_f^\tau = \langle \sigma, \tau \rangle$ over A where $\sigma(i) = f(\tau(i))$.*

It is important to notice that this correspondence between signals and timed traces assumes a common vocabulary A. However, the nature of vocabularies over which signals and timed traces are usually defined

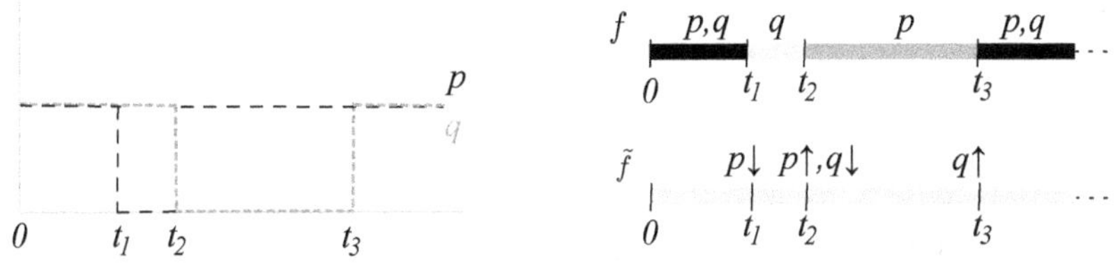

Figure 3.1: A signal f over the set of propositions $\{p, q\}$ and its conversion into a signal $\tilde{f}$ over the set of actions $\{p\uparrow, p\downarrow, q\uparrow, q\downarrow\}$.

is not necessarily the same. As mentioned before, in timed traces, the vocabulary represents the actions that a component or a system can execute. In contrast, signals are usually defined over a set of propositions that characterise a (finite) state space and, in this case, they have no point discontinuities.

In Fig.3.1 we illustrate how certain signals can be converted into signals over a set of actions in a straightforward way. More generally, given a signal f over A with no point discontinuities, $\tilde{f}$ is the signal over $\tilde{A} = \{a\uparrow, a\downarrow : a \in A\}$ such that, for every $t \in \mathbb{R}_{\geq 0}$,

$$\tilde{f}(t) = \begin{cases} \{a\uparrow : a \notin f(t^-) \wedge a \in f(t^+)\} \cup \{a\downarrow : a \in f(t^-) \wedge a \notin f(t^+)\} \\ \qquad\qquad\qquad\qquad\qquad \text{if } t \neq 0 \text{ and } f \text{ has a discontinuity in } t \\ \emptyset \qquad\qquad\qquad\qquad\quad\; \text{otherwise} \end{cases}$$

Notice that, because signals have only finitely many discontinuities in any finite amount of time, $f(t^-)$ and $f(t^+)$ are well defined ($f(t^-) = \lim_{\epsilon\to 0} f(t-\epsilon)$ and $f(t^+) = \lim_{\epsilon\to 0} f(t+\epsilon)$).

Signals obtained in this way have only point discontinuities and map any other time instant to the empty set. In fact, these properties hold for every signal defined over a vocabulary representing a set of actions, including every signal defined by a timed trace.

The following result relates this type of signals and timed traces:

Proposition 3.4 *Let $\tilde{F}(A)$ be the set of signals f over A such that, for every discontinuity t, $f(t^-) = f(t^+)$ and $f(t) = \emptyset$ otherwise.*

1. *Let $\lambda = \langle \sigma, \tau \rangle$ be a timed trace. We have that $\boldsymbol{f}_\lambda \in \tilde{F}(A)$ and $\boldsymbol{\lambda}^{\tau}_{(\boldsymbol{f}_\lambda)} = \lambda$.*

2. *Let f be a signal in $\tilde{F}(A)$ and τ a time sequence including all discontinuities of f. We have that $\boldsymbol{f}_{(\boldsymbol{\lambda}^{\tau}_{f})} = f$.*

3.3 Time Refinement

Hybrid systems may have components with different time granularities such as physical components sampled at different rates or software components running in execution platforms with different clock periods. In order to model such systems, we need a time-related refinement relation through which behaviours observed at different components can be brought down to a finer time granularity. In this section, we present a notion of time refinement and a time-related closure operator that can be used to characterise the fact that a component is open to be integrated as a part of a system with heterogeneous granularities.

Definition 3.5 (Time refinement) *Let $\rho : \mathbb{N} \to \mathbb{N}$ be a monotonically increasing function that satisfies $\rho(0) = 0$.*

- *Let τ, τ' be two time sequences. We say that τ'* refines *τ through ρ, which we denote by $\tau' \preceq_\rho \tau$, iff, for every $i \in \mathbb{N}$, $\tau(i) = \tau'(\rho(i))$. We say that τ'* refines *τ, which we denote by $\tau' \preceq \tau$, iff $\tau' \preceq_\rho \tau$ for some ρ.*

- *Let $\lambda = \langle\sigma, \tau\rangle$, $\lambda' = \langle\sigma', \tau'\rangle$ be two timed traces. We say that λ' refines λ through ρ, which we denote by $\lambda' \preceq_\rho \lambda$, iff $\tau' \preceq_\rho \tau$ and, for every $i \in \mathbb{N}$ and $\rho(i)<j<\rho(i+1)$, $\sigma(i) = \sigma'(\rho(i))$ and $\sigma'(j) = \emptyset$. We also say that λ' refines λ, which we denote by $\lambda' \preceq \lambda$, iff $\lambda' \preceq_\rho \lambda$ for some ρ.*

That is, a time sequence refines another if the former interleaves time observations between any two time observations of the latter. Refinement extends to timed traces by requiring that no actions be observed in the finer trace between two consecutive times of the coarser. Fig. 3.2 presents an example of a timed trace with periodic observations every four time units in which actions a and b are executed at instant 8 and action c at instant 12. This trace is refined by the timed trace λ', in which periodic observations are made every two time units, through $\rho(i) = 2 * i$.

We are interested in sets of timed traces as models of the behaviour of systems and their components. In this context, the behaviour of a system is given by the intersection of the behaviour of its components and, hence, composition corresponds to set-theoretic intersection of sets of timed traces.

Consider again the timed traces λ and λ' represented in Fig. 3.2 and assume that λ is an admissible behaviour of a given component C. If λ' is not an admissible behaviour of C, C cannot be integrated in a system that, for instance, has a physical component sampled at rate 2. More

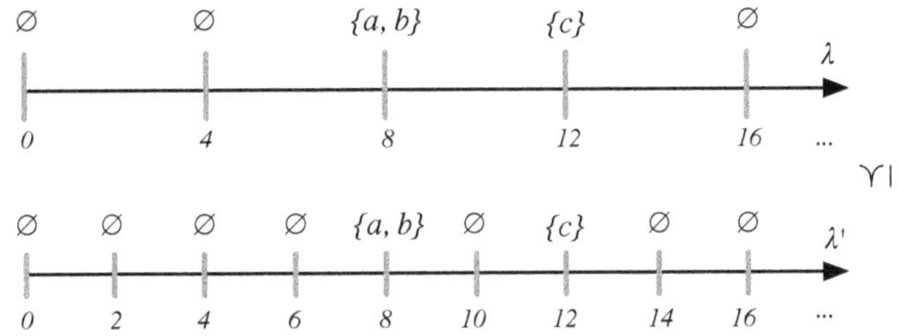

Figure 3.2: An example of time refinement: $\lambda' \preceq \lambda$.

generally, in order to be possible to integrate C in a system with heterogeneous granularities, C cannot constrain the timing of observations made by its environment, i.e., every time refinement of an admissible behaviour of C has also to be an admissible behaviour of C. This property can be characterised in terms of the following closure operator over timed traces.

Definition 3.6 (r-closure) *The r-closure of a set of timed traces Λ is the set $\Lambda^r = \{\lambda' : \exists\lambda \in \Lambda | \lambda' \preceq \lambda\}$. We say that Λ is* closed under time refinement, *or r-closed, iff $\Lambda^r \subseteq \Lambda$.*

The r-closure of a set of a timed traces adds all possible interleavings of empty observations to its traces, capturing its behaviour in any possible environment. This is, to some extent, similar to stuttering [AL91] in the sense that it ensures that components do not constrain their environment. A component whose behaviour is given by an r-closed set of timed traces is *open* to be integrated in a system with heterogeneous granularities.

Proposition 3.7 *The following properties of time refinement are useful:*

1. *If two sets of timed traces Λ_1 and Λ_2 are r-closed, so is their intersection $\Lambda_1 \cap \Lambda_2$.*

2. *Given two timed traces λ and λ', $\lambda' \preceq \lambda$ implies $\boldsymbol{f}_{\lambda'} = \boldsymbol{f}_{\lambda}$.*

It follows from the first proposition that the behaviour of a system whose components are r-closed is still r-closed. The second result means that all refinements of a given timed trace λ define the same signal as the trace λ.

3.4 Time-model Invariance

In this section, we consider MTL and its two main semantics: a discrete, point-based, semantics, and a continuous one [HOW13]. We start by presenting two functions that connect the continuous and pointwise time domains. Then, we show how, through these functions, we can identify a fragment of MTL whose formulas have the same meaning under both semantics. We also show that the formulas of this sub-logic of MTL specify r-closed sets of timed traces.

The formulas of MTL are built from a set of atomic propositions A using Boolean connectives and time-constrained versions of the until operator of the form $\mathcal{U}_I$ where $I \subseteq [0, \infty)$ is an interval with endpoints in $\mathbb{Q}_{\geq 0} \cup \{\infty\}$:

$$\phi ::= a \mid \neg\phi \mid \phi \supset \phi \mid \phi\, \mathcal{U}_I\, \phi$$

with $a \in A$.

The pointwise semantics of MTL evaluates sentences over the indexes of timed traces as follows.

Definition 3.8 (Pointwise semantics)
For every timed trace $\lambda = \langle \sigma, \tau \rangle$ over A and $i \in \mathbb{N}$:

- $\lambda, i \models_{pw} a$ *iff* $a \in \sigma(\tau(i))$
- $\lambda, i \models_{pw} \neg\phi$ *iff* $\lambda, i \not\models_{pw} \phi$
- $\lambda, i \models_{pw} \phi_1 \supset \phi_2$ *iff if* $\lambda, i \models_{pw} \phi_1$ *then* $\lambda, i \models_{pw} \phi_2$
- $\lambda, i \models_{pw} \phi_1\, \mathcal{U}_I\, \phi_2$ *iff there exists* $k \geq i$ *such that* $(\tau(k) - \tau(i)) \in I$, $\lambda, k \models_{pw} \phi_2$ *and, for all* $i \leq j < k$, $\lambda, j \models_{pw} \phi_1$

We say that λ satisfies ϕ in the pointwise semantics, denoted by $\lambda \models_{pw} \phi$, iff $\lambda, 0 \models_{pw} \phi$. We use $[\![\phi]\!]^{pw}$ to denote $\{\lambda \in \Lambda(A) : \lambda \models_{pw} \phi\}$.

Notice that linear temporal logic (LTL) is captured by the fragment of MTL whose formulas are built exclusively with $I = [0, \infty)$. For simplicity, in this case we omit the interval and write $\phi_1\, \mathcal{U}\, \phi_2$. Notice also that this version of MTL is, according to the terminology used in [FR07], non-strict on the left argument and non-matching: $\phi_1\, \mathcal{U}\, \phi_2$ does not constrain ϕ_2 to be true strictly in the future (i.e, excluding the present), and does not require ϕ_1 to hold when ϕ_2 becomes true.

The continuous semantics of MTL is usually defined in terms of signals as follows.

Definition 3.9 (Continuous semantics over signals)
For every signal f over A and $t \in \mathbb{R}_{\geq 0}$:

- $f, t \models a$ *iff* $a \in f(t)$
- $f, t \models \neg\phi$ *iff* $f, t \not\models \phi$
- $f, t \models \phi_1 \supset \phi_2$ *iff if* $f, t \models \phi_1$ *then* $f, t \models \phi_2$
- $f, t \models \phi_1 \; \mathcal{U}_I \; \phi_2$ *iff there exists* $u \geq t$ *s.t.* $(u - t) \in I$, $f, u \models \phi_2$ *and, for all* $t \leq r < u$, $f, r \models \phi_1$

We say that f *satisfies* ϕ *in the continuous semantics, denoted by* $f \models \phi$, *iff* $f, 0 \models \phi$.

In order to compare the meaning of MTL sentences over the pointwise and continuous semantics, it is more convenient to consider instead the continuous interpretation over timed traces proposed in [SP07] and defined as follows.

Definition 3.10 (Continuous semantics over timed traces) *For every timed trace* $\lambda = \langle \sigma, \tau \rangle$ *and* $t \in \mathbb{R}_{\geq 0}$:

- $\lambda, t \models_c a$ *iff exists* $i \in \mathbb{N}$ *such that* $\tau(i) = t$ *and* $a \in \sigma(i)$
- $\lambda, t \models_c \neg\phi$ *iff* $\lambda, t \not\models_c \phi$
- $\lambda, t \models_c \phi_1 \supset \phi_2$ *iff if* $\lambda, t \models_c \phi_1$ *then* $\lambda, t \models_c \phi_2$
- $\lambda, t \models_c \phi_1 \; \mathcal{U}_I \; \phi_2$ *iff there exists* $u \geq t$ *such that* $(u - t) \in I$, $\lambda, u \models_c \phi_2$ *and, for all* $t \leq r < u$, $\lambda, r \models_c \phi_1$

We say that λ *satisfies* ϕ *in the continuous semantics, denoted by* $\lambda \models_c \phi$, *iff* $\lambda, 0 \models_c \phi$. *We use* $[\![\phi]\!]^c$ *to denote* $\{\lambda : \lambda \models_c \phi\}$.

An important observation is that, because all refinements of a given timed trace define the same signal (c.f. Prop. 3.7), the continuous interpretation of a formula is the same over any of the refinements of a timed trace.

Proposition 3.11 *Given timed traces* λ *and* λ' *such that* $\lambda' \preceq \lambda$, $\lambda \models_c \phi$ *iff* $\lambda' \models_c \phi$. *It follows that, for every set of formulas* Φ, $[\![\Phi]\!]^c$ *is r-closed.*

The two continuous semantics of MTL are equivalent in the following sense:

Proposition 3.12 *For every timed trace* λ *and* $t \in \mathbb{R}_{\geq 0}$, $\lambda, t \models_c \phi$ *iff* $\boldsymbol{f}_\lambda, t \models \phi$. *Moreover,* $\{f \in \tilde{F}(A) : f \models \phi\} = \{\boldsymbol{f}_\lambda : \lambda \in \Lambda(A)$ *and* $\lambda \models_c \phi\}$.

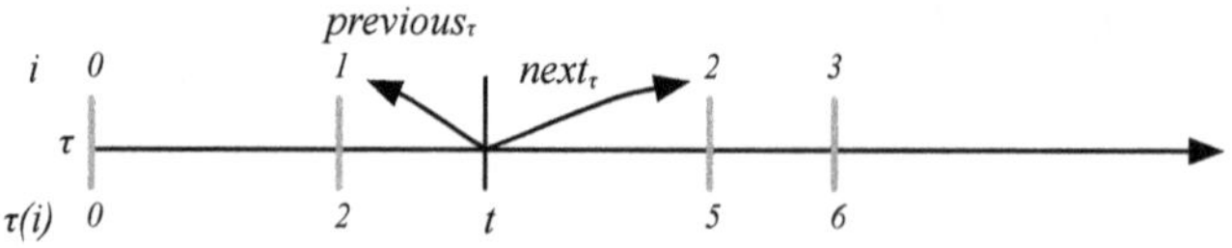

Figure 3.3: Next and previous functions.

That is to say, the two semantics are equivalent when we consider formulas expressed over a vocabulary that represents the set of all possible actions. Recall that, in this case, as discussed in Sec. 3.2, admissible signals have only point discontinuities and take the value $\emptyset$ everywhere else. Given that we now have a continuous and pointwise semantics of MTL defined over a common set of models (timed traces), our problem can be reduced to identifying the formulas ϕ for which $[\![\phi]\!]^c = [\![\phi]\!]^{pw}$.

It is easy to confirm that this property holds for some formulas but not for others. For example, in both semantics, a trace $\lambda = \langle\sigma, \tau\rangle$ satisfies the formula $(true\ \mathcal{U}\ a)$ if there exists $i \in \mathbb{N}$ such that $a \in \sigma(\tau(i))$. If we consider instead the formula $\neg(true\ \mathcal{U}\ \neg a)$, we have that no timed trace satisfies this formula in the continuous semantics whereas, in the pointwise semantics, it is satisfied by every timed trace λ such that for all $i \in \mathbb{N}$, $a \in \sigma(\tau(i))$. Proceeding in a systematic way, we are able to identify a non-trivial fragment of MTL with the same continuous and pointwise semantics.

We start by observing that the following two functions can be used for connecting the continuous and the pointwise time domains:

Definition 3.13 (next, previous) *Given a time sequence τ, we define the functions $next_\tau : \mathbb{R}_{\geq 0} \to \mathbb{N}$ and $previous_\tau : \mathbb{R}_{\geq 0} \to \mathbb{N}$ as follows:*

$$next_\tau(t) = \min\{i \in \mathbb{N} : \tau(i) \geq t\} \quad previous_\tau(t) = \max\{i \in \mathbb{N} : t \geq \tau(i)\}$$

For simplicity, given a timed trace $\lambda = \langle\sigma, \tau\rangle$, we use $next_\lambda$ and $previous_\lambda$ to mean $next_\tau$ and $previous_\tau$, respectively.

These functions, as shown in Fig. 3.3, map time instances to indexes of a given timed trace. It is easy to see that both functions map the time instants $\tau(i)$ into i. We use these functions to identify formulas of MTL that, in the continuous semantics, are satisfied by all the timed traces that satisfy them under the pointwise semantics.

We proceed as follows. We start by considering formulas written in terms of *true*, *false*, $\wedge$, $\vee$ and $\mathcal{R}_I$. The pointwise and continuous semantics of these formulas is derived from Defs. 3.8 and 3.10, respectively, considering the usual abbreviations (e.g., $\phi_1\mathcal{R}_I\phi_2 \triangleq \neg(\neg\phi_1\mathcal{U}_I\neg\phi_2)$).

Because some common real-time properties are easily expressible in terms of the strict and non-matching variant of the until and the release operators, we also consider formulas written in terms of these operators. As shown in [FR07], the formulas written in terms of the strict non-matching until are expressible in terms of the non-strict variant. However, since the encodings presented therein are far too complex to be used in practice, it is convenient to find directly sufficient conditions for formulas written with these variants of until and release operators to be time-invariant.

The continuous and pointwise semantics of the non-strict non-matching until, which we represent by $\tilde{\mathcal{U}}_I$, is as follows:

- $\lambda, t \models_c \phi_1 \, \tilde{\mathcal{U}}_I \, \phi_2$ iff there exists $u \geq t$ such that $(u - t) \in I$, $\lambda, u \models_c \phi_2$ and, for all $t < r < u$, $\lambda, r \models_c \phi_1$

- $\lambda, i \models_{pw} \phi_1 \, \tilde{\mathcal{U}}_I \, \phi_2$ iff there exists $k \geq i$ such that $(\tau(k) - \tau(i)) \in I$, $\lambda, k \models_{pw} \phi_2$ and, for all $i < j < k$, $\lambda, j \models_{pw} \phi_1$

The semantics of the non-strict non-matching relase operator is derived considering the abbreviation $\phi_1 \tilde{\mathcal{R}}_I \phi_2 \triangleq \neg(\neg\phi_1 \tilde{\mathcal{U}}_I \neg\phi_2)$.

We now look for formulas ϕ° whose pointwise satisfaction in a point i of a trace is preserved when we move to the continuous semantics and evaluate it at the instant $\tau(i)$; and we look for formulas $\phi^\bullet$ with the inverse property.

By examining the structure of formulas we identify, for each case, additional properties that, once required of its sub-formulas, ensure that the whole formula has the desired property. Then, for identifying formulas that meet each of these properties, we proceed in the same way.

More concretely, in the case of formulas ϕ°, we can identify three important properties defining three additional classes of formulas: formulas ϕ^- (resp., ϕ^+) whose pointwise satisfaction in a point $previous_\tau(t)$ (resp., $next_\tau(t)$) ensures the continuous satisfaction at instant t, and formulas $\phi^\diamond$ that are satisfied when evaluated at instants not represented in the timed trace.

Definition 3.14 (MTL^{c}_{pw}) *The language* MTL^{c}_{pw} *is defined as follows:*

$$
\begin{aligned}
\phi^{\circ} &::= \mathit{true} \mid \mathit{false} \mid a \mid \neg a \mid \phi^{\circ} \wedge \phi^{\circ} \mid \phi^{\circ} \vee \phi^{\circ} \\
&\quad \mid \phi^{\circ}\ \hat{\mathcal{R}}_I\ \phi^{\diamond} \mid \phi^{\circ}\ \hat{\mathcal{R}}_{I^<}\ \phi^{-} \mid \phi^{\circ}\ \hat{\mathcal{R}}_{I^+}\ \phi^{+} \mid \phi^{-}\ \mathcal{U}_I\ \phi^{\circ} \mid \phi^{\diamond}\ \tilde{\mathcal{U}}_I\ \phi^{\circ} \\
\phi^{-} &::= \mathit{true} \mid \mathit{false} \mid \neg a \mid \phi^{-} \wedge \phi^{-} \mid \phi^{-} \vee \phi^{-} \mid \phi^{-} \vee \phi^{\diamond} \mid \phi^{\diamond} \vee \phi^{-} \\
&\quad \mid \phi^{\circ}\ \mathcal{R}\ \phi^{-} \mid \phi^{-}\ \mathcal{U}_{I^-}\ \phi^{-} \mid \phi^{-}\ \mathcal{U}_{I_0^-}\ \phi^{\circ} \mid \phi^{\diamond}\ \tilde{\mathcal{U}}_{I_0^-}\ \phi^{\circ} \\
\phi^{+} &::= \mathit{true} \mid \mathit{false} \mid \neg a \mid \phi^{+} \wedge \phi^{+} \mid \phi^{+} \vee \phi^{+} \mid \phi^{+} \vee \phi^{\diamond} \mid \phi^{\diamond} \vee \phi^{+} \\
&\quad \mid \phi^{\circ}\ \hat{\mathcal{R}}\ \phi^{+} \mid \phi^{\diamond}\ \mathcal{U}_{I^+}\ \phi^{\circ} \\
\phi^{\diamond} &::= \mathit{true} \mid \neg a \mid \phi^{\diamond} \wedge \phi^{\diamond} \mid \phi^{\diamond} \vee \phi^{\circ} \mid \phi^{\circ} \vee \phi^{\diamond}
\end{aligned}
$$

where $\hat{\mathcal{R}}_I$ *represents both* $\mathcal{R}_I$ *and* $\tilde{\mathcal{R}}_I$, I^- *is of the form* $[0,t)$ *or* $[0,t]$, $I^<$ *is of the form* I^- *or* $[0,\infty)$, I_0^- *is of the form* $(0,t)$ *or* $(0,t]$ *or* $(0,\infty)$, I^+ *is of the form* $[t,\infty)$ *or* (t,∞) *and* I *is any interval.*

These classes of formulas satisfy the following properties:

Lemma 3.15 *Let* $\lambda = \langle \sigma, \tau \rangle$ *be a timed trace,* $i \in \mathbb{N}$ *and* $t \in \mathbb{R}_{\geq 0}$.

1. *If* $t \notin \tau(\mathbb{N})$ *then* $\lambda, t \vDash_c \phi^{\diamond}$
2. *If* $\lambda, next_{\lambda}(t) \vDash_{pw} \phi^{+}$ *then* $\lambda, t \vDash_c \phi^{+}$.
3. *If* $\lambda, previous_{\lambda}(t) \vDash_{pw} \phi^{-}$ *then* $\lambda, t \vDash_c \phi^{-}$.
4. *If* $\lambda, i \vDash_{pw} \phi^{\circ}$ *then* $\lambda, \tau(i) \vDash_c \phi^{\circ}$.

Proposition 3.16 *If* $\lambda \vDash_{pw} \phi^{\circ}$ *then* $\lambda \vDash_c \phi^{\circ}$.

That is, formulas of the form ϕ° are satisfied in the continuous semantics if they are satisfied in the pointwise semantics.

The other fragment of MTL that is of interest consists of the formulas $\phi^{\bullet}$ that in the pointwise semantics are satisfied by all timed traces that satisfy them under the continuous semantics. In this case, we identify two important properties defining two additional classes of formulas: formulas $\phi^{\star}$ that are only true at instants of the form $\tau(i)$ and formulas $\phi^{\bullet\star}$ that, in addition, have their satisfaction preserved when they are evaluated at point i.

Definition 3.17 (MTL^{pw}_{c}) MTL^{pw}_{c} *has the following language:*

$$
\begin{aligned}
\phi^{\bullet} &::= \mathit{true} \mid \mathit{false} \mid a \mid \neg a \mid \phi^{\bullet} \wedge \phi^{\bullet} \mid \phi^{\bullet} \vee \phi^{\bullet} \mid \phi^{\bullet\star}\ \hat{\mathcal{R}}_I\ \phi^{\bullet} \mid \phi^{\bullet}\ \hat{\mathcal{U}}_I\ \phi^{\bullet\star} \\
\phi^{\star} &::= \mathit{false} \mid a \mid \phi^{\star} \wedge \phi \mid \phi \wedge \phi^{\star} \mid \phi^{\star} \vee \phi^{\star} \mid \phi\ \hat{\mathcal{R}}_{I^<}\ \phi^{\star} \mid \phi^{\star}\ \mathcal{U}_I\ \phi^{\star} \mid \phi^{\star}\ \mathcal{U}_{I_0}\ \phi \\
\phi^{\bullet\star} &::= \mathit{false} \mid a \mid \phi^{\bullet} \wedge \phi^{\bullet\star} \mid \phi^{\bullet\star} \wedge \phi^{\bullet} \mid \phi^{\bullet\star} \vee \phi^{\bullet\star} \mid \phi^{\bullet\star}\ \hat{\mathcal{R}}_{I^<}\ \phi^{\bullet\star} \mid \phi^{\bullet\star}\ \mathcal{U}_I\ \phi^{\bullet\star}
\end{aligned}
$$

where $\hat{\mathcal{U}}_I$ represents both $\mathcal{U}_I$ and $\tilde{\mathcal{U}}_I$, $\hat{\mathcal{R}}_I$ represents both $\mathcal{R}_I$ and $\tilde{\mathcal{R}}_I$, ϕ is any formula of MTL, *$I^<$ is of the form $[0,t)$ or $[0,t]$ or $[0,\infty)$, I_0 is any interval not including 0 and I is any interval.*

Lemma 3.18 *Let $\lambda = \langle \sigma, \tau \rangle$ be a timed trace, $i \in \mathbb{N}$ and $t \in \mathbb{R}_{\geq 0}$.*

1. *If $\lambda, t \vDash_c \phi^\star$ then there exists i such that $\tau(i) = t$.*
2. *If $\lambda, t \vDash_c \phi^{\bullet\star}$ then there exists i such that $\tau(i) = t$ and $\lambda, i \vDash_{pw} \phi^{\bullet\star}$.*
3. *If $\lambda, \tau(i) \vDash_c \phi^\bullet$ then $\lambda, i \vDash_{pw} \phi^\bullet$.*

Proposition 3.19 *If $\lambda \vDash_c \phi^\bullet$ then $\lambda \vDash_{pw} \phi^\bullet$.*

That is, formulas of the form $\phi^\bullet$ are satisfied in the pointwise semantics if they are satisfied in the continuous semantics (i.e., $\bullet$ is the dual of $\circ$).

Therefore, formulas that are both of the form ϕ° (in the language of MTL^c_{pw}) and $\phi^\bullet$ (in the language of MTL^{pw}_c) have the same continuous and pointwise semantics. We refer to the language of these formulas as HMTL, the syntax of which can be found in the Appendix.

Theorem 3.20 *Let Φ be a set of sentences and ϕ a sentence of* HMTL. *We have that $\lambda \vDash_c \phi$ iff $\lambda \vDash_{pw} \phi$ and, hence, $[\![\Phi]\!] \triangleq [\![\Phi]\!]^{pw} (= [\![\Phi]\!]^c)$ is r-closed.*

This means that HMTL is appropriate for describing the behaviour of hybrid, real-time systems. On the one hand, we can put together formulas describing properties of physical components that are meant to be interpreted in a continuous time model and formulas describing properties of software components that are meant to be interpreted in a discrete time domain. On the other hand, properties expressed in HMTL are guaranteed not to constrain the timing of environment observations. Because the behaviour of any component described in HMTL is r-closed, components specified in HMTL are necessarily open to be integrated in a system with heterogeneous granularities.

In terms of expressivity, we cannot claim that HMTL is complete, i.e., that it includes all MTL formulas in negation normal form that have the same continuous and pointwise semantics. From a practical point of view, the ability of HMTL to express relevant properties of hybrid systems is in fact more important than completeness. Although a study of the expressiveness of HMTL with respect to the properties of hybrid systems is out of the scope of this paper, we can confirm that HMTL allows us to express several types of properties that are recurrently used for the description of the behaviour of real-time systems.

Some examples of properties presented in different sources, namely [Koy90, OW08, FL14], are presented below. We use $\Diamond_I \phi$ and $\Box_I \phi$ to abbreviate ($true\ \mathcal{U}_I\ \phi$) and ($false\ \mathcal{R}_I\ \phi$), respectively.

- $\Box(\neg a \vee \Diamond_{(0,3)} b)$ – Every execution of a is followed by an execution of b within three time units (a bounded response time requirement).
- $\Box(\neg a \vee (\neg a\ \tilde{\mathcal{U}}_{(0,3)} b))$ – Every time a is executed it became disable until b is executed and that will happen within 3 time units (a bounded response time requirement with an additional constraint).
- $\Box(\neg a \vee \Diamond_{\{7\}} b)$ – Every execution of a is followed by an execution of b in exactly seven time units (e.g., setting of a timer and its timeout).
- $\Box(\neg a \vee (\Diamond_{(0,10)} c \vee \Diamond_{\{10\}} b))$ – Every execution of a is followed by an execution of b in exactly ten time units unless c is executed first.
- $\Box(\neg a \vee (\Box_{(0,5)} \neg a \wedge \Diamond_{[5,\infty)} a))$ – If a is ever executed then it is executed infinitely often and its consecutive executions are at least 5 time units apart (e.g., an assumption about the rate of input from the environment).
- $\Box(\neg a \vee (\Box_{(0,4)} \neg a \wedge \Diamond_{\{4\}} a))$ – After the first execution of a, it is executed regularly with a period of 4 time units.
- $a\ \mathcal{R}(\neg a \vee \Diamond_{(0,4)} (b \vee c))$ – Either b or c is executed within four time units from the first execution of a.
- $\Box(\neg a \vee (b\ \mathcal{R}_{[0,5]} (\neg b \vee \Diamond_{[0,3]} c)))$ – The first execution of b within 5 time units of every execution of a is followed by an execution of c within three time units.
- $\Box(\neg p\!\uparrow \vee\ (\neg p\!\downarrow \mathcal{U}\ q\!\uparrow) \vee\ \Box \neg p\!\downarrow)$ — p does not change to false until q became true (which may never happen).

All these formulas are both in the language of MTL^{c}_{pw} and MTL^{pw}_{c}(see the Appendix for proofs).

3.5 Related Work

The formal description and analysis of real-time properties of hybrid systems has become an increasingly important topic, which been addressed considering different formalisms [AHL+00, Pla08, LPFL13]. In particular, a wide variety of logics able to express quantitative time requirements have been developed (e.g. [Koy90, AH91, MMG92, AH94, Pla08]), taking very different perspectives [DN00]. Among these, the discrete and continuous versions of MTL are the most popular logics. The fact that the version of MTL most amenable to automated analysis is also considered the less natural led Reynolds in [Rey13] to propose the new metric temporal logic 1CMTL. All the examples used by Reynolds to illustrate why he considers the discrete version of MTL less natural do not have the same meaning under the continuous and discrete semantics and, hence, are not expressible in HMTL. For instance, the formula $\Diamond_{(0,5)}\Diamond_{(0,5)}a$ is considered to have un-intuitive meaning under discrete semantics because it is not equivalent to $\Diamond_{(0,10)}a$: the formula $\Diamond_{(0,5)}\Diamond_{(0,5)}a$ is not true even if a is executed within ten time units, if no observation has been made in between. Reynolds considers 1CMTL particularly suitable for hybrid systems because it allows for imprecision in metric constraints, i.e., metric constraints can be defined to any desired precision.

In the context of logic-based approaches, the problem of describing, in a uniform manner, the behaviour of systems that encompass components with different time models has been addressed directly or indirectly by several authors. In [HMP92], a sufficient criterion for the real-time verification problem to be invariant is investigated under two different time models, none of which is truly continuous: one is a real-time model capturing the use of analog clocks and modelled by timed traces, and the other is an integral-time model capturing the use of digital clocks and represented by timed traces with only integer timestamps. This work led to the notion of *digitization* of timed traces and digitizable systems. These systems, whose behaviour is a set of timed traces closed under digitization, were proved to have the same integral-time and digitized integral-time semantics. Two classes of MTL formulas were shown to be preserved under digitization: formulas that express bounded invariance, such as $\Box(p \supset \Box_{[0,3)}\neg q)$, and formulas that express bounded response, such as $\Box(p \supset \Diamond_{[0,5]}q)$. Many of these formulas, such as $\Box(p \supset \Box_{[0,3)}\neg q)$ and $\Box(p \supset \Diamond_{[0,5]}q)$, are also in HMTL.

Furia and Rossi addressed in [FR06] the problem of integrating specifications described using different time models in mind, so that they have a global semantics and can be analyzed. Their approach is based on the notion of *sampling* of continuous-time behaviours, which involves making observations at some periodic rate, obtaining discrete-time behaviours. They established that properties in the TRIO metric temporal logic [GMM90] are *sampling invariant* for any sampling period, i.e., have a consistent truth value when moving from a continuous to a discrete-time semantics and vice-versa. However, this requires that continuous behaviour be restricted by making use of complex constraints and that time bounds in formulas be rewritten. The results presented therein do not seem to be transposable to MTL, mainly because the complex constraints imposed on signals over sets of actions exclude most signals.

Another closely related approach is the translation of MTL formulas presented in [SP07], which was developed to show that MTL with the continuous semantics over finite timed words is strictly more expressive than with the pointwise semantics. This translation converts properties over a discrete time model into equivalent properties under a continuous time model. However, this translation does not extend to timed traces, which are strictly more expressive than timed words (they allow the simultaneous execution of two or more actions to be modelled). This happens because there is no MTL formula that characterises the time instants in a timed trace in which observations are made (recall that $[\![\phi]\!]^c$ is r-closed). Other works that have also analysed the expressiveness of different variants and fragments of MTL include [AH93, OW07, FR07, SP07, HOW13].

3.6 Concluding Remarks

In this paper we addressed a problem that arises in the specification of real-time properties of systems encompassing components that are best described considering a continuous time model and others that are more naturally described considering a discrete time model: reasoning about such systems requires that properties expressed in two real-time logics with different time models be handled together. To address this issue, we presented in this paper a sub-language of MTL that allows us to specify real-time properties of the different components without have to be concerned with which time model they are meant to be interpreted in.

For example, consider the scenario depicted in Fig. 3.4: a hybrid system with a component specified by a set of formulas Φ_1 meant to be interpreted in terms of signals and another component specified by a set of formulas Φ_2 meant to be interpreted in terms of timed traces. If Φ_1 and Φ_2 are in HMTL, then we can deduce the properties of the system by reasoning about $\Phi_1 \cup \Phi_2$ in MTL with the pointwise semantics or the continuous semantics. On the one hand, we know that all properties deduced from $\Phi_1 \cup \Phi_2$ considering the continuous semantics over signals are also deducible considering instead the pointwise semantics over timed traces. On the other hand, all properties deduced from $\Phi_1 \cup \Phi_2$ considering the pointwise semantics over timed traces are also deducible in the continuous semantics if we consider only signals in $\tilde{F}(A)$(i.e., the signals

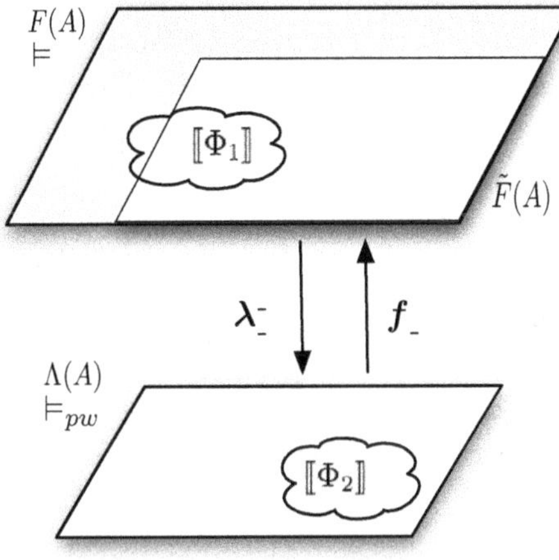

Figure 3.4: Application scenario.

that are admissible when the vocabulary over which they are defined represents a set of actions). Given that MTL with discrete semantics is more amenable to automated analysis, it will be more convenient to use $\models_{pw}$ for reasoning about the system.

Another motivation for the development of this fragment of MTL concerns heterogeneous timed systems, i.e., systems in which different components may compute over platforms with different clock granularities. The fact that the formulas of HMTL are closed in relation to a timed notion of refinement shows that HMTL is suitable for the specification of components that can be integrated in a system with heterogeneous time granularities. More on this aspect can be found in [FL14].

Bibliography

[AFH96] Rajeev Alur, Tomás Feder, and Thomas A. Henzinger. The benefits of relaxing punctuality. *J. ACM*, 43(1):116–146, 1996.

[AH91] Rajeev Alur and Thomas A. Henzinger. Logics and models of real time: A survey. In *REX Workshop*, volume 600 of *LNCS*, pages 74–106. Springer, 1991.

[AH93] Rajeev Alur and Thomas A. Henzinger. Real-time logics: Complexity and expressiveness. *Inf. Comput.*, 104(1):35–77, 1993.

[AH94] Rajeev Alur and Thomas A. Henzinger. A really temporal logic. *J. ACM*, 41(1):181–204, 1994.

[AHL+00] Rajeev Alur, Tom Henzinger, Gerardo Lafferriere, George, and George J. Pappas. Discrete abstractions of hybrid systems. In *Proceedings of the IEEE*, pages 971–984, 2000.

[AL91] Martín Abadi and Leslie Lamport. The existence of refinement mappings. *Theor. Comput. Sci.*, 82(2):253–284, 1991.

[AS87] Bowen Alpern and Fred B. Schneider. Recognizing safety and liveness. *Distributed Computing*, 2(3):117–126, 1987.

[DN00] Jennifer Daroven and Anil Nerode. Logics for hybrid systems. *Proceedings of the IEEE*, 88(7):985–1010, 2000.

[FL14] José Luiz Fiadeiro and Antónia Lopes. Heterogeneous and asynchronous networks of timed systems. In Stefania Gnesi and Arend Rensink, editors, *FASE*, volume 8411 of *LNCS*, pages 79–93. Springer, 2014.

[FR06] Carlo A. Furia and Matteo Rossi. Integrating discrete- and continuous-time metric temporal logics through sampling. In Eugene Asarin and Patricia Bouyer, editors, *FORMATS*, volume 4202 of *LNCS*, pages 215–229. Springer, 2006.

[FR07] Carlo A. Furia and Matteo Rossi. On the expressiveness of MTL variants over dense time. In Jean-François Raskin and P. S. Thiagarajan, editors, *FORMATS*, volume 4763 of *LNCS*, pages 163–178. Springer, 2007.

[GMM90] C. Ghezzi, D. Mandrioli, and A. Morzenti. Trio: A logic language for executable specifications of real-time systems. *J. Syst. Softw.*, 12(2):107–123, May 1990.

[HMP92] Thomas A. Henzinger, Zohar Manna, and Amir Pnueli. What good are digital clocks? In Werner Kuich, editor, *ICALP'92*, volume 623 of *LNCS*, pages 545–558. Springer, 1992.

[HOW13] Paul Hunter, Joël Ouaknine, and James Worrell. Expressive completeness for metric temporal logic. In *LICS*, pages 349–357. IEEE Computer Society, 2013.

[HR04] Yoram Hirshfeld and Alexander Rabinovich. Logics for real time: Decidability and complexity. *Fundam. Inf.*, 62(1):1–28, January 2004.

[Koy90] Ron Koymans. Specifying real-time properties with metric temporal logic. *Real-Time Systems*, 2(4):255–299, 1990.

[LPFL13] Thi Thieu Hoa Le, Roberto Passerone, Uli Fahrenberg, and Axel Legay. Tag machines for modeling heterogeneous systems. In *ACSD*, pages 186–195. IEEE Computer Society, 2013.

[MMG92] Angelo Morzenti, Dino Mandrioli, and Carlo Ghezzi. A model parametric real-time logic. *ACM Trans. Program. Lang. Syst.*, 14(4):521–573, October 1992.

[OW07] Joël Ouaknine and James Worrell. On the decidability and complexity of metric temporal logic over finite words. *Logical Methods in Computer Science*, 3(1), 2007.

[OW08] Joël Ouaknine and James Worrell. Some recent results in metric temporal logic. In *FORMATS*, volume 5215 of *LNCS*, pages 1–13, 2008.

[Pla08] André Platzer. Differential dynamic logic for hybrid systems. *J. Autom. Reas.*, 41(2):143–189, 2008.

[Rey13] Mark Reynolds. A new metric temporal logic for hybrid systems. In *Proceedings of the 2013 20th International Symposium on Temporal Representation and Reasoning*, TIME '13, pages 73–80, Washington, DC, USA, 2013. IEEE Computer Society.

[SP07] Deepak Souza and Pavithra Prabhakar. On the expressiveness of MTL in the pointwise and continuous semantics. *Int. J. Softw. Tools Technol. Transf.*, 9(1):1–4, February 2007.

Appendix

Proof of Prop. 3.4

1. Let $\lambda = \langle\sigma,\tau\rangle$ be a timed trace. We have that $\boldsymbol{f}_\lambda \in \tilde{F}(A)$ and $\boldsymbol{\lambda}^\tau_{(\boldsymbol{f}_\lambda)} = \lambda$.
2. Let f be a signal in $\tilde{F}(A)$ and τ a time sequence including all discontinuities of f. We have that $\boldsymbol{f}_{(\boldsymbol{\lambda}^\tau_f)} = f$.

Proof.

1. The fact that $\boldsymbol{f}_\lambda \in \tilde{F}(A)$ follows immediately from Def. 3.3. Also by Def. 3.3, we have that $\boldsymbol{\lambda}^\tau_{(\boldsymbol{f}_\lambda)} = \langle\sigma',\tau\rangle$ with $\sigma'(i) = \boldsymbol{f}_\lambda(\tau(i)) = \sigma(i)$ for every i and, hence, $\boldsymbol{\lambda}^\tau_{(\boldsymbol{f}_\lambda)} = \lambda$.
2. Let $\boldsymbol{\lambda}^\tau_f = \langle\sigma',\tau\rangle$. We have that:
   (a) For every discontinuity t of f, there is i such that $\tau(i) = t$ and hence, by Def. 3.3, $\boldsymbol{f}_{(\boldsymbol{\lambda}^\tau_f)}(t) = \sigma'(i)$. Since $\sigma'(i) = f(\tau(i))$, we have that $\boldsymbol{f}_{(\boldsymbol{\lambda}^\tau_f)}(t) = f(t)$.
   (b) For every t that is not a discontinuity of f, since $f \in \tilde{F}(A)$, $f(t) = \emptyset$. If there is i such that $\tau(i) = t$, as in the previous case, we can conclude that $\boldsymbol{f}_{(\boldsymbol{\lambda}^\tau_f)}(t) = f(t)$. Otherwise, by Def. 3.3, $\boldsymbol{f}_{(\boldsymbol{\lambda}^\tau_f)}(t) = \emptyset$ and, hence, also in this case $\boldsymbol{f}_{(\boldsymbol{\lambda}^\tau_f)}(t) = f(t)$.

□

Proof of Prop. 3.11

Given timed traces λ and λ' such that $\lambda' \preceq \lambda$, $\lambda \models_c \phi$ iff $\lambda' \models_c \phi$. It follows that, for every set of formulas Φ, $[\![\Phi]\!]^c$ is r-closed.

Proof. Let $\lambda = \langle\sigma,\tau\rangle$ and $\lambda' = \langle\sigma',\tau'\rangle$ be two timed traces. We prove by induction on the structure of the formula that, if $\lambda' \preceq_\rho \lambda$, for every $t \in \mathbb{R}_{\geq 0}$, $\lambda, t \models_c \phi$ iff $\lambda', t \models_c \phi$:

a : By definition, $\lambda, t \models_c \phi$ iff there exists $i \in \mathbb{N}$ such that $\tau(i) = t$ and $a \in \sigma(i)$ and $\lambda', t \models_c \phi$ iff there exists $j \in \mathbb{N}$ such that $\tau'(j) = t$ and $a \in \sigma'(j)$.

$\Rightarrow$ If there exists $i \in \mathbb{N}$ such that $\tau(i) = t$ and $a \in \sigma(i)$, we take $j = \rho(i)$ since $\sigma'(\rho(i)) = \sigma(i)$.

$\Leftarrow$ If there exists $j \in \mathbb{N}$ such that $\tau'(j) = t$ and $a \in \sigma'(j)$, then $\sigma'(j)$ is not empty. Because $\lambda' \preceq \lambda$, then there exists $i \in \mathbb{N}$ such that $\rho(i) = j$ and, hence, the result follows immediately.

$\neg\phi$: It follows straightforwardly from the induction hypothesis.

$\phi_1 \supset \phi_2$: It follows straightforwardly from the induction hypothesis.

$\phi_1 \; \mathcal{U}_I \; \phi_2$: It follows straightforwardly from the induction hypothesis.

□

Proof of Prop. 3.12

For every timed trace λ and $t \in \mathbb{R}_{\geq 0}$, $\lambda, t \models_c \phi$ iff $\boldsymbol{f}_\lambda, t \models \phi$. Moreover, $\{f \in \tilde{F}(A) : f \models \phi\} = \{\boldsymbol{f}_\lambda : \lambda \in \Lambda(A) \wedge \lambda \models_c \phi\}$.

Proof. Let $\lambda = \langle \sigma, \tau \rangle$ be a timed trace. By Def 3.3, $\boldsymbol{f}_\lambda(t) = \sigma(i)$ if $t = \tau(i)$ and $\boldsymbol{f}_\lambda(t) = \emptyset$ if $t \notin \tau(\mathbb{N})$. We prove by induction on the structure of the formula that, for every $t \in \mathbb{R}_{\geq 0}$, $\lambda, t \models_c \phi$ iff $\boldsymbol{f}_\lambda, t \models \phi$:

a : By definition, $\lambda, t \models_c \phi$ iff there exists $i \in \mathbb{N}$ such that $\tau(i) = t$ and $a \in \sigma(i)$ and $\boldsymbol{f}_\lambda, t \models \phi$ iff $a \in \boldsymbol{f}_\lambda(t)$ which is equivalent to $\tau(i) = t$ and $a \in \sigma(i)$.

$\neg\phi$: It follows straightforwardly from the induction hypothesis.

$\phi_1 \supset \phi_2$: It follows straightforwardly from the induction hypothesis.

$\phi_1 \; \mathcal{U}_I \; \phi_2$: It follows straightforwardly from the induction hypothesis.

It remains to prove that $\{f \in \tilde{F}(A) : f \models \phi\} = \{\boldsymbol{f}_\lambda : \lambda \in \Lambda(A) \wedge \lambda \models_c \phi\}$.
$\subseteq$ Let $f \in \tilde{F}(A)$ such that $f \models \phi$. By Prop. 3.4, we have that $\boldsymbol{f}_{(\boldsymbol{\lambda}_f^\mathcal{T})} = f$. The result follows from the fact that , as proved below, $\boldsymbol{f}_{(\boldsymbol{\lambda}_f^\mathcal{T})} \models \phi$ iff $\boldsymbol{\lambda}_f^\mathcal{T} \models_c \phi$.
$\supseteq$ Let λ such that $\lambda \models_c \phi$. As proved above, $\lambda \models_c \phi$ iff $\boldsymbol{f}_\lambda \models \phi$. Since, also by Prop. 3.4, $\boldsymbol{f}_\lambda \in \tilde{F}(A)$, we can conclude that $\boldsymbol{f}_\lambda$ is in $\{f \in \tilde{F}(A) : f \models \phi\}$. □

Proof of Lemma 3.15

Let $\lambda = \langle \sigma, \tau \rangle$ be a timed trace, $i \in \mathbb{N}$ and $t \in \mathbb{R}_{\geq 0}$.

1. If $t \notin \tau(\mathbb{N})$ then $\lambda, t \vDash_c \phi^\diamond$
2. If $\lambda, next_\lambda(t) \vDash_{pw} \phi^+$ then $\lambda, t \vDash_c \phi^+$.
3. If $\lambda, previous_\lambda(t) \vDash_{pw} \phi^-$ then $\lambda, t \vDash_c \phi^-$.
4. If $\lambda, i \vDash_{pw} \phi^\circ$ then $\lambda, \tau(i) \vDash_c \phi^\circ$.

Proof. We prove (1)-(4) by simultaneous induction on the number of connectives that occur in the formulas $\phi^\diamond, \phi^+, \phi^-$ and ϕ°.

Base case:

$\phi^\diamond$ **is** *true*: Follows trivially from the fact that, for every t, $\lambda, t \vDash_c$ *true*.

ϕ^+ **is** *true*: Similar to the previous case.

ϕ^+ **is** *false*: By Def. 3.8, $\lambda, next_\lambda(t) \nvDash_{pw}$ *false*.

ϕ^- **is** *true*: Similar to the first case.

ϕ^- **is** *false*: By Def. 3.8, $\lambda, previous_\lambda(t) \nvDash_{pw}$ *false*.

ϕ° **is** *true*: Similar to the first case.

ϕ° **is** *false*: By Def. 3.8, $\lambda, i \nvDash_{pw}$ *false*.

ϕ° **is** a: By Def. 3.8, $\lambda, i \vDash_{pw} a$ implies that $a \in \sigma(i)$ and, hence, by Def. 3.10, $\lambda, \tau(i) \vDash_c a$.

Inductive step:

$\phi^\diamond$ **is** $\neg a$: By Def. 3.10, $t \notin \tau(\mathbb{N})$ implies $\lambda, t \vDash_c \neg a$.

$\phi^\diamond$ **is** $\phi_1^\diamond \wedge \phi_2^\diamond$: If $t \notin \tau(\mathbb{N})$, then by the induction hypothesis of (1) we have that $\lambda, t \vDash_c \phi_1^\diamond$ and $\lambda, t \vDash_c \phi_2^\diamond$ and, hence, by Def. 3.10, $\lambda, t \vDash_c \phi^\diamond$.

$\phi^\diamond$ **is** $\phi_1^\diamond \vee \phi_2^\circ$: If $t \notin \tau(\mathbb{N})$, then by the induction hypothesis of (1) we have that $\lambda, t \vDash_c \phi_1^\diamond$ and, hence, by Def. 3.10, $\lambda, t \vDash_c \phi^\diamond$.

$\phi^\diamond$ **is** $\phi_1^\circ \vee \phi_2^\diamond$: Similar to the previous case.

ϕ^+ **is** $\neg a$: By Def. 3.8, $\lambda, next_\lambda(t) \vDash_{pw} \neg a$ implies that $a \notin \sigma(next_\lambda(t))$. If $\tau(next_\lambda(t)) = t$ then, by Def. 3.10, $\lambda, t \vDash_c \neg a$. If $\tau(next_\lambda(t)) \neq t$, then $t \notin \tau(\mathbb{N})$ and it follows that $\lambda, t \vDash_c \neg a$.

ϕ^+ is $\phi_1^+ \wedge \phi_2^+$: By Def. 3.8, $\lambda, next_\lambda(t) \vDash_{pw} \phi^+$ implies that $\lambda, next_\lambda(t) \vDash_{pw} \phi_1^+$ and $\lambda, next_\lambda(t) \vDash_{pw} \phi_2^+$. By the induction hypothesis of (2), we have that $\lambda, t \vDash_c \phi_1^+$ and $\lambda, t \vDash_c \phi_2^+$ and, hence, $\lambda, t \vDash_c \phi^+$.

ϕ^+ is $\phi_1^+ \vee \phi_2^+$: By Def. 3.8, $\lambda, next_\lambda(t) \vDash_{pw} \phi^+$ implies that $\lambda, next_\lambda(t) \vDash_{pw} \phi_1^+$ or $\lambda, next_\lambda(t) \vDash_{pw} \phi_2^+$. By the induction hypothesis of (2), we have that $\lambda, t \vDash_c \phi_1^+$ or $\lambda, t \vDash_c \phi_2^+$ and, hence, $\lambda, t \vDash_c \phi^+$.

ϕ^+ is $\phi_1^\circ \mathcal{R} \phi_2^+$: By Def. 3.8, $\lambda, next_\lambda(t) \vDash_{pw} \phi^+$ implies that, for all $k \geq next_\lambda(t)$, either $\lambda, k \vDash_{pw} \phi_2^+$ or there exists j such that $next_\lambda(t) \leq j < k$ and $\lambda, j \vDash_{pw} \phi_1^\circ$. Let $t' \geq t$ and $k = next_\lambda(t')$. Since $k \geq next_\lambda(t)$, we have that either $\lambda, next_\lambda(t') \vDash_{pw} \phi_2^+$ or there exists j such that $next_\lambda(t) \leq j < next_\lambda(t')$ and $\lambda, j \vDash_{pw} \phi_1^\circ$. In the first case, by the induction hypothesis of (2), we have that $\lambda, t' \vDash_c \phi_2^+$. In the second case, by the induction hypothesis of (4), we have that $\lambda, \tau(j) \vDash_c \phi_1^\circ$. Let $r = \tau(j)$. Since $next_\lambda(t) \leq j < next_\lambda(t')$ implies $t \leq \tau(j) < t'$, then we have $t \leq r < t'$ and $\lambda, r \vDash_c \phi_1^\circ$. We can then conclude that $\lambda, t \vDash_c \phi^+$.

ϕ^+ is $\phi_1^\circ \tilde{\mathcal{R}} \phi_2^+$: By Def. 3.8, $\lambda, next_\lambda(t) \vDash_{pw} \phi^+$ implies that, for all $k \geq next_\lambda(t)$, either $\lambda, k \vDash_{pw} \phi_2^+$ or there exists j such that $next_\lambda(t) < j < k$ and $\lambda, j \vDash_{pw} \phi_1^\circ$. Let $t' \geq t$ and $k = next_\lambda(t')$. Since $k \geq next_\lambda(t)$, we have that either $\lambda, next_\lambda(t') \vDash_{pw} \phi_2^+$ or there exists j such that $next_\lambda(t) < j < next_\lambda(t')$ and $\lambda, j \vDash_{pw} \phi_1^\circ$. In the first case, by the induction hypothesis of (2), we have that $\lambda, t' \vDash_c \phi_2^+$. In the second case, by the induction hypothesis of (4), we have that $\lambda, \tau(j) \vDash_c \phi_1^\circ$. Let $r = \tau(j)$. Since $next_\lambda(t) < j < next_\lambda(t')$ implies $t < \tau(j) < t'$, then we have $t < r < t'$ and $\lambda, r \vDash_c \phi_1^\circ$. We can then conclude that $\lambda, t \vDash_c \phi^+$.

ϕ^+ is $\phi_1^\diamond \mathcal{U}_{I^+} \phi_2^\circ$: By Def. 3.8, $\lambda, next_\lambda(t) \vDash_{pw} \phi^+$ implies that, there exists $k \geq next_\lambda(t)$ such that $(\tau(k) - \tau(next_\lambda(t)) \in I^+$, $\lambda, k \vDash_{pw} \phi_2^\circ$ and, for all j such that $next_\lambda(t) \leq j < k$, $\lambda, j \vDash_{pw} \phi_1^\diamond$. Let $t' = \tau(k)$. We have that $t' \geq \tau(next_\lambda(t)) \geq t$ and $(t' - t) \geq (\tau(k) - \tau(next(_\lambda(t))$. The fact that I^+ is of the form $[u, \infty)$ or (u, ∞) ensures that $(t' - t) \in I^+$. By the induction hypothesis of (4), we have that $\lambda, t' \vDash_c \phi_2^\circ$. Let r be such that $t \leq r < t'$. There are two cases: (a) $r \in \tau(\mathbb{N})$: let j such that $r = \tau(j)$. On the one hand, we have that $next_\lambda(t) \leq j < k$ and, hence, $\lambda, j \vDash_{pw} \phi_1^\diamond$. It is easy to see that all $\diamond$-formulas are also $\circ$-formulas and, hence, we can apply the induction hypothesis of (4) and conclude that $\lambda, r \vDash_c \phi_1^\diamond$. (b) $r \notin \tau(\mathbb{N})$: By the induction hypothesis of (1), we conclude that $\lambda, r \vDash_c \phi_1^\diamond$.

ϕ^- is $\neg a$: By Def. 3.8, if $\lambda, previous_\lambda(t) \vDash_{pw} \neg a$, $a \notin \sigma(previous_\lambda(t))$. If $\tau(previous_\lambda(t)) = t$, then $\lambda, t \vDash_c \neg a$. If $\tau(next_\lambda(t)) \neq t$, then $t \notin \tau(\mathbb{N})$ and it follows that $\lambda, t \vDash_c \neg a$.

ϕ^- is $\phi_1^- \wedge \phi_2^-$: By Def. 3.8, we have that $\lambda, previous_\lambda(t) \vDash_{pw} \phi^-$ implies that $\lambda, previous_\lambda(t) \vDash_{pw} \phi_1^-$ and $\lambda, previous_\lambda(t) \vDash_{pw} \phi_2^-$. By the induction hypothesis of (3), we have that $\lambda, t \vDash_c \phi_1^-$ and $\lambda, t \vDash_c \phi_2^-$ and, hence, $\lambda, t \vDash_c \phi^-$.

ϕ^- is $\phi_1^- \vee \phi_2^-$: By Def. 3.8, we have that $\lambda, previous_\lambda(t) \vDash_{pw} \phi^-$ implies that $\lambda, previous_\lambda(t) \vDash_{pw} \phi_1^-$ or $\lambda, previous_\lambda(t) \vDash_{pw} \phi_2^-$. By the induction hypothesis of (3), we have that $\lambda, t \vDash_c \phi_1^-$ or $\lambda, t \vDash_c \phi_2^-$ and, hence, $\lambda, t \vDash_c \phi^-$.

ϕ^- is $\phi_1^\circ \mathcal{R} \phi_2^-$: By Def. 3.8, $\lambda, previous_\lambda(t) \vDash_{pw} \phi^-$ implies that, for all $k \geq previous_\lambda(t)$, either $\lambda, k \vDash_{pw} \phi_2^-$ or there exists j such that $previous_\lambda(t) \leq j < k$ and $\lambda, j \vDash_{pw} \phi_1^\circ$. Let $t' \geq t$ and $k = previous_\lambda(t')$. Since $k \geq previous_\lambda(t)$, either λ, $previous_\lambda(t') \vDash_{pw} \phi_2^-$ or there exists j such that $previous_\lambda(t) \leq j < previous_\lambda(t')$ and $\lambda, j \vDash_{pw} \phi_1^\circ$. In the first case, by the induction hypothesis of (3), we have that $\lambda, t' \vDash_c \phi_2^-$. In the second case, by the induction hypothesis of (4), we have that $\lambda, \tau(j) \vDash_c \phi_1^\circ$. Let $r = \tau(j)$. Since $previous_\lambda(t) \leq j < previous_\lambda(t')$ implies $t \leq \tau(j) < t'$, then we have $t \leq r < t'$ and $\lambda, r \vDash_c \phi_1^\circ$. We can then conclude that $\lambda, t \vDash_c \phi^-$.

ϕ^- is $\phi_1^- \mathcal{U}_{I^-} \phi_2^-$: By Def. 3.8, $\lambda, previous_\lambda(t) \vDash_{pw} \phi^-$ implies that there is $k \geq previous_\lambda(t)$ such that $(\tau(k) - \tau(previous_\lambda(t))) \in I^-$, $\lambda, k \vDash_{pw} \phi_2^-$ and, for all j such that $previous_\lambda(t) \leq j < k$, $\lambda, j \vDash_{pw} \phi_1^-$. Let $t' = \tau(k)$. We have two cases: (a) $0 \in I^-$ and $k = previous_\lambda(t)$. In such case, by the induction hypothesis of (3), $\lambda, t \vDash_{pw} \phi_2^-$ and, hence, $\lambda, t \vDash_{pw} \phi^-$. (b) $0 \notin I^-$ and, the result follows as in the previous case, noticing that $-$-formulas are also $\circ$-formulas.

ϕ^- is $\phi_1^- \mathcal{U}_{I_0^-} \phi_2^\circ$: By Def. 3.8, $\lambda, previous_\lambda(t) \vDash_{pw} \phi^-$ implies that there is $k \geq previous_\lambda(t)$ such that $(\tau(k) - \tau(previous_\lambda(t))) \in I_0^-$, $\lambda, k \vDash_{pw} \phi_2^\circ$ and, for all j such that $previous_\lambda(t) \leq j < k$, $\lambda, j \vDash_{pw} \phi_1^-$. Let $t' = \tau(k)$. Since $0 \notin I_0^-$, $k > previous_\lambda(t)$ and, hence, $t' > t$. We also have that $\tau(previous_\lambda(t)) \leq t$

and, hence, $(t' - t) \leq (t' - \tau(previous_\lambda(t)))$. The fact that I_0^- is of the form $(0, u)$ or $(0, u]$ allows us to conclude that $(t' - t) \in I_0^-$. By the induction hypothesis of (4), we additionally have that $\lambda, t' \models_c \phi_2^\circ$. Let r be such that $t \leq r < t'$. We have that $previous_\lambda(t) \leq previous_\lambda(r) < k$. Hence, we know that $\lambda, previous_\lambda(r) \models_{pw} \phi_1^-$ and we can apply the induction hypothesis of (3), which ensures that $\lambda, r \models_c \phi_1^-$.

ϕ^- is $\phi_1^\diamond \tilde{\mathcal{U}}_{I_0^-} \phi_2^\circ$: By Def. 3.8, $\lambda, previous_\lambda(t) \models_{pw} \phi^-$ implies that there exists $k \geq previous_\lambda(t)$ such that $(\tau(k) - \tau(previous_\lambda(t))) \in I_0^-$, $\lambda, k \models_{pw} \phi_2^\circ$ and, for all j such that $previous_\lambda(t) < j < k$, $\lambda, j \models_{pw} \phi_1^\diamond$. Let $t' = \tau(k)$. Since $0 \notin I_0^-$, $k > previous_\lambda(t)$ and, hence, $t' > t$. We also have that $\tau(previous_\lambda(t)) \leq t$ and, hence, $(t' - t) \leq (t' - \tau(previous_\lambda(t)))$. The fact that I_0^- is of the form $(0, u)$ or $(0, u]$ allows us to conclude that $(t' - t) \in I_0^-$. By the induction hypothesis of (4), we additionally have that $\lambda, t' \models_c \phi_2^\circ$. Let r be such that $t < r < t'$. We have that $previous_\lambda(t) \leq previous_\lambda(r) < k$. We have that either $previous_\lambda(t) < previous_\lambda(r)$ or $previous_\lambda(t) = previous_\lambda(r)$. In the first case, we know that $\lambda, previous_\lambda(r) \models_{pw} \phi_1^\diamond$. Since all $\diamond$-formulas are also $\circ$-formulas, we can apply the induction hypothesis of (4), which ensures that $\lambda, r \models_c \phi_1^\diamond$. In the second case, since $t < r$ we know that $r \notin \tau(\mathbb{N})$ and, hence, we can apply the induction hypothesis of (1), which ensures that $\lambda, r \models_c \phi_1^\diamond$.

ϕ° is $\neg a$: By Def. 3.8, if $\lambda, i \models_{pw} \neg a$, we have that $a \notin \sigma(i)$. Then, by Def. 3.10, $\lambda, \tau(i) \not\models_c a$ and, hence, $\lambda, \tau(i) \models_c \neg a$.

ϕ° is $\phi_1^\circ \wedge \phi_2^\circ$: By Def. 3.8, we have that $\lambda, i \models_{pw} \phi^\circ$ implies that $\lambda, i \models_{pw} \phi_1^\circ$ and $\lambda, i \models_{pw} \phi_2^\circ$. By the induction hypothesis of (4), we have that $\lambda, \tau(i) \models_c \phi_1^\circ$ and $\lambda, \tau(i) \models_c \phi_2^\circ$ and, hence, $\lambda, \tau(i) \models_c \phi^\circ$.

ϕ° is $\phi_1^\circ \vee \phi_2^\circ$: By Def. 3.8, we have that $\lambda, i \models_{pw} \phi^\circ$ implies that $\lambda, i \models_{pw} \phi_1^\circ$ or $\lambda, i \models_{pw} \phi_2^\circ$. By the induction hypothesis of (4), we have that $\lambda, \tau(i) \models_c \phi_1^\circ$ or $\lambda, \tau(i) \models_c \phi_2^\circ$ and, hence, $\lambda, \tau(i) \models_c \phi^\circ$.

ϕ° is $\phi_1^\circ \mathcal{R}_I \phi_2^\diamond$: By Def. 3.8, $\lambda, i \models_{pw} \phi^\circ$ implies that, for all $k \geq i$ such that $(\tau(k) - \tau(i)) \in I$, either $\lambda, k \models_{pw} \phi_2^\diamond$ or there exists j such that $i \leq j < k$ and $\lambda, j \models_{pw} \phi_1^\circ$. Let $t \geq \tau(i)$ such that $(t - \tau(i)) \in I$. If $t \notin \tau(\mathbb{N})$, by the induction hypothesis of (1), $\lambda, t \models_c \phi_2^\diamond$. If $t \in \tau(\mathbb{N})$, then let k be such that $t = \tau(k)$. We have that either $\lambda, k \models_{pw} \phi_2^{\diamond}$ or there exists j such that $i \leq j < k$ and $\lambda, j \models_{pw} \phi_1^{\circ}$. Since all $\diamond$-formulas are also $\circ$-formula, in the first case we can apply the induction hypothesis of (4) to conclude that $\lambda, t \models_c \phi_2^{\diamond}$. In the second case, also by the induction hypothesis of (4), we have that $\lambda, \tau(j) \models_c \phi_1^{\circ}$. Let $r = \tau(j)$. Since $i \leq j < k$ implies $\tau(i) \leq \tau(j) < t$, then we have $\tau(i) \leq r < t$ and $\lambda, r \models_c \phi_1^{\circ}$. We can then conclude that $\lambda, \tau(i) \models_c \phi^{\circ}$.

ϕ° is $\phi_1^{\circ}\tilde{\mathcal{R}}_I\phi_2^{\diamond}$: By Def. 3.8, $\lambda, i \models_{pw} \phi^{\circ}$ implies that, for all $k \geq i$ such that $(\tau(k) - \tau(i)) \in I$, either $\lambda, k \models_{pw} \phi_2^{\diamond}$ or there exists j such that $i < j < k$ and $\lambda, j \models_{pw} \phi_1^{\circ}$. Let $t \geq \tau(i)$ such that $(t - \tau(i)) \in I$. If $t \notin \tau(\mathbb{N})$, by the induction hypothesis of (1), $\lambda, t \models_c \phi_2^{\diamond}$. If $t \in \tau(\mathbb{N})$, then let k be such that $t = \tau(k)$. We have that either $\lambda, k \models_{pw} \phi_2^{\diamond}$ or there exists j such that $i < j < k$ and $\lambda, j \models_{pw} \phi_1^{\circ}$. Since all $\diamond$-formulas are also $\circ$-formulas, in the first case we can apply the induction hypothesis of (4) to conclude that $\lambda, t \models_c \phi_2^{\diamond}$. In the second case, also by the induction hypothesis of (4), we have that $\lambda, \tau(j) \models_c \phi_1^{\circ}$. Let $r = \tau(j)$. Since $i < j < k$ implies $\tau(i) < \tau(j) < t$, then we have $\tau(i) < r < t$ and $\lambda, r \models_c \phi_1^{\circ}$. We can then conclude that $\lambda, \tau(i) \models_c \phi^{\circ}$.

ϕ° is $\phi_1^{\circ}\mathcal{R}_{I^<}\phi_2^{-}$: By Def. 3.8, $\lambda, i \models_{pw} \phi^{\circ}$ implies that, for all $k \geq i$ such that $(\tau(k) - \tau(i)) \in I^<$, either $\lambda, k \models_{pw} \phi_2^{-}$ or there exists j such that $i \leq j < k$ and $\lambda, j \models_{pw} \phi_1^{\circ}$. Let $t \geq \tau(i)$ such that $(t - \tau(i)) \in I^<$ and $k = previous_{\lambda}(t)$. It follows that $k \geq i$ and that $\tau(k) \leq t$, which implies $(\tau(k) - \tau(i)) \leq (t - \tau(i))$. Since $I^<$ is any interval starting in 0, $(\tau(k) - \tau(i)) \in I^<$. Then, we have that either $\lambda, previous_{\lambda}(t) \models_{pw} \phi_2^{-}$ or there exists j such that $i \leq j < previous_{\lambda}(t)$ and $\lambda, j \models_{pw} \phi_1^{\circ}$. In the first case, by the induction hypothesis of (3), we have that $\lambda, t \models_c \phi_2^{-}$. In the second case, by the induction hypothesis of (4), we have that $\lambda, \tau(j) \models_c \phi_1^{\circ}$. Let $r = \tau(j)$. Since $i \leq j < previous_{\lambda}(t)$ implies $\tau(i) \leq \tau(j) < t$, then we have $\tau(i) \leq r < t$ and $\lambda, r \models_c \phi_1^{\circ}$. We can then conclude that $\lambda, \tau(i) \models_c \phi^{\circ}$.

ϕ° is $\phi_1^{\circ}\tilde{\mathcal{R}}_{I^<}\phi_2^{-}$: By Def. 3.8, $\lambda, i \models_{pw} \phi^{\circ}$ implies that, for all $k \geq i$ such that $(\tau(k) - \tau(i)) \in I^<$, either $\lambda, k \models_{pw} \phi_2^{-}$ or there exists j such that $i < j < k$ and $\lambda, j \models_{pw} \phi_1^{\circ}$. Let $t \geq \tau(i)$ such that $(t - \tau(i)) \in I^<$ and $k = previous_{\lambda}(t)$. It follows that $k \geq i$ and that $\tau(k) \leq t$, which implies $(\tau(k) - \tau(i)) \leq (t - \tau(i))$. Since $I^<$ is any interval starting in 0, $(\tau(k) - \tau(i)) \in I^<$. Then, we have that either $\lambda, previous_{\lambda}(t) \models_{pw} \phi_2^{-}$ or there exists j such that $i < j < previous_\lambda(t)$ and $\lambda, j \models_{pw} \phi_1^\circ$. In the first case, by the induction hypothesis of (3), we have that $\lambda, t \models_c \phi_2^-$. In the second case, by the induction hypothesis of (4), we have that $\lambda, \tau(j) \models_c \phi_1^\circ$. Let $r = \tau(j)$. Since $i < j < previous_\lambda(t)$ implies $\tau(i) < \tau(j) < t$, then we have $\tau(i) < r < t$ and $\lambda, r \models_c \phi_1^\circ$. We can then conclude that $\lambda, \tau(i) \models_c \phi^\circ$.

ϕ° is $\phi_1^\circ \mathcal{R}_{I^+} \phi_2^+$: By Def. 3.8, $\lambda, i \models_{pw} \phi^\circ$ implies that, for all $k \geq i$ such that $(\tau(k) - \tau(i)) \in I^+$, either $\lambda, k \models_{pw} \phi_2^+$ or there exists j such that $i \leq j < k$ and $\lambda, j \models_{pw} \phi_1^\circ$. Let $t \geq \tau(i)$ such that $(t - \tau(i)) \in I^+$ and $k = next_\lambda(t)$. It follows that $k \geq i$ and $\tau(k) \geq t$, which implies $(\tau(k) - \tau(i)) \geq (t - \tau(i))$. Since I^+ is an unbounded interval, $(\tau(k) - \tau(i)) \in I^+$. Then, we have that either $\lambda, next_\lambda(t) \models_{pw} \phi_2^+$ or there exists j such that $i \leq j < next_\lambda(t)$ and $\lambda, j \models_{pw} \phi_1^\circ$. In the first case, by the induction hypothesis of (2), we have that $\lambda, t \models_c \phi_2^+$. In the second case, by the induction hypothesis of (4), we have that $\lambda, \tau(j) \models_c \phi_1^\circ$. Let $r = \tau(j)$. Since $i \leq j < next_\lambda(t)$ implies $\tau(i) \leq \tau(j) < t$, then we have $\tau(i) \leq r < t$ and $\lambda, r \models_c \phi_1^\circ$. We can then conclude that $\lambda, \tau(i) \models_c \phi^\circ$.

ϕ° is $\phi_1^\circ \tilde{\mathcal{R}}_{I^+} \phi_2^+$: By Def. 3.8, $\lambda, i \models_{pw} \phi^\circ$ implies that, for all $k \geq i$ such that $(\tau(k) - \tau(i)) \in I^+$, either $\lambda, k \models_{pw} \phi_2^+$ or there exists j such that $i < j < k$ and $\lambda, j \models_{pw} \phi_1^\circ$. Let $t \geq \tau(i)$ such that $(t - \tau(i)) \in I^+$ and $k = next_\lambda(t)$. It follows that $k \geq i$ and $\tau(k) \geq t$, which implies $(\tau(k) - \tau(i)) \geq (t - \tau(i))$. Since I^+ is an unbounded interval, $(\tau(k) - \tau(i)) \in I^+$. Then, we have that either $\lambda, next_\lambda(t) \models_{pw} \phi_2^+$ or there exists j such that $i < j < next_\lambda(t)$ and $\lambda, j \models_{pw} \phi_1^\circ$. In the first case, by the induction hypothesis of (2), we have that $\lambda, t \models_c \phi_2^+$. In the second case, by the induction hypothesis of (4), we have that $\lambda, \tau(j) \models_c \phi_1^\circ$. Let $r = \tau(j)$. Since $i < j < next_\lambda(t)$ implies $\tau(i) < \tau(j) < t$, then we have $\tau(i) < r < t$ and $\lambda, r \models_c \phi_1^\circ$. We can then conclude that $\lambda, \tau(i) \models_c \phi^\circ$.

ϕ° is $\phi_1^- \mathcal{U}_I \phi_2^\circ$: By Def. 3.8, $\lambda, i \models_{pw} \phi^\circ$ implies that there exists $k \geq i$ such that $(\tau(k) - \tau(i)) \in I$, $\lambda, k \models_{pw} \phi_2^\circ$ and, for all j such that $i \leq j < k$, $\lambda, j \models_{pw} \phi_1^-$. Let $t' = \tau(k)$. We have that $t' \geq \tau(i)$ and, by the induction hypothesis of (4), $\lambda, t' \models_c \phi_2^\circ$. Let r be such that $\tau(i) \leq r < t' = \tau(k)$. We have that $i \leq previous_\lambda(r) < k$. Hence, we can apply the induction hypothesis of (3), which ensures that $\lambda, r \models_c \phi_1^-$. We can then conclude that $\lambda, \tau(i) \models_c \phi^\circ$.

ϕ° is $\phi_1^\diamond \tilde{\mathcal{U}}_I \phi_2^\circ$: By Def. 3.8, $\lambda, i \vDash_{pw} \phi^\circ$ implies that there exists $k \geq i$ such that $(\tau(k) - \tau(i)) \in I$, $\lambda, k \vDash_{pw} \phi_2^\circ$ and, for all j such that $i < j < k$, $\lambda, j \vDash_{pw} \phi_1^\diamond$. Let $t' = \tau(k)$. We have that $t' \geq \tau(i)$ and, by the induction hypothesis of (4), $\lambda, t' \vDash_c \phi_2^\circ$. Let r be such that $\tau(i) < r < t' = \tau(k)$. We have that either $i < previous_\lambda(r) < k$ or $i = previous_\lambda(r) < k$. In the first case, we know that $\lambda, previous_\lambda(r) \vDash_{pw} \phi_1^\diamond$. Since all $\diamond$-formulas are also $\circ$-formulas, we can apply the induction hypothesis of (4), which ensures that $\lambda, r \vDash_c \phi_1^\diamond$. In the second case, since $\tau(i) < r$ we know that $r \notin \tau(\mathbb{N})$ and, hence, we can apply the induction hypothesis of (1), which ensures that $\lambda, r \vDash_c \phi_1^\diamond$.

□

Proof of Prop. 3.16

If $\lambda \vDash_{pw} \phi^\circ$ then $\lambda \vDash_c \phi^\circ$.

Proof. This follows from (4) of Lemma 3.15. By Def. 3.8, $\lambda \vDash_{pw} \phi^\circ$ iff $\lambda, 0 \vDash_{pw} \phi^\circ$. By definition of timed trace $\tau(0) = 0$ and, hence, $\lambda, 0 \vDash_c \phi^\circ$. By Def. 3.10, we then conclude $\lambda \vDash_c \phi^\circ$. □

Proof of Lemma 3.18

Let $\lambda = \langle \sigma, \tau \rangle$ be a timed trace, $i \in \mathbb{N}$ and $t \in \mathbb{R}_{\geq 0}$.

1. If $\lambda, t \vDash_c \phi^\star$ then there exists i such that $\tau(i) = t$.
2. If $\lambda, t \vDash_c \phi^{\bullet\star}$ then there exists i such that $\tau(i) = t$ and $\lambda, i \vDash_{pw} \phi^{\bullet\star}$.
3. If $\lambda, \tau(i) \vDash_c \phi^\bullet$ then $\lambda, i \vDash_{pw} \phi^\bullet$.

Proof. We prove (1)-(3) by simultaneous induction on the number of connectives that occur in the formulas $\phi^\star, \phi^{\bullet\star}$ and $\phi^\bullet$.

Base case:

$\phi^\star$ is *false*: The result follows from the fact that, by Def. 3.10, $\lambda, t \nvDash_c$ *false* for every t.

$\phi^\star$ is a: The result follows from the fact that, by Def. 3.10, $\lambda, t \vDash_c a$ implies that $t \in \tau(\mathbb{N})$.

$\phi^{\bullet\star}$ **is** *false*: Similar to the case of $\phi^\star$.

$\phi^{\bullet\star}$ **is** a: The result follows from the fact that, by Def. 3.8, $\lambda, t \models_c a$ implies that there exists i such that $\tau(i) = t$ and $a \in \sigma(i)$. Hence, by Def. 3.8, $\lambda, i \models_{pw} a$.

$\phi^\bullet$ **is** *false*: Similar to the case of $\phi^\star$.

$\phi^\bullet$ **is** *true*: The result follows from the fact that, by Def. 3.8, for every t, $\lambda, t \models_c$ *true*.

$\phi^\bullet$ **is** a: By Def. 3.10, $\lambda, \tau(i) \models_c a$ implies that $a \in \sigma(i)$ and, hence, by Def. 3.8, $\lambda, i \models_{pw} a$.

Inductive step:

$\phi^\star$ **is** $\phi_1^\star \wedge \phi_2$: By Def. 3.10, $\lambda, t \models_c \phi_1^\star$. By the induction hypothesis of (1), we have that $t \in \tau(\mathbb{N})$.

$\phi^\star$ **is** $\phi_1 \wedge \phi_2^\star$: Similar to the previous case.

$\phi^\star$ **is** $\phi_1^\star \vee \phi_2^\star$: By Def. 3.10, either $\lambda, t \models_c \phi_1^\star$ or $\lambda, t \models_c \phi_2^\star$. By the induction hypothesis of (1), $t \in \tau(\mathbb{N})$ holds in both cases.

$\phi^\star$ **is** $\phi_1 \mathcal{R}_{I^<} \phi_2^\star$: By Def. 3.10, $\lambda, t \models_c \phi^\star$ implies that, for all $u \geq t$ such that $(u - t) \in I^<$, either $\lambda, u \models_c \phi_2^\star$ or there exists r such that $t \leq r < u$ and $\lambda, r \models_c \phi_1$. Because $0 \in I^<$, this also holds for $u = t$, i.e., either $\lambda, t \models_c \phi_2^\star$ or there exists r such that $t \leq r < t$ and $\lambda, r \models_c \phi_1$. Because there is no r such that $t \leq r < t$, it follows that $\lambda, t \models_c \phi_2^\star$. By the induction hypothesis of (1), $t \in \tau(\mathbb{N})$ holds.

$\phi^\star$ **is** $\phi_1 \tilde{\mathcal{R}}_{I^<} \phi_2^\star$: By Def. 3.10, $\lambda, t \models_c \phi^\star$ implies that, for all $u \geq t$ such that $(u - t) \in I^<$, either $\lambda, u \models_c \phi_2^\star$ or there exists r such that $t < r < u$ and $\lambda, r \models_c \phi_1$. Because $0 \in I^<$, this also holds for $u = t$, i.e., either $\lambda, t \models_c \phi_2^\star$ or there exists r such that $t < r < t$ and $\lambda, r \models_c \phi_1$. Because there is no r such that $t < r < t$, it follows that $\lambda, t \models_c \phi_2^\star$. By the induction hypothesis of (1), $t \in \tau(\mathbb{N})$ holds.

$\phi^\star$ **is** $\phi_1^\star \mathcal{U}_I \phi_2^\star$: By Def. 3.10, $\lambda, t \models_c \phi^\star$ implies that there exists $u \geq t$ such that $(u - t) \in I$, $\lambda, u \models_c \phi_2^\star$ and, for all r such that $t \leq r < u$, $\lambda, r \models_c \phi_1^\star$. We have two cases: (a) $u = t$ and, hence, by induction hypothesis of (1), $t \in \tau(\mathbb{N})$ holds. (b) $u > t$ and, hence, $\lambda, t \models_c \phi_1^\star$. Again by induction hypothesis of (1), it follows that $t \in \tau(\mathbb{N})$.

$\phi^\star$ **is** $\phi_1^\star \mathcal{U}_{I_0} \phi_2$: By Def. 3.10, $\lambda, t \models_c \phi^\star$ implies that there exists $u \geq t$ such that $(u - t) \in I$, $\lambda, u \models_c \phi_2^\star$ and, for all r such that

$t \leq r < u$, $\lambda, r \vDash_c \phi_1^\star$. Since $0 \notin I_0$ we have that $u > t$ and, hence, $\lambda, t \vDash_c \phi_1^\star$. By induction hypothesis of (1), it follows that $t \in \tau(\mathbb{N})$.

$\phi^{\bullet\star}$ is $\phi_1^\bullet \wedge \phi_2^{\bullet\star}$: By Def. 3.10, $\lambda, t \vDash_c \phi^{\bullet\star}$ implies that $\lambda, t \vDash_c \phi_1^\bullet$ and $\lambda, t \vDash_c \phi_2^{\bullet\star}$. By the induction hypothesis of (2), we have that there exists i such that $t = \tau(i)$ and $\lambda, i \vDash_{pw} \phi_2^{\bullet\star}$. Then, we are also in the conditions of applying the induction hypothesis of (3), and can conclude that $\lambda, i \vDash_{pw} \phi_1^\bullet$. Hence, there exists i such that $t = \tau(i)$ and $\lambda, i \vDash_c \phi^{\bullet\star}$.

$\phi^{\bullet\star}$ is $\phi_1^{\bullet\star} \wedge \phi_2^\bullet$: Similar to the previous case.

$\phi^{\bullet\star}$ is $\phi_1^{\bullet\star} \vee \phi_2^{\bullet\star}$: By Def. 3.10, either $\lambda, t \vDash_c \phi_1^{\bullet\star}$ or $\lambda, t \vDash_c \phi_2^{\bullet\star}$. By the induction hypothesis of (2), we have that there exists i such that $t = \tau(i)$ and either $\lambda, i \vDash_{pw} \phi_1^{\bullet\star}$ or $\lambda, i \vDash_{pw} \phi_2^{\bullet\star}$. Therefore, $\lambda, i \vDash_{pw} \phi^{\bullet\star}$.

$\phi^{\bullet\star}$ is $\phi_1^{\bullet\star}\mathcal{R}_{I^<}\phi_2^{\bullet\star}$: By Def. 3.10, $\lambda, t \vDash_c \phi^{\bullet\star}$ implies that, for all $u \geq t$ such that $(u - t) \in I^<$, either $\lambda, u \vDash_c \phi_2^{\bullet\star}$ or there exists r such that $t \leq r < u$ and $\lambda, r \vDash_c \phi_1^{\bullet\star}$ (†). In particular, since $0 \in I^<$, this is also true for $u = t$. Because it does not exist a r such that $t \leq r < t$, it follows that $\lambda, t \vDash_c \phi_2^{\bullet\star}$. By the induction hypothesis of (2), we have that there exists i such that $t = \tau(i)$. Let $k \geq i$ such that $(\tau(k) - \tau(i)) \in I^<$. We are in the conditions of applying (†) to $u = \tau(k)$ and, hence, either $\lambda, \tau(k) \vDash_c \phi_2^{\bullet\star}$ or there exists r such that $\tau(i) \leq r < \tau(k)$ and $\lambda, r \vDash_c \phi_1^{\bullet\star}$. In the first case, by the induction hypothesis of (2), we have that $\lambda, k \vDash_{pw} \phi_2^{\bullet\star}$. In the second case, by the induction hypothesis of (2) we have that there exists j such that $\tau(j) = r$ and $\lambda, j \vDash_{pw} \phi_1^{\bullet\star}$. To conclude that $\lambda, i \vDash_{pw} \phi^{\bullet\star}$, it remains to ensure that $i \leq j < k$, which follows from the fact that $\tau(i) \leq r < \tau(k)$.

$\phi^{\bullet\star}$ is $\phi_1^{\bullet\star}\tilde{\mathcal{R}}_{I^<}\phi_2^{\bullet\star}$: By Def. 3.10, $\lambda, t \vDash_c \phi^{\bullet\star}$ implies that, for all $u \geq t$ such that $(u - t) \in I^<$, either $\lambda, u \vDash_c \phi_2^{\bullet\star}$ or there exists r such that $t < r < u$ and $\lambda, r \vDash_c \phi_1^{\bullet\star}$ (†). In particular, since $0 \in I^<$, this is also true for $u = t$. Because it does not exist a r such that $t < r < t$, it follows that $\lambda, t \vDash_c \phi_2^{\bullet\star}$. By the induction hypothesis of (2), we have that there exists i such that $t = \tau(i)$. Let $k \geq i$ such that $(\tau(k) - \tau(i)) \in I^<$. We are in the conditions of applying (†) to $u = \tau(k)$ and, hence, either $\lambda, \tau(k) \vDash_c \phi_2^{\bullet\star}$ or there exists r such that $\tau(i) < r < \tau(k)$ and $\lambda, r \vDash_c \phi_1^{\bullet\star}$. In the first case, by the induction hypothesis of

(2), we have that $\lambda, k \models_{pw} \phi_2^{\bullet\star}$. In the second case, by the induction hypothesis of (2) we have that there exists j such that $\tau(j) = r$ and $\lambda, j \models_{pw} \phi_1^{\bullet\star}$. To conclude that $\lambda, i \models_{pw} \phi^{\bullet\star}$, it remains to ensure that $i < j < k$, which follows from the fact that $\tau(i) < r < \tau(k)$.

$\phi^{\bullet\star}$ is $\phi_1^{\bullet\star}\mathcal{U}_I\phi_2^{\bullet\star}$: By Def. 3.10, $\lambda, t \models_c \phi^{\bullet\star}$ implies that there exists $u \geq t$ such that $(u - t) \in I$, $\lambda, u \models_c \phi_2^{\bullet\star}$ and, for all r such that $t \leq r < u$, $\lambda, r \models_c \phi_1^{\bullet\star}$ (†). We have two cases: (a) $u = t$ and, hence, by induction hypothesis of (2), there exists i such that $t = \tau(i)$ and $\lambda, i \models_{pw} \phi_2^{\bullet\star}$. In this case, $\lambda, i \models_{pw} \phi^{\bullet\star}$. (b) $u > t$ and, hence, $\lambda, t \models_c \phi_1^{\bullet\star}$. Again by induction hypothesis of (2), it follows there exists i such that $t = \tau(i)$ and $\lambda, i \models_{pw} \phi_1^{\bullet\star}$. Also we have by induction hypothesis of (2) that there exists k such that $u = \tau(k)$ and $\lambda, k \models_{pw} \phi_2^{\bullet\star}$. Since $u = \tau(k)$, $u \geq t$ and $(u - t) \in I$, we have that $k \geq i$ and $(\tau(k) - \tau(i)) \in I$. Let j be such that $i \leq j < k$. Since $\tau(i) \leq \tau(j) < \tau(k)$, we are in the conditions of applying (†) to $r = \tau(j)$ and, hence, $\lambda, \tau(j) \models_c \phi_1^{\bullet\star}$. Since all $\bullet\star$-formulas are also $\bullet$-formulas, we can use the induction hypothesis of (3), $\lambda, j \models_{pw} \phi_1^{\bullet\star}$. Hence, we can conclude that also in this case $\lambda, i \models_{pw} \phi^{\bullet\star}$.

$\phi^{\bullet}$ is $\neg a$: By Def. 3.10, $\lambda, \tau(i) \models_c \neg a$ implies that, $a \notin \sigma(i)$. By Def. 3.8, $\lambda, i \models_{pw} \neg a$.

$\phi^{\bullet}$ is $\phi_1^{\bullet} \wedge \phi_2^{\bullet}$: By Def. 3.10, $\lambda, \tau(i) \models_c \phi^{\bullet}$ implies that $\lambda, \tau(i) \models_c \phi_1^{\bullet}$ and $\lambda, \tau(i) \models_c \phi_2^{\bullet}$. By the induction hypothesis of (3), if follows that $\lambda, i \models_{pw} \phi_1^{\bullet}$ and $\lambda, i \models_{pw} \phi_2^{\bullet}$. Therefore, $\lambda, i \models_c \phi^{\bullet}$.

$\phi^{\bullet}$ is $\phi_1^{\bullet} \vee \phi_2^{\bullet}$: By Def. 3.10, either $\lambda, \tau(i) \models_c \phi_1^{\bullet}$ or $\lambda, \tau(i) \models_c \phi_2^{\bullet}$. By the induction hypothesis of (3), it follows that $\lambda, i \models_{pw} \phi_1^{\bullet}$ or $\lambda, i \models_{pw} \phi_2^{\bullet}$. Therefore, $\lambda, i \models_{pw} \phi^{\bullet}$.

$\phi^{\bullet}$ is $\phi_1^{\bullet\star}\mathcal{R}_I\phi_2^{\bullet}$: By Def. 3.10, $\lambda, \tau(i) \models_c \phi^{\bullet}$ implies that, for all $u \geq \tau(i)$ such that $(u - \tau(i)) \in I$, either $\lambda, u \models_c \phi_2^{\bullet}$ or there exists r such that $\tau(i) \leq r < u$ and $\lambda, r \models_c \phi_1^{\bullet\star}$ (†). Let $k \geq i$ such that $(\tau(k) - \tau(i)) \in I$. We are in the conditions of applying (†) to $u = \tau(k)$ and, hence, either $\lambda, \tau(k) \models_c \phi_2^{\bullet}$ or there exists r such that $\tau(i) \leq r < \tau(k)$ and $\lambda, r \models_c \phi_1^{\bullet\star}$. In the first case, by the induction hypothesis of (3), we have that $\lambda, k \models_{pw} \phi_2^{\bullet}$. In the second case, by the induction hypothesis of (2) we have that there exists j such that $\tau(j) = r$ and $\lambda, j \models_{pw} \phi_1^{\bullet\star}$. To conclude that $\lambda, i \models_{pw} \phi^{\bullet\star}$, it remains to ensure that $i \leq j < k$, which follows from the fact that $\tau(i) \leq r < \tau(k)$.

$\phi^\bullet$ is $\phi_1^{\bullet\star}\tilde{\mathcal{R}}_I\phi_2^\bullet$: By Def. 3.10, $\lambda, \tau(i) \vDash_c \phi^\bullet$ implies that, for all $u \geq \tau(i)$ such that $(u - \tau(i)){\in}I$, either $\lambda, u \vDash_c \phi_2^\bullet$ or there exists r such that $\tau(i) < r < u$ and $\lambda, r \vDash_c \phi_1^{\bullet\star}$ (†). Let $k \geq i$ such that $(\tau(k) - \tau(i)) \in I$. We are in the conditions of applying (†) to $u = \tau(k)$ and, hence, either $\lambda, \tau(k) \vDash_c \phi_2^\bullet$ or there exists r such that $\tau(i) < r < \tau(k)$ and $\lambda, r \vDash_c \phi_1^{\bullet\star}$. In the first case, by the induction hypothesis of (3), we have that $\lambda, k \vDash_{pw} \phi_2^\bullet$. In the second case, by the induction hypothesis of (2) we have that there exists j such that $\tau(j) = r$ and $\lambda, j \vDash_{pw} \phi_1^{\bullet\star}$. To conclude that $\lambda, i \vDash_{pw} \phi^{\bullet\star}$, it remains to ensure that $i < j < k$, which follows from the fact that $\tau(i) < r < \tau(k)$.

$\phi^\bullet$ is $\phi_1^\bullet\mathcal{U}_I\phi_2^{\bullet\star}$: By Def. 3.10, $\lambda, \tau(i) \vDash_c \phi^\bullet$ implies that there exists $u \geq \tau(i)$ such that $(u - \tau(i)) \in I$, $\lambda, u \vDash_c \phi_2^{\bullet\star}$ and, for all r such that $\tau(i) \leq r < u$, $\lambda, r \vDash_c \phi_1^\bullet$ (†). By the induction hypothesis of (2), there exists k such that $\tau(k) = u$ and $\lambda, k \vDash_{pw} \phi_2^{\bullet\star}$. For this k we have that $(\tau(k) - \tau(i)) \in I$ and $\lambda, k \vDash_{pw} \phi_2^{\bullet\star}$. Let j be such that $i \leq j < k$. Since $\tau(i) \leq \tau(j) < \tau(k)$, we are in the conditions of applying (†) to $r = \tau(j)$ and, hence, $\lambda, \tau(j) \vDash_c \phi_1^\bullet$. By the induction hypothesis of (3), $\lambda, j \vDash_{pw} \phi_1^\bullet$. Hence, we can conclude that $\lambda, i \vDash_{pw} \phi^\bullet$.

$\phi^\bullet$ is $\phi_1^\bullet\tilde{\mathcal{U}}_I\phi_2^{\bullet\star}$: By Def. 3.10, $\lambda, \tau(i) \vDash_c \phi^\bullet$ implies that there exists $u \geq \tau(i)$ such that $(u - \tau(i)) \in I$, $\lambda, u \vDash_c \phi_2^{\bullet\star}$ and, for all r such that $\tau(i) < r < u$, $\lambda, r \vDash_c \phi_1^\bullet$ (†). By the induction hypothesis of (2), there exists k such that $\tau(k) = u$ and $\lambda, k \vDash_{pw} \phi_2^{\bullet\star}$. For this k we have that $(\tau(k) - \tau(i)) \in I$ and $\lambda, k \vDash_{pw} \phi_2^{\bullet\star}$. Let j be such that $i < j < k$. Since $\tau(i) < \tau(j) < \tau(k)$, we are in the conditions of applying (†) to $r = \tau(j)$ and, hence, $\lambda, \tau(j) \vDash_c \phi_1^\bullet$. By the induction hypothesis of (3), $\lambda, j \vDash_{pw} \phi_1^\bullet$. Hence, we can conclude that $\lambda, i \vDash_{pw} \phi^\bullet$.

□

Proof of Prop. 3.19

If $\lambda \vDash_c \phi^\bullet$ then $\lambda \vDash_{pw} \phi^\bullet$.

Proof. This result is an immediate consequence of (3) of Lemma 3.18. By Def. 3.10, $\lambda \vDash_c \phi^\bullet$ iff $\lambda, 0 \vDash_c \phi^\bullet$. By definition of timed trace $\tau(0) = 0$ and, hence, $\lambda, 0 \vDash_{pw} \phi^\bullet$. By Def. 3.8, we then conclude $\lambda \vDash_{pw} \phi^\bullet$. □

Language of HMTL

The language HMTL can be defined as follows:

$$
\begin{aligned}
\phi^{\circ\bullet} ::= {} & \mathit{true} \mid \mathit{false} \mid a \mid \neg a \mid \phi^{\circ\bullet} \wedge \phi^{\circ\bullet} \mid \phi^{\circ\bullet} \vee \phi^{\circ\bullet} \mid \phi^{\circ\bullet\star}\, \hat{\mathcal{R}}_I\, \phi^{\diamond\bullet} \\
& \mid \phi^{\circ\bullet\star}\, \hat{\mathcal{R}}_{I^<}\, \phi^{-\bullet} \mid \phi^{\circ\bullet\star}\, \hat{\mathcal{R}}_{I^+}\, \phi^{+\bullet} \mid \phi^{-\bullet}\, \mathcal{U}_I\, \phi^{\circ\bullet\star} \mid \phi^{\diamond\bullet}\, \tilde{\mathcal{U}}_I\, \phi^{\circ\bullet\star}
\end{aligned}
$$

where

$$
\begin{aligned}
\phi^{\diamond\bullet} ::= {} & \mathit{true} \mid \neg a \mid \phi^{\diamond\bullet} \wedge \phi^{\diamond\bullet} \mid \phi^{\diamond\bullet} \vee \phi^{\circ\bullet} \mid \phi^{\circ\bullet} \vee \phi^{\diamond\bullet} \\
\phi^{-\bullet} ::= {} & \mathit{true} \mid \mathit{false} \mid \neg a \mid \phi^{-\bullet} \wedge \phi^{-\bullet} \mid \phi^{-\bullet} \vee \phi^{-\bullet} \mid \phi^{-\bullet} \vee \phi^{\diamond\bullet} \mid \phi^{\diamond\bullet} \vee \phi^{-\bullet} \\
& \mid \phi^{\circ\bullet\star}\, \mathcal{R}\, \phi^{-\bullet} \mid \phi^{-\bullet}\, \mathcal{U}_{I^-}\, \phi^{-\bullet\star} \mid \phi^{-\bullet}\, \mathcal{U}_{I_0^-}\, \phi^{\circ\bullet\star} \mid \phi^{\diamond\bullet}\, \tilde{\mathcal{U}}_{I_0^-}\, \phi^{\circ\bullet\star} \\
\phi^{+\bullet} ::= {} & \mathit{true} \mid \mathit{false} \mid \neg a \mid \phi^{+\bullet} \wedge \phi^{+\bullet} \mid \phi^{+\bullet} \vee \phi^{+\bullet} \mid \phi^{+\bullet} \vee \phi^{\diamond\bullet} \\
& \mid \phi^{\diamond\bullet} \vee \phi^{+\bullet} \mid \phi^{\circ\bullet\star}\, \hat{\mathcal{R}}\, \phi^{+\bullet} \mid \phi^{\diamond\bullet}\, \mathcal{U}_{I^+}\, \phi^{\circ\bullet\star} \\
\phi^{\circ\bullet\star} ::= {} & \mathit{false} \mid a \mid \phi^{\circ\bullet\star} \wedge \phi^{\circ\bullet} \mid \phi^{\circ\bullet} \wedge \phi^{\circ\bullet\star} \mid \phi^{\circ\bullet\star} \vee \phi^{\circ\bullet\star} \\
& \mid \phi^{\circ\bullet\star}\, \hat{\mathcal{R}}_{I^<}\, \phi^{-\bullet\star} \mid \phi^{\circ\bullet\star}\, \hat{\mathcal{R}}\, \phi^{+\bullet\star} \mid \phi^{-\bullet\star}\, \mathcal{U}_I\, \phi^{\circ\bullet\star} \\
\phi^{-\bullet\star} ::= {} & \mathit{false} \mid \phi^{-\bullet} \wedge \phi^{-\bullet\star} \mid \phi^{-\bullet\star} \wedge \phi^{-\bullet} \mid \phi^{-\bullet\star} \vee \phi^{-\bullet\star} \\
& \mid \phi^{\circ\bullet\star}\, \mathcal{R}\, \phi^{-\bullet\star} \mid \phi^{-\bullet\star}\, \mathcal{U}_{I_0^-}\, \phi^{\circ\bullet\star} \mid \phi^{-\bullet\star}\, \mathcal{U}_{I^-}\, \phi^{-\bullet\star} \\
\phi^{+\bullet\star} ::= {} & \mathit{false} \mid \phi^{+\bullet} \wedge \phi^{+\bullet\star} \mid \phi^{+\bullet\star} \wedge \phi^{+\bullet} \mid \phi^{+\bullet\star} \vee \phi^{+\bullet\star} \mid \phi^{\circ\bullet\star}\, \hat{\mathcal{R}}\, \phi^{+\bullet\star}
\end{aligned}
$$

where I^- is of the form $[0,t)$ or $[0,t]$, $I^<$ is of the form I^- or $[0,\infty)$, I_0^- is of the form $(0,t)$ or $(0,t]$, I^+ is of the form $[t,\infty)$ or (t,∞) and I is any interval.

Proof of Theorem 3.20

Let Φ be a set of sentences and ϕ a sentence of HMTL. We have that $\lambda \models_c \phi$ iff $\lambda \models_{pw} \phi$ and, hence, $[\![\Phi]\!] \triangleq [\![\Phi]\!]^{pw} (= [\![\Phi]\!]^c)$ is r-closed.

Proof. The first part of the proof follows from Propositions 3.16 and 3.19. Then, the fact that $[\![\Phi]\!]$ is r-closed is a direct consequence of Proposition 3.11. □

Examples of properties in HMTL

In the figures below we sketch the proofs that some of the properties presented in the end of Section 3.4 are indeed in the language of HMTL.

$$false \; \mathcal{R}(\neg a \vee (true \; \mathcal{U}_{(0,3)}b))$$

$$\cfrac{\cfrac{\cfrac{\overline{\phi^-} \qquad \overline{\phi^\circ}}{\phi^\circ} \qquad \cfrac{}{\phi^\circ}}{\cfrac{\overline{\phi^\circ} \qquad\qquad \phi^\circ}{\phi^\circ}}}{\phi^\circ}$$

$$\phi^\circ ::= \phi^- \; \mathcal{U}_I \; \phi^\circ$$
$$\phi^\circ ::= \phi^\circ \vee \phi^\circ$$
$$\phi^\circ ::= \phi^\circ \; \hat{\mathcal{R}}_I \; \phi^\circ$$

$$false \; \mathcal{R}(\neg a \vee (true \; \mathcal{U}_{(0,3)}b))$$

$$\cfrac{\cfrac{\cfrac{\overline{\phi^\bullet} \qquad \overline{\phi^{\bullet\star}}}{\phi^\bullet} \qquad \cfrac{}{\phi^\bullet}}{\cfrac{\overline{\phi^{\bullet\star}} \qquad\qquad \phi^\bullet}{\phi^\bullet}}}{\phi^\bullet}$$

$$\phi^\bullet ::= \phi^\bullet \; \hat{\mathcal{U}}_I \; \phi^{\bullet\star}$$
$$\phi^\bullet ::= \phi^\bullet \vee \phi^\bullet$$
$$\phi^\bullet ::= \phi^{\bullet\star} \; \hat{\mathcal{R}}_I \; \phi^\bullet$$

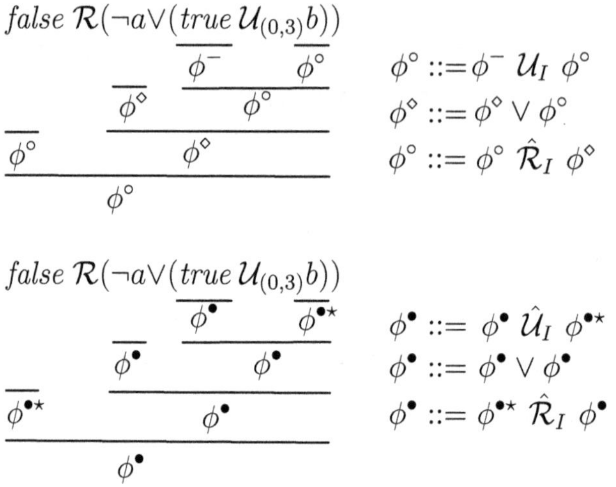

Figure 3.5: Proof that $\Box(\neg a \vee \Diamond_{(0,3)}b)$ is in HMTL.

$$false \; \mathcal{R} \; (\neg a \vee \neg a \; \tilde{\mathcal{U}}_{(0,3)}b)$$

$$\cfrac{\cfrac{\cfrac{\overline{\phi^\circ} \qquad \overline{\phi^\circ}}{\phi^\circ} \qquad \cfrac{}{\phi^\circ}}{\cfrac{\overline{\phi^\circ} \qquad\qquad \phi^\circ}{\phi^\circ}}}{\phi^\circ}$$

$$\phi^\circ ::= \phi^\circ \; \tilde{\mathcal{U}}_I \; \phi^\circ$$
$$\phi^\circ ::= \phi^\circ \vee \phi^\circ$$
$$\phi^\circ ::= \phi^\circ \; \hat{\mathcal{R}}_I \; \phi^\circ$$

$$false \; \mathcal{R} \; (\neg a \vee \neg a \; \tilde{\mathcal{U}}_{(0,3)}b)$$

$$\cfrac{\cfrac{\cfrac{\overline{\phi^\bullet} \qquad \overline{\phi^{\bullet\star}}}{\phi^\bullet} \qquad \cfrac{}{\phi^\bullet}}{\cfrac{\overline{\phi^{\bullet\star}} \qquad\qquad \phi^\bullet}{\phi^\bullet}}}{\phi^\bullet}$$

$$\phi^\bullet ::= \phi^\bullet \; \hat{\mathcal{U}}_I \; \phi^{\bullet\star}$$
$$\phi^\bullet ::= \phi^\bullet \vee \phi^\bullet$$
$$\phi^\bullet ::= \phi^{\bullet\star} \; \hat{\mathcal{R}}_I \; \phi^\bullet$$

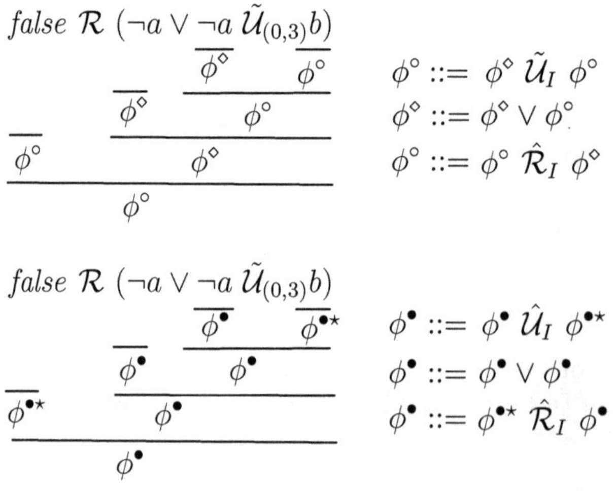

Figure 3.6: Proof that $\Box(\neg a \vee \neg a \; \tilde{\mathcal{U}}_{(0,3)}b)$ is in HMTL.

false $\mathcal{R}\ (\neg a \vee (b\ \mathcal{R}_{[0,5]}(\neg b \vee (true\mathcal{U}_{[0,3]}c))))$

$$\frac{\overline{\phi^-} \qquad \overline{\phi^\circ}}{\frac{\overline{\phi^\circ} \qquad \qquad \phi^\circ}{\frac{\overline{\phi^\circ} \qquad \qquad \qquad \phi^\circ}{\frac{\overline{\phi^\circ} \qquad \qquad \qquad \qquad \phi^\circ}{\frac{\overline{\phi^\circ} \qquad \qquad \qquad \qquad \qquad \phi^\circ}{\phi^\circ}}}}}$$

$$\phi^\circ ::= \phi^- \ \mathcal{U}_I \ \phi^\circ$$
$$\phi^\circ ::= \phi^\circ \vee \phi^\circ$$
$$\phi^\circ ::= \phi^\circ \ \hat{\mathcal{R}}_I \ \phi^\circ$$
$$\phi^\circ ::= \phi^\circ \vee \phi^\circ$$
$$\phi^\circ ::= \phi^\circ \ \hat{\mathcal{R}}_I \ \phi^\circ$$

false $\mathcal{R}\ (\neg a \vee (b\ \mathcal{R}_{[0,5]}(\neg b \vee (true\mathcal{U}_{[0,3]}c))))$

$$\frac{\overline{\phi^\bullet} \qquad \overline{\phi^{\bullet\star}}}{\frac{\overline{\phi^\bullet} \qquad \qquad \phi^\bullet}{\frac{\overline{\phi^{\bullet\star}} \qquad \qquad \qquad \phi^\bullet}{\frac{\overline{\phi^\bullet} \qquad \qquad \qquad \qquad \phi^\bullet}{\frac{\overline{\phi^{\bullet\star}} \qquad \qquad \qquad \qquad \qquad \phi^\bullet}{\phi^\bullet}}}}}$$

$$\phi^\bullet ::= \phi^\bullet \ \hat{\mathcal{U}}_I \ \phi^{\bullet\star}$$
$$\phi^\bullet ::= \phi^\bullet \vee \phi^\bullet$$
$$\phi^\circ ::= \phi^{\bullet\star} \ \hat{\mathcal{R}}_I \ \phi^\bullet$$
$$\phi^\bullet ::= \phi^\bullet \vee \phi^\bullet$$
$$\phi^\bullet ::= \phi^{\bullet\star} \ \hat{\mathcal{R}}_I \ \phi^\bullet$$

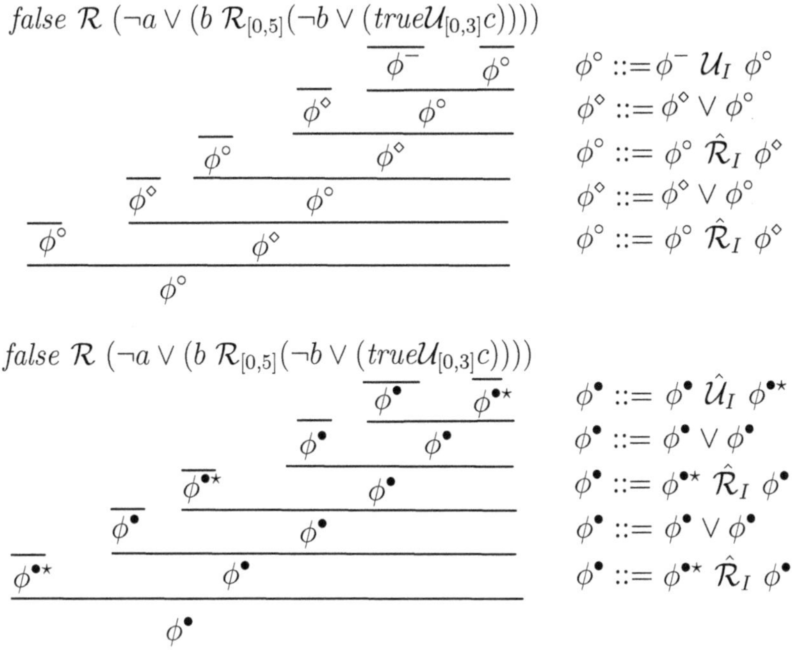

Figure 3.7: Proof that $\Box(\neg a \vee (b\ \mathcal{R}_{[0,5]}(\neg b \vee \Diamond_{[0,3]}c)))$ is in HMTL.

4

A TABLEAUX-BASED DECISION PROCEDURE FOR DISTRIBUTED TEMPORAL LOGIC

Carlos Caleiro, Paula Gouveia, Jaime Ramos
Dep. Matemática, Instituto Superior Técnico, Universidade de Lisboa, Portugal
SQIG, Instituto de Telecomunicações, Portugal
{carlos.caleiro, paula.gouveia, jaime.ramos}@tecnico.ulisboa.pt

Luca Viganò
Department of Informatics, King's College London, UK
luca.vigano@kcl.ac.uk

Abstract

The distributed temporal logic DTL is a logic for reasoning about temporal properties of distributed systems from the local point of view of the system's agents, which are assumed to execute sequentially and to interact by means of synchronous event sharing. Different versions of DTL have been given over the years for a number of different applications, reflecting different perspectives on how non-local information can be accessed by each agent. In this paper, we propose a decidable Fisher-Ladner-style tableaux system for an anchored version of DTL. Our tableaux system is built on top of a tableaux system for LTL and integrates in a smooth way both the usual rules for the temporal operators and rules for tackling the specific communication features of DTL. We endow our tableaux system with a specific decision procedure for dealing with entailment and we show that our system can be coupled with a model-checking-like feature for deciding global DTL properties.

4.1 Introduction

The distributed temporal logic DTL was introduced in [7] as a logic for specifying and reasoning about distributed information systems. DTL allows one to reason about temporal properties of distributed systems from the local point of view of the system's agents, which are assumed to execute sequentially and to interact by means of synchronous event sharing. In DTL, distribution is implicit and properties of entire systems are formulated in terms of the local properties of the agents and their interaction. The logic was shown to be decidable, as well as trace-consistent, which makes it suitable for model-checking tasks.

Different versions of distributed temporal logic have been given over the years for a number of different applications, reflecting different perspectives on how non-local information can be accessed by each agent. In particular, DTL has proved to be useful in the context of security protocol analysis in order to reason about the interplay between protocol models and security properties [5, 6, 3]. However, most of the results for security protocol analysis and for other case studies were obtained directly by semantic arguments.[1] To overcome this problem, a labeled tableaux system for DTL was proposed in [1, 2]. The main goal was to have a usable deductive system in which deductions followed closely semantic arguments, also thanks to the labeling of the formulas along with a labeling algebra capturing the different semantic properties.

The labeled tableaux system was proved to be sound and complete, but decidability was not considered in [1, 2] and the system included an infinite closure rule to capture eventualities that are always delayed. Hence, the labeled system proved to be quite hard to use in practice although several properties can still be proved using only the tableaux system. For instance, the correctness of the *two-phase commit protocol* is one of such examples where a decision procedure is not needed. The DTL specification for a simplified version of the protocol as well as a proof of correctness using labelled tableaux can be found in [2]. .

Nevertheless, DTL was shown to be decidable via a translation to linear temporal logic (LTL). However, when translating DTL specifications into LTL specifications, we lose one of the main advantages of DTL, namely the naturalness of the distributed nature of DTL, which allows for more natural and simpler specifications.

Hence, in this paper, we propose a *decidable tableaux system for DTL.*

[1]DTL is closely related to the family of temporal logics whose semantics are based on the models of true concurrency introduced and developed in [10, 11, 13]. In particular, the semantics of these logics are based on a conflict-free version of Winskel's event structures [14], enriched with information about sequential agents.

Our tableaux system is built on top of a tableaux system for LTL as presented in [8]. Similar systems for LTL have also been proposed, e.g., [9]. Our tableaux system integrates in a smooth way both the usual rules for the temporal operators and rules for tackling the specific communication features of DTL. As we do not consider labeled formulas, deductions are no longer guided by semantic arguments.

More specifically, we propose a Fisher-Ladner-style tableaux system for DTL and show how it can be used to decide DTL properties. The idea is to use the usual fixpoint properties of temporal operators (from a local perspective) in order to explore the global state space, as is done in [8] for the verification of LTL properties of state systems. In fact, similar to [8], in which an anchored version of LTL is considered, in this paper we consider an *anchored version of DTL*. Although such an anchored version of DTL is less expressive in terms of global reasoning, a suitable labeling discipline can be used to control the distribution dimension.

In this setting, the proposed tableaux system can only be used to decide validity of DTL formulas. However, the notion of entailment is a key ingredient for verification of specifications. Since DTL does not include global temporal operators, we cannot use the usual correspondence between anchored and floating entailment. Hence, we endow the tableaux system with a specific decision procedure for dealing with entailment. We focus on the case $\Delta \models \delta$, where Δ is a set of formulas (i.e., a specification) and δ is a formula with no temporal operators.[2] This is enough to capture the usual notion of invariant. Moreover, as distributed system specifications should be built from local specification blocks (see, e.g., [7]), the focus on an anchored version is not a major restriction. In the end, we also show how the tableaux system proposed here can be coupled with a model-checking-like feature for deciding global DTL properties.

We proceed as follows. In Section 4.2, we briefly introduce DTL and in Section 4.3, we present our tableaux system for DTL. In Section 4.4, we prove the soundness of the proposed system and then, in Section 4.5, we prove its completeness. In Section 4.6, we introduce rules for connectives defined as abbreviations. In Section 4.7, we present a procedure for checking entailment. Finally, in Section 4.8, we conclude and discuss future work. In the appendix, we present all the proofs that for the sake of simplicity and readability were omitted from the body of the paper.

---

[2]In general, when δ involves only one agent, then it is possible to use the usual correspondence between anchored and floating entailment and therefore the decision procedure is not needed.

4.2 The Distributed Temporal Logic $\mathrm{DTL}_\emptyset$

As we mentioned above, a number of variants of DTL have been considered in the past, especially to adapt it to specific applications and case studies. In this paper, we consider an anchored variant of DTL that we call $\mathrm{DTL}_\emptyset$ and that has the following syntax and semantics.

4.2.1 Syntax

The logic is defined over a *distributed signature*

$$\Sigma = \langle Id, \{Prop\}_{i \in Id} \rangle ,$$

where Id is a finite non-empty set (of *agent identifiers*) and, for each agent $i \in Id$, $Prop_i$ is a set of *local state propositions*, which, intuitively characterize the current local states of the agents.

The *local language* $\mathcal{L}_i$ of each agent $i \in Id$ is defined by

$$\mathcal{L}_i ::= Prop_i \mid \bot \mid \mathcal{L}_i \Rightarrow \mathcal{L}_i \mid \mathsf{X}\,\mathcal{L}_i \mid \mathsf{G}\,\mathcal{L}_i \mid ©_j[\mathcal{L}_j]$$

with $j \in Id$. We will denote such *local formulas* by the letters φ and ψ. As the names suggests, local formulas hold locally for the different agents. For instance, locally for an agent i, the operators X and G are the usual *next (tomorrow)* and *always in the future* temporal operators, whereas the *communication formula* $©_j[\psi]$ means that agent i has just communicated (synchronized) with agent j, for whom ψ held.

Other logical connectives (conjunction $\wedge$, disjunction $\vee$, true $\top$, etc.) and temporal operators (sometime in the future F, for instance) can be defined as abbreviations as is standard; note that we postpone the introduction of the *until* operator U to Section 4.6.

The *global language* $\mathcal{L}$ is defined by

$$\mathcal{L} ::= @_i[\mathcal{L}_i] \mid \bot \mid \mathcal{L} \Rightarrow \mathcal{L}$$

with $i \in Id$. We will denote the *global formulas* by α, β and δ. A *global formula* $@_i[\varphi]$ means that the local formula φ holds for agent i.

A formula is said to be

- an *atomic global formula* if it is $\bot$ or $@_i[\varphi]$ for some $\varphi \in \mathcal{L}_i$ and $i \in Id$;
- *atomic* if it is $\bot$, $@_i[\bot]$ or $@_i[p]$, for some $i \in Id$ and $p \in Prop_i$.

A *state formula* is either an atomic formula or a formula $@_i[\mathsf{X}\,\varphi]$, for some $i \in Id$ and $\varphi \in \mathcal{L}_i$.

A formula $@_i[\varphi] \in \mathcal{L}$, for some $i \in Id$, is an *i-formula*. Given $i \in Id$ and $\Delta \subseteq \mathcal{L}$, we denote by $\Delta{\downharpoonleft_i}$ the set of all i-formulas in Δ.

4.2.2 Semantics

The interpretation structures of $\mathcal{L}$ are labeled distributed life-cycles, built upon a simplified form of Winskel's *event structures* [14].

A *local life-cycle* of an agent $i \in Id$ is a countable infinite, discrete, and well-founded total order $\lambda_i = \langle Ev_i, \leq_i \rangle$, where Ev_i is the set of *local events* and $\leq_i$ the *local order of causality*. We define the corresponding *local successor relation* $\rightarrow_i \subseteq Ev_i \times Ev_i$ to be the relation such that $e \rightarrow_i e'$ if $e <_i e'$ and there is no e'' such that $e <_i e'' <_i e'$. As a consequence, we have that $\leq_i = \rightarrow_i^*$, i.e., $\leq_i$ is the reflexive and transitive closure of $\rightarrow_i$.

A *distributed life-cycle* is a family $\lambda = \{\lambda_i\}_{i \in Id}$ of local life-cycles such that $\leq = (\bigcup_{i \in Id} \leq_i)^*$ defines a partial order of *global causality* on the set of all events $Ev = \bigcup_{i \in Id} Ev_i$.

Communication is modeled by event sharing, and thus for some event e we may have $e \in Ev_i \cap Ev_j$, with $i \neq j$. In that case, requiring $\leq$ to be a partial order amounts to requiring that the local orders are globally compatible, thus excluding the existence of another $e' \in Ev_i \cap Ev_j$ such that $e <_i e'$ but $e' <_j e$. We denote by $Ids(e)$ the set $\{i \in Id \mid e \in Ev_i\}$, for each $e \in Ev$.

We can check the progress of an agent by collecting all the local events that have occurred up to a given point. This yields the notion of the *local state* of agent i, which is a finite set $\xi_i \subseteq Ev_i$ down-closed for local causality, i.e., if $e \leq_i e'$ and $e' \in \xi_i$ then also $e \in \xi_i$. The set Ξ_i of all local states of an agent i is totally ordered by inclusion and has $\emptyset$ as the minimal element.

Each non-empty local state ξ_i is reached, by the occurrence of an event that we call $last(\xi_i)$, from the local state $\xi_i \setminus \{last(\xi_i)\}$. The local states of each agent are totally ordered, as a consequence of the total order on local events. Since they are discrete and well-founded, we can enumerate them as follows: $\emptyset$ is the 0th state; $\{e\}$, where e is the minimum of $\langle Ev_i, \leq_i \rangle$, is the 1st state; and if ξ_i is the kth state of agent i and $last(\xi_i) \rightarrow_i e$, then $\xi_i \cup \{e\}$ is agent i's $(k+1)^{\text{th}}$ state.

We will denote by ξ_i^k the kth state of agent i, so $\xi_i^0 = \emptyset$ is the initial state and ξ_i^k is the state reached from the initial state after the occurrence of the first k events. In fact, ξ_i^k is the only state of agent i that contains k elements, i.e., where $|\xi_i^k| = k$. Given $e \in Ev_i$, $e{\downarrow_i} = \{e' \in Ev_i \mid e' \leq_i e\}$ is always a local state. Moreover, if ξ_i is non-empty, then $last(\xi_i){\downarrow_i} = \xi_i$.

We can also define the notion of a *global state*: a finite set $\xi \subseteq Ev$ closed for global causality, i.e. if $e \leq e'$ and $e' \in \xi$, then also $e \in \xi$. The set Ξ of all global states constitutes a lattice under inclusion and has $\emptyset$ as the minimal element. Clearly, every global state ξ includes the local state $\xi|_i = \xi \cap Ev_i$ of each agent i. Given $e \in Ev$, $e{\downarrow} = \{e' \in Ev \mid e' \leq e\}$ is always a global state.

An *interpretation structure* $\mu = \langle \lambda, \vartheta \rangle$ consists of a distributed life-cycle λ and a family $\vartheta = \{\vartheta_i\}_{i \in Id}$ of local *labeling functions*, where, for each $i \in Id$, $\vartheta_i : \Xi_i \to \wp(Prop_i)$ associates a set of local state propositions to each local state. We denote the tuple $\langle \lambda_i, \vartheta_i \rangle$ also by μ_i.

We can the define a *global satisfaction relation* as follows. Given a global interpretation structure μ and a global state ξ then

- $\mu, \xi \not\Vdash \bot$;
- $\mu, \xi \Vdash \alpha \Rightarrow \beta$ if $\mu, \xi \not\Vdash \alpha$ or $\mu, \xi, \Vdash \beta$;
- $\mu, \xi \Vdash @_i[\varphi]$ if $\mu_i, \xi|_i \Vdash_i \varphi$.

The *local satisfaction relations* at local states are defined by

- $\mu_i, \xi_i \Vdash_i p$ if $p \in \vartheta_i(\xi_i)$;
- $\mu_i, \xi_i \not\Vdash_i \bot$;
- $\mu_i, \xi_i \Vdash_i \varphi \Rightarrow \psi$ if $\mu_i, \xi_i \not\Vdash_i \varphi$ or $\mu_i, \xi_i \Vdash_i \psi$;
- $\mu_i, \xi_i \Vdash_i \mathsf{X}\,\varphi$ if there is $e \in Ev_i \setminus \xi_i$ such that $\xi_i \cup \{e\} \in \Xi_i$ and $\mu_i, \xi_i \cup \{e\} \Vdash_i \varphi$;
- $\mu_i, \xi_i \Vdash_i \mathsf{G}\,\varphi$ if $\mu_i, \xi_i' \Vdash_i \varphi$, for every $\xi_i' \in \Xi_i$ such that $\xi_i \subseteq \xi_i'$;
- $\mu_i, \xi_i \Vdash_i ©_j[\varphi]$ if $\xi_i \neq \emptyset$, $last(\xi_i) \in Ev_j$ and $\mu_j, last(\xi_i){\downarrow_j} \Vdash_j \varphi$.

We say that μ *(globally) satisfies* α, written $\mu \Vdash \alpha$, whenever $\mu, \emptyset \Vdash \alpha$. As expected, α is said to be *satisfiable* whenever there is a μ such that $\mu \Vdash \alpha$. We say that

- μ is a *model* of $\Delta \subseteq \mathcal{L}$ if $\mu \Vdash \delta$, for every $\delta \in \Delta$;
- α is *valid*, in symbols $\vDash \alpha$, whenever $\mu \Vdash \alpha$ for every interpretation structure μ;
- Δ *entails* $\alpha \in \mathcal{L}$, in symbols $\Delta \vDash \alpha$, if every global model of Δ is also a model of α.

We proceed similarly for the local languages. We say that μ_i *(locally) satisfies* φ, written $\mu_i \Vdash_i \varphi$ if $\mu_i, \emptyset \Vdash_i \varphi$. We say that μ_i is a *(local) model* of $\Psi \subseteq \mathcal{L}_i$ if $\mu_i \Vdash_i \psi$, for every $\psi \in \Psi$.

4.3 A tableaux system for $\mathrm{DTL}_\emptyset$

In this section we present a tableaux system for $\mathrm{DTL}_\emptyset$ and some examples.

4.3.1 The rules of the tableaux systems for $\mathrm{DTL}_\emptyset$

In order to define our tableaux systems for $\mathrm{DTL}_\emptyset$, let us first introduce some useful terminology and notation.

A *state tuple* is a tuple $Q = \langle Q^+, Q^-, Q^{\bowtie}, Q^{\#}, Q^{\times}\rangle$, where $Q^+, Q^- \subseteq \mathcal{L}$ are finite sets, and $Q^{\bowtie}, Q^{\#}, Q^{\times} \subseteq Id$. A state tuple is intended to characterize a global state of a model:

- Formulas in Q^+ are intended to be *true* and formulas in Q^- are intended to be *false* at that state.
- $Q^{\bowtie}$, $Q^{\#}$ and $Q^{\times}$ impose *synchronization requirements* on how the state was reached: the identifiers in $Q^{\times}$ indicate which are the agents that synchronize to reach the current state, whereas the identifiers in $Q^{\bowtie}$ and $Q^{\#}$ impose, respectively, so called *forced synchronizations* (identifiers that must be in $Q^{\times}$) and *conflicts* (identifiers that must be outside $Q^{\times}$).

We will write $\mathcal{S}$ to denote the set of all state tuples (over a given signature Σ).

We say that Q is a state tuple with no synchronization requirements when, as expected, $Q^{\bowtie} = Q^{\#} = Q^{\times} = \emptyset$. Given state tuples Q_1 and Q_2, we denote by $Q_1 \setminus Q_2$ the state tuple $\langle Q_1^+ \setminus Q_2^+, Q_1^- \setminus Q_2^-, Q_1^{\bowtie} \setminus Q_2^{\bowtie}, Q_1^{\#} \setminus Q_2^{\#}, Q_1^{\times} \setminus Q_2^{\times}\rangle$.

Let us state rigorously the intended semantics of state tuples. We say that an interpretation structure μ and a global state ξ *meet the synchronization requirements* in Q when all the following conditions hold:

(S1) if $Q^{\times} = \emptyset$ then $\xi = \emptyset$;

(S2) if $Q^{\times} \neq \emptyset$ then there is a maximal event e_Q with respect to ξ such that $\{i \in Id \mid e_Q \in Ev_i\} = Q^{\times}$;

(S3) $Q^{\bowtie} \subseteq Q^{\times}$;

(S4) $Q^{\#} \cap Q^{\times} = \emptyset$.

By maximal event e_Q with respect to ξ we mean an event $e_Q \in \xi$ such that there is no other $e \in \xi$ with $e_Q < e$. Note that the above conditions are actually quite redundant. Rather than aiming for a minimal specification, we opted for conditions that are quite intuitive and easy to understand and apply.

We say that μ and ξ *satisfy* the state tuple Q, in symbols $\mu, \xi \Vdash Q$, if μ and ξ meet the synchronization requirements in Q and also $\mu, \xi \Vdash \widehat{Q}$, where $\widehat{Q}$ is the formula

$$(\bigwedge_{\alpha \in Q^+} \alpha) \wedge (\bigwedge_{\beta \in Q^-} \neg \beta),$$

where the empty conjunctions are identified with $\top$.

Our tableaux system is based on state tuples, with rules that allow one to take advantage of the fixpoint properties of the temporal operators in order to unfold the formulas in Q^+ and Q^- into a full model whenever possible. Synchronization constraints allow one to control the distributed dimension of the logic. To this end, it is useful to consider the following notion: a state tuple is said to be *contradictory* whenever one of the following conditions holds:

($\perp$1) $\perp \in Q^+$ or $@_i[\perp] \in Q^+$, for some $i \in Id$;

($\perp$2) $Q^+ \cap Q^- \neq \emptyset$;

($\perp$3) $j \notin Q^{\mathsf{X}}$ for some $j \in Q^{\bowtie}$;

($\perp$4) $j \in Q^{\mathsf{X}}$ for some $j \in Q^{\#}$.

A contradictory state tuple corresponds to an unfeasible situation. Conditions ($\perp$1) and ($\perp$2) are straightforward, given that $\perp$ cannot be true and no formula can be simultaneously true and false. Conditions ($\perp$3) and ($\perp$4) are impossible given the above definition of satisfaction of a state tuple at a global state of an interpretation structure.

Before we proceed with the definition of our tableaux system, we define some auxiliary functions. For each $I \subseteq Id$, let $\sigma_I^+, \sigma_I^- : \mathcal{S} \to 2^{\mathcal{L}}$ be functions such that, for each $Q \in \mathcal{S}$:

$$\begin{aligned} \sigma_I^+(Q) &= \{@_i[\varphi] \mid @_i[\mathsf{X}\,\varphi] \in Q^+ \text{ and } i \in I\} \\ \sigma_I^-(Q) &= \{@_i[\varphi] \mid @_i[\mathsf{X}\,\varphi] \in Q^- \text{ and } i \in I\} \end{aligned}$$

Function σ_I^+, when applied to Q, returns the set of formulas of the agents in I that must hold in the next state. Similarly, function σ_I^-, when applied to Q, returns the set of formulas of the agents in I that cannot hold in the next state.

For each $i, j \in Id$, let $\eta^{+}_{i,j}, \eta^{-}_{i,j}, \eta^{\bowtie}_{i,j}, \eta^{\#}_{i,j} : \mathcal{S} \to 2^{\mathcal{L}}$ be functions such that, for each $Q \in \mathcal{S}$:

$$\begin{aligned} \eta^{+}_{i,j}(Q) &= \{@_j[\varphi] \mid @_i[©_j[\varphi]] \in Q^+\} \\ \eta^{-}_{i,j}(Q) &= \{@_j[\varphi] \mid @_i[©_j[\varphi]] \in Q^-\} \\ \eta^{\bowtie}_{i,j}(Q) &= \{@_i[©_j[\varphi]] \mid @_i[©_j[\varphi]] \in Q^+\} \\ \eta^{\#}_{i,j}(Q) &= \{@_i[©_j[\varphi]] \mid @_i[©_j[\varphi]] \in Q^-\} \end{aligned}$$

Function $\eta^{+}_{i,j}$, when applied to Q, returns the set of formulas that must hold for agent j forced by the interaction triggered by agent i. Similarly, function $\eta^{-}_{i,j}$, when applied to Q, returns the set of formulas that cannot hold for agent j, if agent j is synchronized with agent i. Observe that, while positive interaction forces synchronization between the agents, negative interaction does not. It only requires that if the agents are synchronized then the formulas must not hold. This will be reflected in the rules of the tableaux.

Function $\eta^{\bowtie}_{i,j}$, when applied to Q, returns the set of formulas that force the synchronization between agents i and j, triggered by i. Function $\eta^{\#}_{i,j}$, when applied to Q, returns the set of formulas that prevent a possible synchronization between agents i and j, triggered by i.

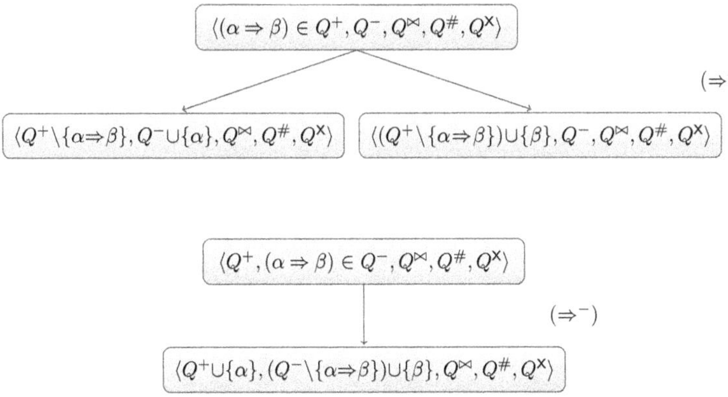

Figure 4.1: Rules for global implication.

We are finally ready to define what is a *tableau* $\mathcal{T}$. Namely, $\mathcal{T}$ is a directed graph whose nodes are state tuples and the (directed) edges are defined according to the rules in Figures 4.1, 4.2, 4.3 and 4.4, with the proviso that exactly one rule is applied to each non-contradictory node and no rule is applied to a contradictory node. A *tableau for* $Q_0 \in \mathcal{S}$ is a directed graph whose construction starts with Q_0 and proceeds with the application of rules until a tableau is obtained.

For simplicity, to denote the state tuple $\langle Q^+, Q^-, Q^{\bowtie}, Q^{\#}, Q^{\times} \rangle$ with α is in Q^+, we will write $\langle \alpha \in Q^+, Q^-, Q^{\bowtie}, Q^{\#}, Q^{\times} \rangle$ when depicting a rule. Similarly for $\langle Q^+, \alpha \in Q^-, Q^{\bowtie}, Q^{\#}, Q^{\times} \rangle$. As terminology, the source node of the edges is the *conclusion* of a rule, whereas the target nodes are the *premises* of that rule.

The rules for global implication in Figure 4.1 are the usual ones. If $(\alpha \Rightarrow \beta)$ is true then either α is false or β is true. Hence, we have two premises in rule $(\Rightarrow^+)$, one for each case. Rule $(\Rightarrow^-)$ for when the formula is false is similar. The other global formulas have no specific global rules. Formulas of the form $@_i[\varphi]$, for some $\varphi \in \mathcal{L}_i$, have specific local rules that are introduced below. The global formula $\perp$ is used for closure.

$$\langle @_i[\varphi \Rightarrow \psi] \in Q^+, Q^-, Q^{\bowtie}, Q^{\#}, Q^{\times} \rangle$$

$(\Rightarrow_i^+)$

$$\langle Q^+ \backslash \{@_i[\varphi \Rightarrow \psi]\}, Q^- \cup \{@_i[\varphi]\}, Q^{\bowtie}, Q^{\#}, Q^{\times} \rangle \qquad \langle (Q^+ \backslash \{@_i[\varphi \Rightarrow \psi]\}) \cup \{@_i[\psi]\}, Q^-, Q^{\bowtie}, Q^{\#}, Q^{\times} \rangle$$

$$\langle Q^+, @_i[\varphi \Rightarrow \psi] \in Q^-, Q^{\bowtie}, Q^{\#}, Q^{\times} \rangle$$

$(\Rightarrow_i^-)$

$$\langle Q^+ \cup \{@_i[\varphi]\}, (Q^- \backslash \{@_i[\varphi \Rightarrow \psi]\}) \cup \{@_i[\psi]\}, Q^{\bowtie}, Q^{\#}, Q^{\times} \rangle$$

$$\langle @_i[\mathsf{G}\,\varphi] \in Q^+, Q^-, Q^{\bowtie}, Q^{\#}, Q^{\times} \rangle$$

(G_i^+)

$$\langle (Q^+ \backslash \{@_i[\mathsf{G}\,\varphi]\}) \cup \{@_i[\varphi], @_i[\mathsf{X}\,\mathsf{G}\,\varphi]\}, Q^-, Q^{\bowtie}, Q^{\#}, Q^{\times} \rangle$$

$$\langle Q^+, @_i[\mathsf{G}\,\varphi] \in Q^-, Q^{\bowtie}, Q^{\#}, Q^{\times} \rangle$$

(G_i^-)

$$\langle Q^+, (Q^- \backslash \{@_i[\mathsf{G}\,\varphi]\}) \cup \{@_i[\varphi]\}, Q^{\bowtie}, Q^{\#}, Q^{\times} \rangle \qquad \langle Q^+, (Q^- \backslash \{@_i[\mathsf{G}\,\varphi]\}) \cup \{@_i[\mathsf{X}\,\mathsf{G}\,\varphi]\}, Q^{\bowtie}, Q^{\#}, Q^{\times} \rangle$$

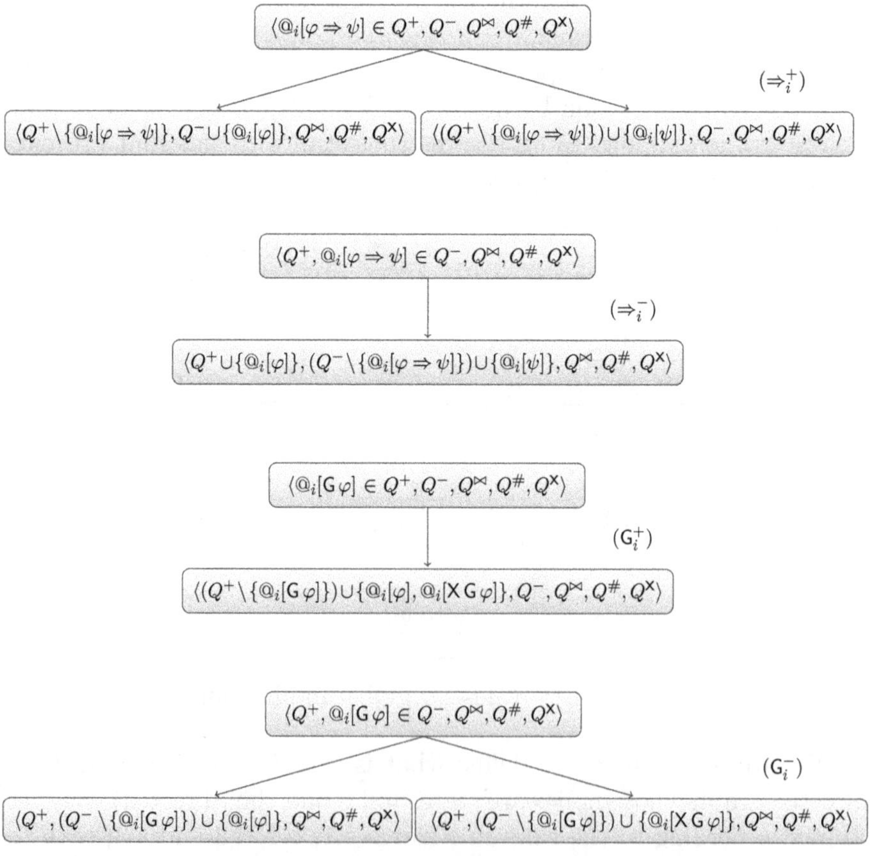

Figure 4.2: Local rules for $\Rightarrow$ and G.

Figure 4.2 gives the rules for local implication and for the temporal operator G. The rules for local implication are similar to the global case. Rule (G_i^+) expresses that if, for some $i \in Id$, $@_i[\mathsf{G}\,\varphi]$ is true, then $@_i[\varphi]$ must hold now and must also hold from the next instant on, i.e., $@_i[\mathsf{X}\,\mathsf{G}\,\varphi]$ must hold. Rule (G_i^-) expresses that if, on the other hand, $@_i[\mathsf{G}\,\varphi]$ is false, then either $@_i[\varphi]$ is false now or it is false in some future instant, i.e., $@_i[\mathsf{X}\,\mathsf{G}\,\varphi]$ is false.

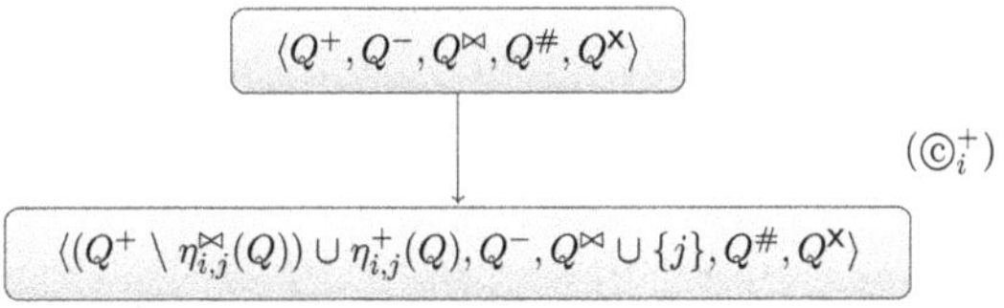

$(©_i^+)$

$$\langle Q^+, Q^-, Q^{\bowtie}, Q^{\#}, Q^{\times}\rangle$$

$$\langle (Q^+ \setminus \eta_{i,j}^{\bowtie}(Q)) \cup \eta_{i,j}^+(Q), Q^-, Q^{\bowtie} \cup \{j\}, Q^{\#}, Q^{\times}\rangle$$

provided that $\eta_{i,j}^{\bowtie}(Q) \neq \emptyset$, for some $j \in Id$

$(©_i^-)$

$$\langle Q^+, Q^-, Q^{\bowtie}, Q^{\#}, Q^{\times}\rangle$$

$$\langle Q^+, (Q^- \setminus \eta_{i,j}^{\#}(Q)) \cup \eta_{i,j}^-(Q), Q^{\bowtie} \cup \{j\}, Q^{\#}, Q^{\times}\rangle \qquad \langle Q^+, Q^- \setminus \eta_{i,j}^{\#}(Q), Q^{\bowtie}, Q^{\#} \cup \{j\}, Q^{\times}\rangle$$

provided that $\eta_{i,j}^{\#}(Q) \neq \emptyset$, for some $j \in Id$

Figure 4.3: Local rules for $©$.

Figure 4.3 gives the rules for communication. Rule $©_i^+$ deals with positive communication for an agent $i \in Id$. If $\eta_{i,j}^{\bowtie}(Q) \neq \emptyset$, for some $j \in Id$, then this means that agent i is executing and needs to synchronize with agent j, i.e., there is at least one formula $@_i[©_j[\varphi]]$ that must hold. Hence, we must require that $@_j[\varphi]$ holds but also that i and j are synchronized by including j in $Q^{\bowtie}$. Note that all the formulas resulting from the synchronization between i and j are given by $\eta_{i,j}^+(Q)$. Rule $©_i^-$ deals with negative communication for an agent $i \in Id$. In this case, if $\eta_{i,j}^{\#}(Q) \neq \emptyset$, for some $j \in Id$, then there is at least one formula $@_i[©_j[\varphi]]$ that cannot hold. There are two possible reasons for this: either i and j synchronize but $@_j[\varphi]$ does not hold, or i and j are in conflict, which is expressed by adding j to $Q^{\#}$.

For each $i \in Id$, rules $(\Rightarrow_i^+)$, $(\Rightarrow_i^-)$, (G_i^+), (G_i^-), $(©_i^+)$ and $(©_i^-)$ are said to be *rules triggered* by i.

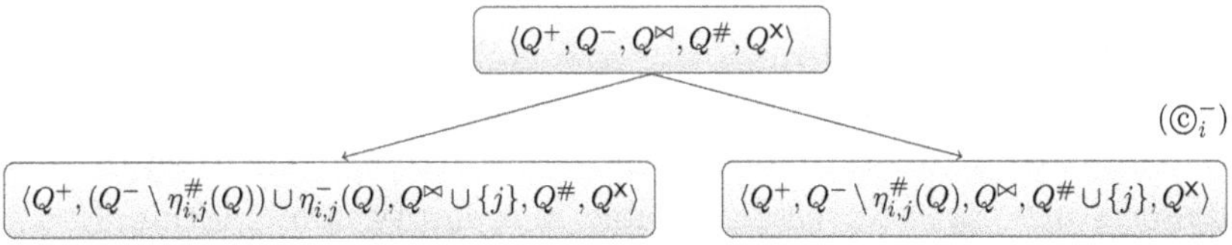

$$\langle Q^+, Q^-, Q^{\bowtie}, Q^{\#}, Q^{\mathsf{X}} \rangle$$

$$\downarrow \qquad (\mathsf{X})$$

$$\begin{gathered}\langle (Q^+ \setminus \{@_k[\varphi] \mid \varphi \in \mathcal{L}_k, \text{ for } k \in K\}) \cup \sigma_K^+(Q), \\ (Q^- \setminus \{@_k[\varphi] \mid \varphi \in \mathcal{L}_k \text{ for } k \in K\}) \cup \sigma_K^-(Q), \\ \emptyset, \emptyset, K \rangle\end{gathered}$$

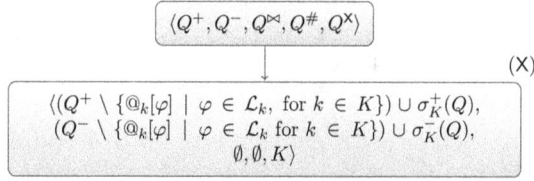

one successor for each $K \subseteq Id$, $K \neq \emptyset$, provided that Q is a state

Figure 4.4: Local rule for X.

Figure 4.4 gives the rule for the temporal operator X. The use of this rule marks a state transition for some agents. In order to allow a state transition, we must first guarantee that all the formulas have been decomposed except for those referring to the next state. That is, in order to be able to change state, all the formulas in Q must be either local atoms or formulas of the form $@_i[\mathsf{X}\,\varphi]$. In this case, there is nothing more to be added to the current state and so we can proceed to analyze the next state. If all the above conditions are met, then we have one successor state tuple for each possible non-empty combination $K \subseteq Id$ of agents. The next state tuple is obtained by removing all the state formulas of the previous state tuple and adding all the formulas that have to be carried on to this new state, i.e., all the formulas $@_i[\varphi]$ such that formulas $@_i[\mathsf{X}\,\varphi]$ were in the previous state. We also eliminate all the forced synchronizations and all the conflicts. Finally, for each successor K, we set Q^{X} to K, stating that the agents that are synchronized to reach this new system state are the ones in K.

One should note that the requirements in the rule (X) impose that before the first state transition, all global rules must have been applied. Since, before the first application of rule (X), global rules and local rules can be applied in an arbitrary order, we will assume that global rules are applied first and only then local rules will be considered.

In general, formulas are decomposed from a tableau node to the next until all that remains are state formulas, which force the change of local state for some agents. When this happens, rule (X) is applied carrying to the next global state of the system only the relevant temporal information. The information contained in the node where rule (X) is applied characterizes a state of the system: the propositional symbols that must be true and false, for each agent, together with the relevant temporal information that must be carried to the next state of the system.

The set $Cl_{\mathcal{T}}$ of *closed nodes* of a tableau $\mathcal{T}$ is inductively defined as follows:

(C1) $Q \in Cl_{\mathcal{T}}$ for every contradictory node Q;

(C2) $Q \in Cl_{\mathcal{T}}$ whenever $Q' \in Cl_{\mathcal{T}}$ for every successor node Q' of Q;

(C3) $Q \in Cl_{\mathcal{T}}$ whenever $@_i[\mathsf{G}\,\varphi] \in Q^-$ for some $i \in Id$ and $\varphi \in \mathcal{L}_i$, and every path from Q to a node Q' with $@_i[\varphi] \in Q'^-$ has a node in $Cl_{\mathcal{T}}$.

(C4) $Q \in Cl_{\mathcal{T}}$ whenever there is an $i \in Id$ such that every path from Q to a node Q' with $i \in Q'^{\mathsf{X}}$ has a node in $Cl_{\mathcal{T}}$.

Figure 4.5: Set of closed nodes.

The *set of closed nodes* of a tableau is defined in Figure 4.5. Herein a node Q' is said to be a *successor* of Q in $\mathcal{T}$ if there is an edge in $\mathcal{T}$ from Q to Q'. Just as contradictory state tuples, closed nodes correspond to unfeasible situations. Conditions $(C1)$ and $(C2)$ are straightforward. Condition $(C3)$ deals with *delayed eventualities*. If a formula $@_i[\mathsf{G}\,\varphi]$ is false then there must be some future state where $@_i[\varphi]$ is also false. But if the paths to all the nodes Q' such that $@_i[\varphi] \in Q'^-$ contain a closed node then that future state is impossible to reach. In condition $(C4)$ we close a node whenever any future transition of an agent i is impossible to reach due to the existence of a closed node. As before, these conditions are not guaranteed to be minimal, but we keep them as they provide for a quite intuitive inductive definition.

A tableau for Q_0 is *unsuccessful* if the node Q_0 is closed, and it is *successful* otherwise.

Proposition 4.1 *A tableau for a finite state tuple is always finite.* □

As a proof sketch, recall that, as we said above, global rules are applied first and always reduce the complexity of the formulas and, since local rules do not produce global formulas, the number of applications of global rules is always finite. The number of applications of local rules is also finite because the set of subformulas that may be generated from the initial set of formulas is finite. In fact, we can inductively define the notion of *local subformula* (*subf*) as follows:

- $subf(@_i[p]) = \{@_i[p]\}$;
- $subf(@_i[\bot]) = \{@_i[\bot]\}$;

- $subf(@_i[\varphi \Rightarrow \psi]) = \{@_i[\varphi \Rightarrow \psi]\} \cup subf_i(@_i[\varphi]) \cup subf(@_i[\psi])$;
- $subf(@_i[\mathsf{X}\,\varphi]) = \{@_i[\mathsf{X}\,\varphi]\} \cup subf(@_i[\varphi])$;
- $subf(@_i[\mathsf{G}\,\varphi]) = \{@_i[\mathsf{G}\,\varphi], @_i[\mathsf{X}\,\mathsf{G}\,\varphi]\} \cup subf(@_i[\varphi])$;
- $subf(@_i[©_j[\varphi]]) = \{@_i[©_j[\varphi]]\} \cup subf(@_j[\varphi])$.

In a strict sense, $@_i[\mathsf{X}\,\mathsf{G}\,\varphi]$ is not a subformula of $@_i[\mathsf{G}\,\varphi]$, but the meaning of subformula herein is to be understood as a traditional subformula or as a state formula.

Now, observe that the set $subf(@_i[\varphi])$ is always finite, for any given formula φ. We refrain from proving this statement but note that *subf* on the right hand side is always applied to a less complex formula. Furthermore, the rules preserve this notion of subformula in the sense that if Q is the conclusion of a rule and Q' is a premise then $Q'^{+} \cup Q'^{-} \subseteq subf(Q^{+} \cup Q^{-})$, which means that the formulas in the premiss are always subformulas of the formulas in the conclusion. In addition, *Id* is finite, which means that the set of possible elements in $Q^{\bowtie}$, $Q^{\#}$ and Q^{X} is also finite. Hence, given a finite state tuple, the set of nodes of the tableau is necessarily finite.

4.3.2 Some examples

We now illustrate the tableaux system at work with some examples. In order to increase readability of the tableaux presented below, we adopt some conventions.

Recall that each node is a state tuple $\langle Q^{+}, Q^{-}, Q^{\bowtie}, Q^{\#}, Q^{\mathsf{X}}\rangle$. For simplicity, when depicting a tableau, we will omit the outer brackets of the state tuples, always maintaining the order of the components. Sometimes, the tuple will be split in two or more lines. In addition, we will label edges with the rule that is being applied. There are two special situations: rule (X) and binary rules. In the case of rule (X) we will label each departing edge with the set of agents (the set K in the rule) that are being considered in that particular edge. In the case of binary rules, we only label one of the edges. So, in some examples, some edges may appear not to have a label, but one just has to look for the other edge leaving that node to learn the rule that is being applied.

Example 4.2 *Consider a DTL-signature with $Id = \{i, j\}$ and $p \in Prop_j$, and consider the state tuple Q such that $Q^{+} = Q^{\bowtie} = Q^{\#} = Q^{\mathsf{X}} = \emptyset$ and $Q^{-} = \{@_i[\mathsf{X}\,©_j[p]] \Rightarrow @_j[p]\}$. Figure 4.6 shows a tableau for Q, where the dark shaded nodes correspond to closed nodes. Node N_9 closes due to condition $(C1)$ since condition $(\bot 3)$ holds. Observe that $j \in N_9^{\bowtie}$ but*

$j \notin N_9^{\mathsf{X}}$. Since node N_9 is the only successor of node N_6, this node is closed due condition $(C2)$. Node N_4 also closes due to condition $(C1)$. In this case, both $(\bot 2)$ and $(\bot 3)$ hold. Node N_1 is closed due $(C2)$ since node N_4 is its only successor. This tableau is successful since node N_0 is not closed. □

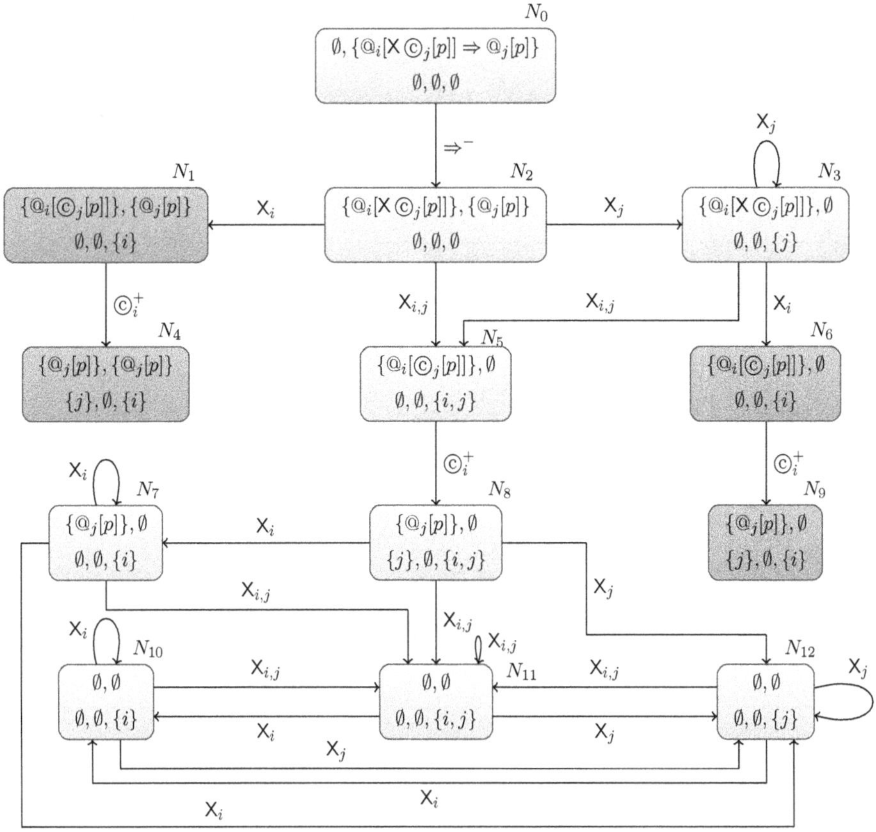

Figure 4.6: Tableau for $\langle \emptyset, \{@_i[\mathsf{X}\,©_j[p]] \Rightarrow @_j[p]\}, \emptyset, \emptyset, \emptyset\rangle$.

Since the tableau in Figure 4.6 is successful, we will see later that, as a consequence of Proposition 4.14, the formula $@_i[\mathsf{X}\,©_j[p]] \Rightarrow @_j[p]$ is not valid. Furthermore, we will be able to extract from the tableau an interpretation structure μ such that $\mu \not\Vdash @_i[\mathsf{X}\,©_j[p]] \Rightarrow @_j[p]$.

Before we proceed, we anticipate some rules for auxiliary connectives. In Section 4.6, we present a more exhaustive list but for now, and to keep the next example simple, we focus on the rules for negation. Recall that $\neg\varphi$ is defined in the standard way as an abbreviation of $\varphi \Rightarrow \bot$. So, we will consider two cases: $@_i[\neg\,\varphi] \in Q^+$ and $@_i[\neg\,\varphi] \in Q^-$.

$$\langle @_i[\varphi \Rightarrow \bot] \in Q^+, Q^-, Q^{\bowtie}, Q^{\#}, Q^{\times}\rangle$$

$(\Rightarrow_i^+)$

$$\langle Q^+\backslash\{@_i[\varphi \Rightarrow \bot]\}, Q^-\cup\{@_i[\varphi]\}, Q^{\bowtie}, Q^{\#}, Q^{\times}\rangle \qquad \langle Q^+\backslash\{@_i[\varphi \Rightarrow \bot]\}\cup\{@_i[\bot]\}, Q^-, Q^{\bowtie}, Q^{\#}, Q^{\times}\rangle$$

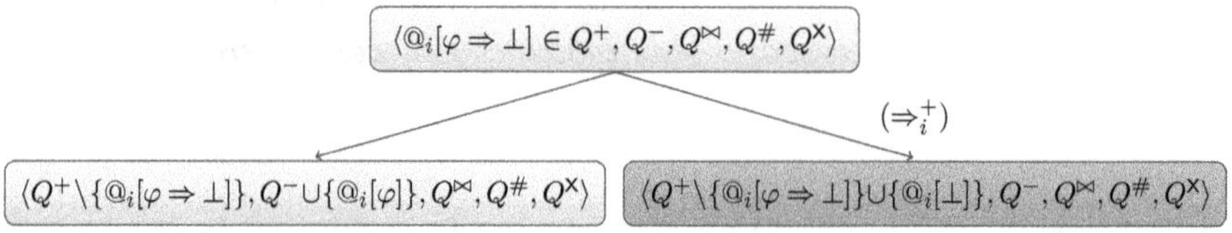

Figure 4.7: Local negation by abbreviation: positive case.

In Figure 4.7, we consider the case when $@_i[\neg\,\varphi] \in Q^+$ and apply the rule $(\Rightarrow_i^+)$. Observe that the right branch is closed and so in this case we can derive the rule $(\neg_i^+)$ given in Figure 4.9.

$$\langle Q^+, @_i[\varphi \Rightarrow \bot] \in Q^-, Q^{\bowtie}, Q^{\#}, Q^{\times}\rangle$$

$(\Rightarrow_i^-)$

$$\langle Q^+ \cup \{@_i[\varphi]\}, Q^- \cup \{@_i[\bot]\}, Q^{\bowtie}, Q^{\#}, Q^{\times}\rangle$$

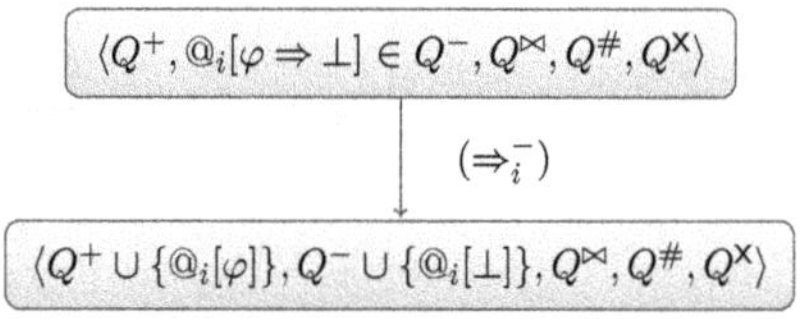

Figure 4.8: Local negation by abbreviation: negative case.

In Figure 4.8, we consider the case when $@_i[\neg\,\varphi] \in Q^-$ and apply the rule $(\Rightarrow_i^-)$. Observe that, in this case, $@_i[\bot] \in Q^-$ is not relevant for the development of the tableau and, consequently, may be omitted. Like in the previous case, we can derive the rule $(\neg_i^-)$ given in Figure 4.9.

$$\langle @_i[\neg\varphi] \in Q^+, Q^-, Q^{\bowtie}, Q^{\#}, Q^{\times}\rangle$$

$(\neg_i^+)$

$$\langle Q^+ \backslash \{@_i[\neg\varphi]\}, Q^- \cup \{@_i[\varphi]\}, Q^{\bowtie}, Q^{\#}, Q^{\times}\rangle$$

$$\langle Q^+, @_i[\neg\varphi] \in Q^-, Q^{\bowtie}, Q^{\#}, Q^{\times}\rangle$$

$(\neg_i^-)$

$$\langle Q^+ \cup \{@_i[\varphi]\}, Q^- \backslash \{@_i[\neg\varphi]\}, Q^{\bowtie}, Q^{\#}, Q^{\times}\rangle$$

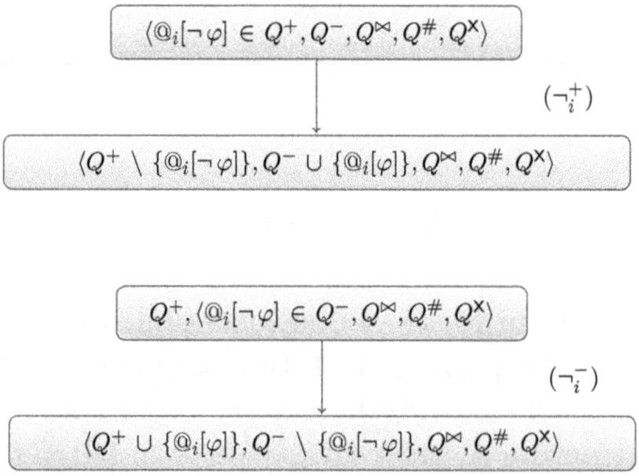

Figure 4.9: Local rules for $\neg$.

Next, we present an example of a closed tableau.

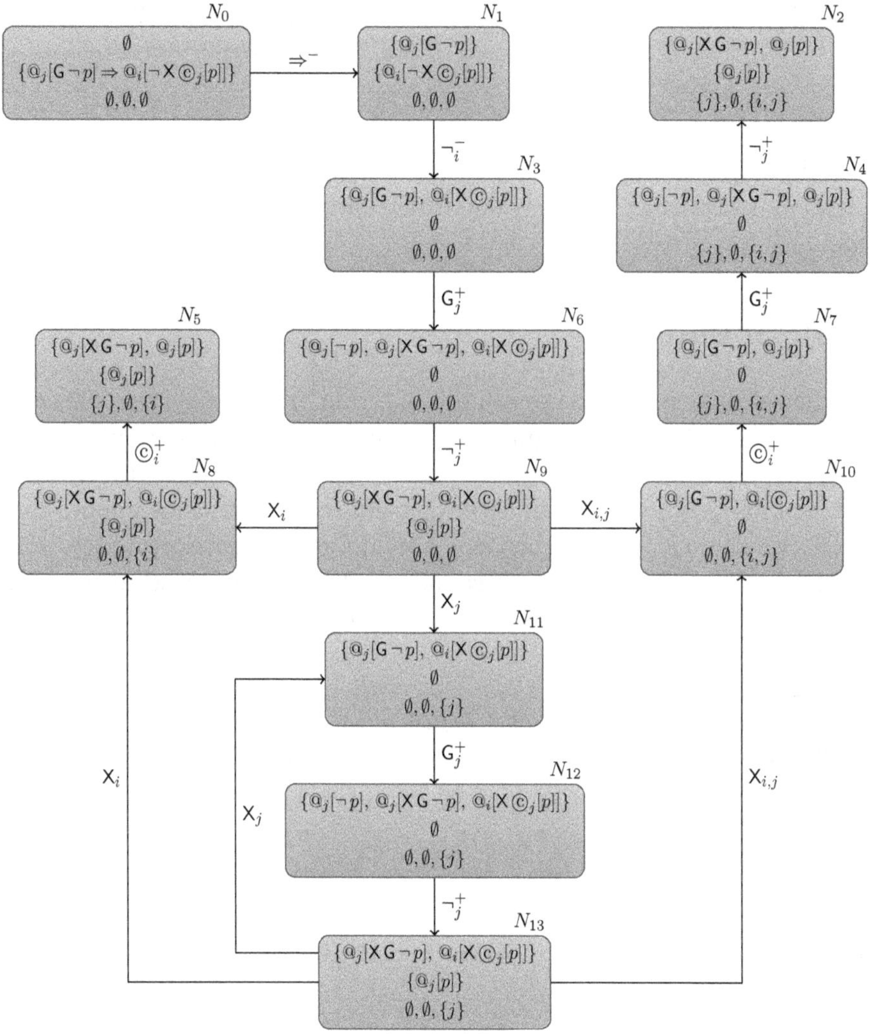

Figure 4.10: Tableau for $\langle \emptyset, \{@_j[\mathsf{G}\neg p] \Rightarrow @_i[\neg \mathsf{X} ©_j[p]]\}, \emptyset, \emptyset, \emptyset \rangle$.

Example 4.3 *Recall the DTL-signature in Example 4.2. Consider the state tuple Q such that $Q^+ = Q^{\bowtie} = Q^{\#} = Q^{\mathsf{X}} = \emptyset$ and $Q^- = \{@_j[\mathsf{G}\neg p] \Rightarrow @_i[\neg \mathsf{X} ©_j[p]]\}$. A tableau for Q is shown in Figure 4.10.*

Node N_2 closes due to $(C1)$ and, in particular, due to condition $(\bot 2)$, as $@_j[p] \in N_2^+ \cap N_2^-$. Node N_5 closes also due to condition $(C1)$. Note that in this case, in addition to $(\bot 2)$, condition $(\bot 3)$ also holds. The fact that these nodes are closed immediately implies that nodes N_4, N_7, N_8 and N_{10} are closed, by condition $(C2)$. Node N_{13} closes due to condition $(C4)$. Observe that every path from N_{13} to a node Q' such that $i \in Q'^{\mathsf{X}}$ must go through either nodes N_8 or N_{10} and these nodes are closed.

Finally, all the remaining nodes close due to condition (C2). Observe that (C4) also holds for these remaining nodes. Hence, this tableau is unsuccessful since node N_0 is closed. □

Since the tableau is unsuccessful, we will later be able to conclude that, by Corollary 4.10, $\widehat{Q}$ is not satisfiable and, consequently, the formula $@_j[\mathsf{G}\,\neg p] \Rightarrow @_i[\neg\,\mathsf{X}\,©_j[p]]$ is valid.

4.4 Soundness

In this section, we prove the soundness of our tableaux system. That is, we prove that for a state tuple Q with no synchronization requirements, if $\widehat{Q}$ is satisfiable, then any tableau for Q is successful. Hence, if a tableau for Q is unsuccessful, then $\widehat{Q}$ is not satisfiable. This implies that if $Q^+ = \emptyset$ and $Q^- = \{\alpha\}$, then α is a valid formula. As expected, we will have to prove that every rule is sound. To this end, we need to establish some technical results along the way, starting with the following lemma.

Lemma 4.4 *Let Q' be a state tuple with no synchronization requirements and $\mathcal{T}$ a tableau for Q'. Let $Q_{s_0}, \ldots, Q_{s_n}$, with $n > 0$, be a path in $\mathcal{T}$ such that Q_{s_n} is not contradictory, and Q_{s_0} is the only state in the path. If some rule triggered by $i \in Id$ is applied at node Q_{s_n} then $i \in Q^{\mathsf{X}}_{s_r}$, for all $r = 1, \ldots, n$.* □

The next proposition states that in any tableau for a state tuple with no synchronization requirements no contradictory state tuple is satisfiable.

Proposition 4.5 *Let Q' be a state tuple with no synchronization requirements and $\mathcal{T}$ a tableau for Q'. If Q is a contradictory node in $\mathcal{T}$, then there is no interpretation structure μ and global state ξ such that $\mu, \xi \Vdash\!\vdash Q$.* □

The next three results establish the soundness of the rules. We consider the use of the rules in the context of a tableau.

Proposition 4.6 *Let μ be an interpretation structure and ξ one of its states. For each global rule, if μ and ξ satisfy the conclusion of rule, then μ and ξ satisfy one of the premisses.*

Proof. ($\Rightarrow^+$) Let Q be the conclusion, Q_1 and Q_2 be the premisses, and assume that $\alpha \Rightarrow \beta \in Q^+$. Observe that $Q^{\bowtie} = Q_1^{\bowtie} = Q_2^{\bowtie}$, $Q^{\#} =$

$Q_1^{\#} = Q_2^{\#}$ and $Q^{\mathsf{X}} = Q_1^{\mathsf{X}} = Q_2^{\mathsf{X}}$ as global rules do not affect any of the synchronization requirements. Hence, if $\mu, \xi \Vdash \alpha \Rightarrow \beta$ holds, then either $\mu, \xi \nVdash \alpha$ or $\mu, \xi \Vdash \beta$ holds. Therefore, $\mu, \xi \Vdash \widehat{Q_1}$ or $\mu, \xi \Vdash \widehat{Q_2}$ hold, implying that $\mu, \xi \Vvdash Q_1$ or $\mu, \xi \Vvdash Q_2$.
($\Rightarrow^-$) Let Q be the conclusion, Q_1 be the premiss, and assume that $\alpha \Rightarrow \beta \in Q^-$. Again, we have that $Q^{\bowtie} = Q_1^{\bowtie}$, $Q^{\#} = Q_1^{\#}$ and $Q^{\mathsf{X}} = Q_1^{\mathsf{X}}$. Hence, if $\mu, \xi \nVdash \alpha \Rightarrow \beta$ holds, then $\mu, \xi \Vdash \alpha$ and $\mu, \xi \nVdash \beta$ also hold. Therefore, $\mu, \xi \Vdash \widehat{Q_1}$ holds, implying that $\mu, \xi \Vvdash Q_1$. □

Proposition 4.7 *Let μ be an interpretation structure and ξ one of its global states. For each local rule for $\Rightarrow$, G and ©$_i$, if μ and ξ satisfy the conclusion of a rule, then μ and ξ satisfy one of its premisses.*

Proof. The proof for ($\Rightarrow_i^+$) and ($\Rightarrow_i^-$) is similar to the global case, so we start by considering the rules for the temporal operator G. As before, we note that the synchronization requirements are not affected by these rules.
(G_i^+) Let Q be the conclusion of the rule, Q_1 its premiss, and assume that $@_i[\mathsf{G}\,\varphi] \in Q^+$. If $\mu, \xi \Vdash @_i[\mathsf{G}\,\varphi]$, then it is easy to conclude that $\mu, \xi \Vdash @_i[\varphi]$ and $\mu, \xi \Vdash @_i[\mathsf{X}\,\mathsf{G}\,\varphi]$. Hence, if $\mu, \xi \Vdash \widehat{Q}$, then $\mu, \xi \Vdash \widehat{Q_1}$ and, thus, if $\mu, \xi \Vvdash Q$ then $\mu, \xi \Vvdash Q_1$.
(G_i^-) Let Q be the conclusion of the rule, Q_1 and Q_2 its premisses, and assume that $@_i[\mathsf{G}\,\varphi] \in Q^-$. Clearly, if $\mu, \xi \nVdash @_i[\mathsf{G}\,\varphi]$, then $\mu, \xi \nVdash @_i[\varphi]$ or $\mu, \xi \nVdash @_i[\mathsf{X}\,\mathsf{G}\,\varphi]$. Hence, if $\mu, \xi \Vdash \widehat{Q}$, then either $\mu, \xi \Vdash \widehat{Q_1}$ or $\mu, \xi \Vdash \widehat{Q_2}$, which implies that if $\mu, \xi \Vvdash Q$, then $\mu, \xi \Vvdash Q_1$ or $\mu, \xi \Vvdash Q_2$.
(©$_i^+$) Let Q be the conclusion of the rule and Q_1 its premiss. Assume that $\mu, \xi \Vvdash Q$. By Proposition 4.5, it follows that Q is not contradictory. Furthermore, $\mu, \xi \Vdash @_i[©_j[\varphi]]$ for every $@_i[©_j[\varphi]] \in \eta_{i,j}^{\bowtie}(Q)$. Then, $\mu_i, \xi|_i \Vdash_i ©_j[\varphi]$ and thus $\xi|_i \neq \emptyset$, $last(\xi|_i) \in Ev_j$ and $\mu_j, last(\xi|_i){\downarrow_j} \Vdash_j \varphi$. Since $Q_1^{\mathsf{X}} = Q^{\mathsf{X}}$ and μ and ξ meet the synchronization requirements in Q, then conditions ($S1$) and ($S2$) also hold for Q_1. Additionally, as $Q_1^{\#} = Q^{\#}$, condition ($S4$) also holds for Q_1. Finally, we show that ($S3$) holds for Q_1. Given that $\xi|_i \neq \emptyset$, then $\xi \neq \emptyset$. By ($S1$), $Q^{\mathsf{X}} \neq \emptyset$. We can then use Lemma 4.4 to conclude that $i \in Q^{\mathsf{X}}$. Moreover, by condition ($S2$), there is a maximal event e_Q with respect to ξ such that $\{i \in Id \mid e_Q \in Ev_i\} = Q^{\mathsf{X}}$, and, therefore, $e_Q \in Ev_i$ and $last(\xi|_i) = e_Q$. Since $last(\xi|_i) \in Ev_j$, then $e_Q \in Ev_j$, which implies that $j \in Q^{\mathsf{X}}$. As a consequence, given that ($S3$) holds for Q, $Q_1^{\bowtie} = Q^{\bowtie} \cup \{j\} \subseteq Q^{\mathsf{X}} = Q_1^{\mathsf{X}}$. Hence, ($S3$) holds for Q_1. Thus, μ and ξ meet the synchronization requirements in Q_1. In order to prove that $\mu, \xi \Vdash \widehat{Q_1}$, it is enough to prove that $\mu, \xi \Vdash @_j[\varphi]$. To this end, given that e_Q is maximal with

respect to ξ and $e_Q \in Ev_j$, it follows that $last(\xi|_i) = last(\xi|_j)$, and therefore, $\mu_j, last(\xi|_j){\downarrow_j}\Vdash_j \varphi$, that is $\mu, \xi \Vdash @_j[\varphi]$. Thus, $\mu, \xi \Vvdash Q_1$.
($©_i^-$) Let Q be the conclusion of the rule, and Q_1 and Q_2 its left and right premisses, respectively. Assume that $\mu, \xi \Vvdash Q$. By Proposition 4.5, Q is not contradictory. Furthermore, as $\mu, \xi \nVdash @_i[©_j[\varphi]]$, for every $@_i[©_j[\varphi]] \in \eta^{\#}_{i,j}(Q) \neq \emptyset$, then either $\xi|_i = \emptyset$, or $last(\xi|_i) \notin Ev_j$, or $\mu_j, last(\xi|_i){\downarrow_j}\nVdash_j \varphi$. Furthermore μ and ξ meet the synchronization requirements in Q. Since $Q_1^{\mathsf{X}} = Q_2^{\mathsf{X}} = Q^{\mathsf{X}}$, conditions ($S1$) and ($S2$) also hold for Q_1 and Q_2. As $Q_1^{\#} = Q^{\#}$, condition ($S4$) holds for Q_1. As $Q_2^{\bowtie} = Q^{\bowtie}$, condition ($S3$) holds for Q_2. And, as $Q_2^+ = Q^+$ and $Q_2^- \subseteq Q^-$, $\mu, \xi \Vdash \widehat{Q_2}$. Let us now consider two cases. If $Q^{\mathsf{X}} = \emptyset$, then we can conclude that $\mu, \xi \Vvdash Q_2$, as $Q_2^{\mathsf{X}} = \emptyset$, and therefore condition ($S4$) holds trivially for Q_2. If, on the other hand, $Q^{\mathsf{X}} \neq \emptyset$, then we can use Lemma 4.4 to conclude that $i \in Q^{\mathsf{X}}$. Reasoning as in ($©_i^+$), it follows that there is $e_Q \in Ev_i \cap \xi$, and so $\xi|_i \neq \emptyset$ and $last(\xi|_i) = e_Q$. If $last(\xi|_i) \notin Ev_j$, then $e_Q \notin Ev_j$, and therefore $j \notin Q^{\mathsf{X}} = Q_2^{\mathsf{X}}$. Since ($S4$) holds for Q, we have that $(Q^{\#} \cup \{j\}) \cap Q_2^{\mathsf{X}} = Q_2^{\#} \cap Q_2^{\mathsf{X}} = \emptyset$, and therefore condition ($S4$) holds again for Q_2. Hence, $\mu, \xi \Vvdash Q_2$. Finally, if $\xi|_i \neq \emptyset$, $last(\xi|_i) \in Ev_j$ and $\mu_j, last(\xi|_i){\downarrow_j}\nVdash_j \varphi$, we can reason as in ($©_i^+$) to conclude that $\mu, \xi \Vvdash Q_1$. □

Proposition 4.8 *Let μ be an interpretation structure and ξ one of its global states. For each $e \in Ev \setminus \xi$ such that $\xi \cup \{e\}$ is also a global state of μ, if μ and ξ satisfy the conclusion of the rule* (X)*, then μ and $\xi \cup \{e\}$ satisfy the K premiss, with $K = Ids(e)$.*

Proof. Let Q be the conclusion and Q_1 the premiss corresponding to $K = Ids(e)$. Assume that $\mu, \xi \Vvdash Q$ holds. Furthermore, let $@_k[\varphi] \in Q_1^+$, for some $k \in Id$.

If $k \notin K$, then $@_k[\varphi]$ was already in Q^+. Furthermore, $\xi|_k = (\xi \cup \{e\})|_k$. Then, since $\mu_k, \xi|_k \Vdash_k \varphi$, it follows that $\mu, \xi \cup \{e\} \Vdash @_k[\varphi]$.

If $k \in K$, then $\xi|_k \cup \{e\} \in \Xi_k$. Furthermore, it must be the case that $@_k[\varphi] \in \sigma_K^+(Q)$, i.e., $@_k[\mathsf{X}\,\varphi] \in Q^+$. Since $\mu, \xi \Vdash @_k[\mathsf{X}\,\varphi]$, we have that $\mu_k, \xi|_k \Vdash_k \mathsf{X}\,\varphi$. Consequently, $\mu_k, \xi|_k \cup \{e\} \Vdash_k \varphi$, which implies that $\mu, \xi \cup \{e\} \Vdash @_k[\varphi]$.

The proof for $@_k[\varphi] \in Q_1^-$ is similar. Hence, we conclude that $\mu, \xi \cup \{e\} \Vdash \widehat{Q_1}$.

Next, we prove that μ and $\xi \cup \{e\}$ meet all the synchronization requirements in Q_1. Condition ($S1$) holds trivially since $Q_1^{\mathsf{X}} = K \neq \emptyset$. Conditions ($S3$) and ($S4$) also hold trivially because $Q_1^{\bowtie} = Q_1^{\#} = \emptyset$. Condition ($S2$) follows by construction as $Q_1^{\mathsf{X}} = \mathit{Ids}(e) = \{i \in \mathit{Id} : e \in \mathit{Ev}_i\}$, $e \in \xi \cup \{e\}$ and, e is maximal with respect to $\xi \cup \{e\}$. □

Let $\mu = \langle \lambda, \vartheta \rangle$ be an interpretation structure and let $\langle \mathit{Ev}, \leq_{\mathit{Ev}} \rangle$ be the underlying global order on events. It is always possible to linearize $\langle \mathit{Ev}, \leq_{\mathit{Ev}} \rangle$, i.e., it is always possible to define a bijection $\ell : \mathbb{N} \to \mathit{Ev}$ such that if $k_1 <_{\mathbb{N}} k_2$, then $\ell(k_1) <_{\mathit{Ev}} \ell(k_2)$, where $<_{\mathbb{N}}$ is the usual ordering on the natural numbers. See, e.g., [4]. From now on, we assume fixed a linearization function (or just *linearization*) ℓ, which induces an enumeration of the global states as follows

- $\xi^0 = \emptyset$;
- $\xi^{k+1} = \xi^k \cup \{\ell(k+1)\}$ for each $k \geq 0$.

Recall that we are assuming a tableau $\mathcal{T}$ for Q. We denote by $\pi_Q^{\mu,\ell}$ the (finite or infinite) path $\pi_0^{\mu,\ell} \cdot \pi_1^{\mu,\ell} \cdot \ldots$ in $\mathcal{T}$ defined as follows:

- $\pi_0^{\mu,\ell} = Q_{0,0}, \ldots, Q_{0,n_0}$ such that:
  - if Q is contradictory or a state, then $Q_{0,0} = Q$ and $n_0 = 0$;
  - otherwise, $Q_{0,0} = Q$, the node Q_{0,n_0} is either a contradictory node or a state, and for each $0 \leq m < n_0$, the node $Q_{0,m}$ is neither a contradictory node nor a state and
    - if it has just one successor node Q', then $Q_{0,m+1} = Q'$;
    - if it has a left successor node Q' and a right successor node Q'', then $Q_{0,m+1} = Q'$ if $\mu, \xi^0 \Vdash Q'$, and $Q_{0,m+1} = Q''$ otherwise;
- if the last node of $\pi_k^{\mu,\ell}$ is contradictory, then $\pi_Q^{\mu,\ell} = \pi_0^{\mu,\ell} \cdot \ldots \cdot \pi_k^{\mu,\ell}$, otherwise $\pi_{k+1}^{\mu,\ell} = Q_{k+1,0}, \ldots, Q_{k+1,n_{k+1}}$ such that $Q_{k+1,0}$ is obtained from the last node of $\pi_k^{\mu,\ell}$ by rule (X) with set $\mathit{Ids}(\ell(k+1))$ and
  - if $Q_{k+1,0}$ is contradictory or a state, then $n_{k+1} = 0$;
  - otherwise, $Q_{k+1,n_{k+1}}$ is either a contradictory node or a state, and for each $0 \leq m < n_{k+1}$, the node $Q_{k+1,m}$ is neither a contradictory node nor a state and
    - if it has just one successor node Q', then $Q_{k+1,m+1} = Q'$;
    - if it has a left successor node Q' and a right successor node Q'', then $Q_{k+1,m+1} = Q'$ if $\mu, \xi^{k+1} \Vdash Q'$, and $Q_{k+1,m+1} = Q''$ otherwise.

Observe that each segment $\pi_k^{\mu,\ell}$ ends with a contradictory node or with a (non-contradictory) state. Moreover, when a contradictory node Q' is found, the path $\pi_Q^{\mu,\ell}$ is finite and its last node is Q'. Otherwise, the path is infinite. Rule (X) is only applied at the last node of each segment, unless a contradictory node is found. As we will prove below, in the case of an infinite path, the information (formulas and synchronization requirements) in the nodes of a segment $\pi_k^{\mu,\ell}$ must hold in the global state ξ^k.

We state without proving that the construction of each of the previous segments is always finite because the sets Q^+,Q^- and Q^{X} of each state tuple are always finite and each rule (other than (G)) always reduces the complexity of the formula. The only one that does not is the rule for G, but in this case the resulting formula with greater complexity cannot be addressed before rule X is applied.

Proposition 4.9 *Let Q be a state tuple with no synchronization requirements and $\mathcal{T}$ a tableau for Q. Let μ an interpretation structure such that $\mu, \emptyset \Vdash \widehat{Q}$ and ℓ be a linearization of μ. Then μ and ξ^k satisfy all the nodes in $\pi_k^{\mu,\ell}$, for all $k \in \mathbb{N}_0$. Furthermore, $\pi_Q^{\mu,\ell}$ is infinite and does not contain any closed node.*

Proof. We prove by induction on $k \in \mathbb{N}_0$ that μ and ξ^k satisfy all the states in $\pi_k^{\mu,\ell}$.

The basis corresponds to proving that μ and ξ^0 satisfy all the states in $\pi_0^{\mu,\ell} = Q_{0,0}, \ldots, Q_{0,n_0}$. Again, the proof follows by induction. As $Q_{0,0} = Q$ and $\mu, \emptyset \Vdash \widehat{Q}$, then μ and ξ^0 satisfy $Q_{0,0}$. Recall that $Q_{0,0}^{\bowtie} = Q_{0,0}^{\#} = Q_{0,0}^{\mathsf{X}} = \emptyset$. If μ and ξ^0 satisfy $Q_{0,m}$, then, using Proposition 4.6 and Proposition 4.7, μ and ξ^0 also satisfy $Q_{0,m+1}$.

We now prove that μ and ξ^{k+1} satisfy all the nodes in $\pi_{k+1}^{\mu,\ell} = Q_{k+1,0}, \ldots, Q_{k+1,n_{k+1}}$. By the induction hypothesis, we know that μ and ξ^k satisfy Q_{k,n_k}. Hence, by Proposition 4.8, if follows that μ and ξ^{k+1} satisfy $Q_{k+1,0}$, by the construction of the path. The rest of the proof follows as in the base case.

This result implies that $\pi_Q^{\mu,\ell}$ is infinite, otherwise, there would be some contradictory node, which cannot be the case because, by Proposition 4.5, no contradictory node is satisfiable.

Next we prove that no node in $\pi_Q^{\mu,\ell}$ is closed. To this end, we will prove by induction on the set of closed nodes $Cl_{\mathcal{T}}$ that every node in $Cl_{\mathcal{T}}$ does not occur in $\pi_Q^{\mu,\ell}$. Recalling ($C1$) in Figure 4.5, the basis of induction corresponds to proving that no contradictory node occurs in $\pi_Q^{\mu,\ell}$. As referred above, this is always the case. The induction step corresponds to proving that if a node is in $Cl_{\mathcal{T}}$ due to ($C2$), ($C3$) or

$(C4)$, then it does not occur in $\pi_Q^{\mu,\ell}$. The proof of this result is presented in the Appendix. □

The intended soundness result now follows as a corollary of the previous proposition.

Corollary 4.10 *Let Q be a state tuple with $Q^{\bowtie} = Q^{\#} = Q^{\mathsf{X}} = \emptyset$, and let $\mathcal{T}$ be a tableau for Q. If $\widehat{Q}$ is satisfiable, then $\mathcal{T}$ is successful.* □

Hence, if a tableau for $Q = \langle \emptyset, \{\alpha\}, \emptyset, \emptyset, \emptyset\rangle$ is unsuccessful, then there is no interpretation structure μ such that $\mu, \emptyset \not\Vdash \alpha$, and therefore α is valid.

4.5 Completeness (and decidability)

We now prove the completeness of the tableaux system by proving that if $\mathcal{T}$ is a successful tableau for a state tuple Q with no synchronization requirements, then $\widehat{Q}$ is satisfiable. Hence, if $\widehat{Q}$ is not satisfiable, then all the tableaux for Q are unsuccessful. In particular, if $Q^+ = \emptyset$ and $Q^- = \{\alpha\}$ and we can then conclude that if α is a valid formula, then every tableau for Q is unsuccessful. We also establish the decidability of the system.

To this end, we need to establish some technical results along the way. We will prove that from any successful tableau for Q it is possible to extract an interpretation structure satisfying $\widehat{Q}$. Recall that we are assuming that $\mathcal{T}$ is a tableau for a state tuple with no synchronization requirements. From now on, we assume that $\mathcal{T}$ is successful. Consider an infinite path $Q_0, Q_1, \ldots$ in $\mathcal{T}$. We say that such a path $Q_0, Q_1, \ldots$ is *complete* if Q_0 is Q and, for every $n \in \mathbb{N}_0$,

- Q_n is not closed;
- if $@_i[\mathsf{G}\,\varphi] \in Q_n^-$, for some $i \in Id$ and $\varphi \in \mathcal{L}_i$, then there is some $m \geq n$ such that $@_i[\varphi] \in Q_m^-$;
- for every $i \in Id$ there is some $m \geq n$ such that $i \in Q_m^{\mathsf{X}}$.

The second condition ensures that there are no delayed eventualities along a complete path. The third condition ensures that, for each agent $i \in Id$, rule (X) with $i \in K$ is applied an infinite number of times along a complete path. This implies that no agent is prevented from progressing. We start by proving that every successful tableau has a complete path. The proof closely follows [8], but it is adapted to encompass the condition on the progress of agents.

Lemma 4.11 *Let Q be a state tuple with no synchronization requirements. If $\mathcal{T}$ is a successful tableau for Q, then $\mathcal{T}$ has a complete path.*

Proof. As it was already observed, $\mathcal{T}$ is a finite graph. Hence, the number of formulas φ such that $@_i[\mathsf{G}\,\varphi] \in Q'^-$ for some node Q' of $\mathcal{T}$ and some $i \in Id$ is finite. Let $\varphi_0, \ldots, \varphi_{q-1}$ be a enumeration of such formulas. Consider the sequence $\pi_0, \pi_1, \ldots$ of finite, non-empty paths in $\mathcal{T}$ that do not contain any closed nodes and such that π_n is a proper prefix of π_{n+1} defined as follows:

- π_0 is Q, which is not closed because $\mathcal{T}$ is successful;
- assume now that $\pi_n = Q_0, \ldots, Q_k$ has been defined and therefore none of its nodes is closed. We consider two cases. If $@_i[\mathsf{G}\,\varphi_{n \bmod q}] \notin Q_k^-$ or $@_i[\varphi_{n \bmod q}] \in Q_k^-$, then we just pick π' to be the sequence constituted by one of the successor nodes Q' of Q_k, that must exist because Q_k is not closed. If $@_i[\mathsf{G}\,\varphi_{n \bmod q}] \in Q_k^-$ and $@_i[\varphi_{n \bmod q}] \notin Q_k^-$, then, as Q_k is not closed, there must exist a finite path π' from Q_k to some node Q' such that $@_i[\varphi_{n \bmod q}] \in Q'^-$ that contains no closed nodes.

  Let j be the $(n \bmod |Id|)$-th agent identifier. If $j \in Q''^{\mathsf{X}}$, for some Q'' in π', then we define π_{n+1} to be the concatenation of π_n with π'. If, on the other hand, $j \notin Q''^{\mathsf{X}}$, for every Q'' in π', then, as Q' (the last node of π') is not closed, there is a finite path π'' from Q' to some Q''' such that $j \in Q'''^{\mathsf{X}}$ that contains no closed nodes. In this case, we define π_{n+1} to be the concatenation of π_n with π' and π''.

Clearly, $\pi_0, \pi_1, \ldots$ determines a complete path in $\mathcal{T}$. □

Given a complete path $\pi = Q_0, Q_1, \ldots,$ we introduce the following notation. We define the function $cnt : \mathbb{N}_0 \to \mathbb{N}_0$ as follows:

$$cnt(n) = |\{k < n \mid Q_k \text{ is a state}\}|\,.$$

Given $n \in \mathbb{N}_0$, $cnt(n)$ counts the number of states that occur in the path until node Q_n. As the path is complete, this function is clearly monotonic and surjective. We define the inverse of cnt to be the function $\rceil : \mathbb{N}_0 \to \mathbb{N}_0$ such that:

$$\rceil(n) = \max\{k \in \mathbb{N}_0 \mid cnt(k) = n\}\,.$$

Given $n \in \mathbb{N}_0$, $\rceil(n)$ is the index of n-th state along the path $Q_0, Q_1, \ldots$

Some considerations are due at this point. First, we observe that $\mid (cnt(n))$ is the index of the first state that occurs after n, if Q_n is not a state. If Q_n is a state, then $\mid (cnt(n)) = n$. Additionally, the following also hold: $cnt(\mid (k)) = k$, $\mid (cnt(k)) \geq k$ and $cnt(\mid (k) + 1) = cnt(\mid (k)) + 1$, for every $k \in \mathbb{N}_0$.

Example 4.12 *Recall the tableau in Example 4.2. The tableau is successful so it is possible to extract a complete path. An example of such a path is*

$$N_0\, N_2\, N_5\, N_8\, N_7\, N_{11}\, N_{12}\, N_{11}\, N_{11}\, N_{11} \ldots$$

Observe that the first node of the path, Q_0, is node N_0, the second node of the path, Q_1, is node N_2, the third node of the path, Q_2, is node N_5, and so on. Furthermore, Q_k is N_{11} for each each $k \geq 7$. The states in this path are N_2, N_7, N_8, N_{11} and N_{12}. Hence, $\mid (0) = 1$ and $\mid (k) = k+2$ for $k > 0$. Observe that $Q_3^{\mathsf{X}} = \{i, j\}$, $Q_4^{\mathsf{X}} = \{i\}$, $Q_5^{\mathsf{X}} = \{i, j\}$, $Q_6^{\mathsf{X}} = \{j\}$ and $Q_k^{\mathsf{X}} = \{i, j\}$, for $k > 6$, and so, $i \notin Q^{\mathsf{X}}_{\mid (4)}$ and $j \notin Q^{\mathsf{X}}_{\mid (2)}$. □

We now establish a technical result that states that between two consecutive states, if the transition did not involve an agent, then nothing about that agent changes. The proof is presented in the Appendix.

Lemma 4.13 *Let π be a complete path, $n \in \mathbb{N}_0$ and $i \in Id$. If $i \notin Q^{\mathsf{X}}_{\mid (n+1)}$ then $Q^{+}_{\mid (n)}\!\downarrow_i = Q^{+}_{\mid (n+1)}\!\downarrow_i$ and $Q^{-}_{\mid (n)}\!\downarrow_i = Q^{-}_{\mid (n+1)}\!\downarrow_i$.* □

We can finally prove the completeness result.

Proposition 4.14 *Let Q be a state tuple with no synchronization requirements. If $\mathcal{T}$ is a successful tableau for Q, then $\widehat{Q}$ is satisfiable.*

Proof. Let $Q_0, Q_1, \ldots$ be a complete path in $\mathcal{T}$, that exists by Lemma 4.11. Let $E = \{e_n \mid n \in \mathbb{N}\}$ and, for each $i \in Id$, let $Ev_i = \{e_k \mid k \in \mathbb{N}$ and $i \in Q^{\mathsf{X}}_{\mid (k)}\}$. Then, define $\lambda_i = \langle Ev_i, \leq_i \rangle$ to be the local life-cycle such that $e_n \leq_i e_m$ if $n \leq m$ and set $\lambda = \{\lambda_i\}_{i \in Id}$ as the corresponding distributed life-cycle.

Next, for each $i \in Id$, let $\vartheta_i : \Xi_i \to \wp(Prop_i)$ be the such that

- if $@_i[p] \in Q^{+}_{\mid (i_k)}$ then $p \in \vartheta_i(\xi_i^k)$ and
- if $@_i[p] \in Q^{-}_{\mid (i_k)}$ then $p \notin \vartheta_i(\xi_i^k)$

for each $\xi_i^k = \{e_{i_1}, \ldots, e_{i_k}\} \in \Xi_i$, where $e_{i_l} \leq e_{i_{l'}}$ whenever $l \leq l'$, and assuming that $i_0 = 0$ and $\xi_i^0 = \emptyset$. Note that ϑ_i is well defined because the path is complete, and so $Q_n^+ \cap Q_n^- = \emptyset$ for every $n \in \mathbb{N}_0$. Set $\vartheta = \{\vartheta_i\}_{i \in Id}$. Then $\mu = \langle \lambda, \vartheta \rangle$ is a global interpretation structure.

Given the way events were identified and ordered above, and how this is closely related to the (sequential) enumeration of the states of the path, we can consider the following enumeration on global states:

- $\xi^0 = \emptyset$;
- $\xi^k = \{e_1, \ldots, e_k\}$, for $k > 0$.

Under these conditions, we prove that $\mu, \xi^{cnt(n)} \Vdash\!\vdash Q_n$, for every $n \in \mathbb{N}_0$, or, equivalently, $\mu, \xi^{cnt(n)} \Vdash \widehat{Q}_n$ and μ and $\xi^{cnt(n)}$ meet the requirements in Q_n. We start by proving the first condition. To this end, we just have to establish that

- if $@_i[\varphi] \in Q_n^+$, then $\mu, \xi^{cnt(n)} \Vdash @_i[\varphi]$;
- if $@_i[\varphi] \in Q_n^-$, then $\mu, \xi^{cnt(n)} \nVdash @_i[\varphi]$.

The proof follows by induction on the structure of the formula φ, simultaneously for all agents, and it is presented in the Appendix, along with the proof of the second condition. Therefore, $\mu, \emptyset \Vdash \widehat{Q}_0$ since $cnt(n) = 0$. Recalling that $Q_0 = Q$, we conclude that $\widehat{Q}$ is satisfiable. □

Hence, if a global formula α is valid, then $Q = \langle \emptyset, \{\alpha\}, \emptyset, \emptyset, \emptyset \rangle$ is not satisfiable and therefore any tableau for Q is unsuccessful.

Example 4.15 *Recall the tableau in Example 4.2 and its complete path*

$$N_0\, N_2\, N_5\, N_8\, N_7\, N_{11}\, N_{12}\, N_{11}\, N_{11}\, N_{11} \ldots$$

given in Example 4.15. We may extract the following interpretation structure μ from this path:

- $\lambda_i = \langle Ev_i, \leq_i \rangle$ *where* $Ev_i = Ev \setminus \{e_4\}$;
- $\vartheta_i : \Xi_i \to \wp(Prop_i)$ *is such that* $\vartheta_i(\xi_i) = \emptyset$;
- $\lambda_j = \langle Ev_j, \leq_j \rangle$ *where* $Ev_j = Ev \setminus \{e_2\}$;
- $\vartheta_j : \Xi_j \to \wp(Prop_j)$ *is such that:*
  - $\vartheta(\emptyset) = \emptyset$ *given that* $@_j[p] \in Q^-_{|(0)}$;
  - $\vartheta(\{e_1\}) = \{p\}$ *given that* $@_j[p] \in Q^+_{|(1)}$;
  - $\vartheta(\xi_j) = \emptyset$ *given that* $@_j[p]$ *does not appear in any other state for* j.

It is not very difficult to conclude that $\mu, \emptyset \nVdash @_i[\mathsf{X}\,\textcircled{C}_j[p]] \Rightarrow @_j[p]$, as expected.   □

The following result is an immediate consequence of Proposition 4.14 and Corollary 4.10 and of the fact that our tableaux are always finite.

Theorem 4.16 *The satisfiability problem for DTL is decidable.*   □

Consequently, the validity problem is also decidable.

4.6   Rules for other connectives

We can extend the above tableaux system to other connectives that are usually introduced as abbreviations.

The rules for conjunction and disjunction are straightforward and are given in Figure 4.11 and Figure 4.12, respectively. The rules for negation have already been introduced (see Figure 4.9). As expected, global rules for conjunction, disjunction and negation are similar.

$$\langle @_i[\varphi \wedge \psi] \in Q^+, Q^-, Q^\bowtie, Q^\#, Q^\times \rangle$$

$$(\wedge_i^+)$$

$$\langle (Q^+ \setminus \{@_i[\varphi \wedge \psi]\}) \cup \{@_i[\varphi], @_i[\psi]\}, Q^-, Q^\bowtie, Q^\#, Q^\times \rangle$$

$$\langle Q^+, @_i[\varphi \wedge \psi] \in Q^-, Q^\bowtie, Q^\#, Q^\times \rangle$$

$$(\wedge_i^-)$$

$$\langle Q^+, (Q^- \setminus \{@_i[\varphi \wedge \psi]\}) \cup \{@_i[\varphi]\}, Q^\bowtie, Q^\#, Q^\times \rangle \quad \langle Q^+, (Q^- \setminus \{@_i[\varphi \wedge \psi]\}) \cup \{@_i[\psi]\}, Q^\bowtie, Q^\#, Q^\times \rangle$$

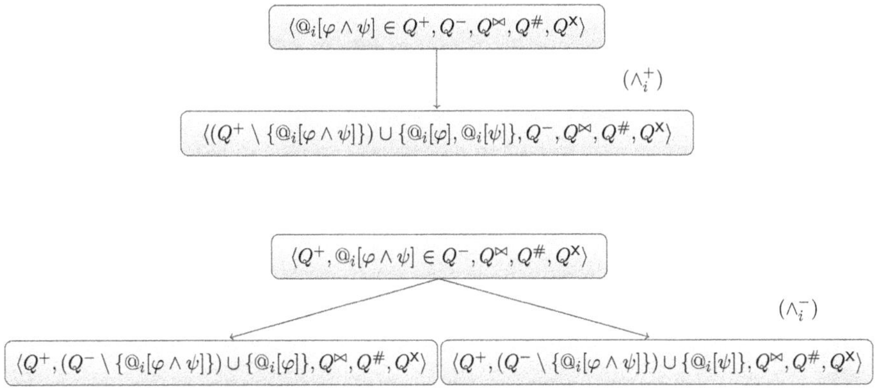

Figure 4.11: Local rules for $\wedge$.

$$\langle @_i[\varphi \vee \psi] \in Q^+, Q^-, Q^\bowtie, Q^\#, Q^\times \rangle$$

$(\vee_i^+)$

$$\langle (Q^+ \setminus \{@_i[\varphi \vee \psi]\}) \cup \{@_i[\varphi]\}, Q^-, Q^\bowtie, Q^\#, Q^\times \rangle \quad \langle (Q^+ \setminus \{@_i[\varphi \vee \psi]\}) \cup \{@_i[\psi]\}, Q^-, Q^\bowtie, Q^\#, Q^\times \rangle$$

$$\langle Q^+, @_i[\varphi \vee \psi] \in Q^-, Q^\bowtie, Q^\#, Q^\times \rangle$$

$(\vee_i^-)$

$$\langle Q^+, (Q^- \setminus \{@_i[\varphi \vee \psi]\}) \cup \{@_i[\varphi], @_i[\psi]\}, Q^\bowtie, Q^\#, Q^\times \rangle$$

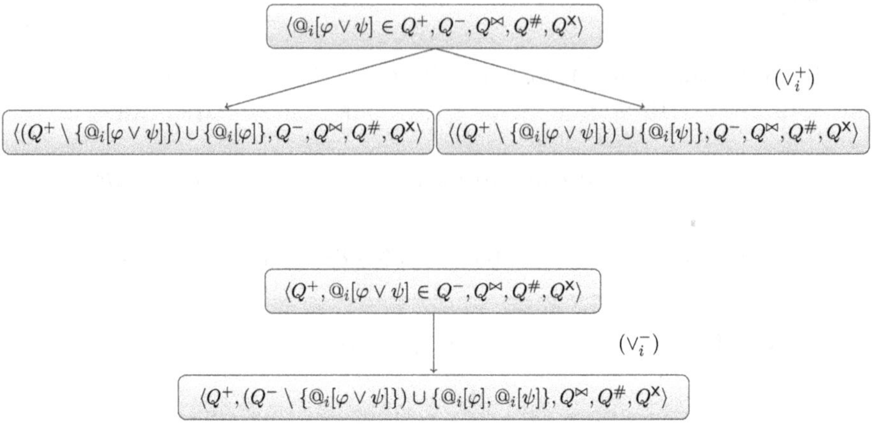

Figure 4.12: Local rules for $\vee$.

The operator $\mathsf{F}\,\varphi \equiv \neg\,\mathsf{G}\,\neg\,\varphi$ has the rules depicted in Figure 4.13. In addition, we also need to extend the definition of the set of closed nodes $Cl_{\mathcal{T}}$ with

(C5) $Q \in Cl_{\mathcal{T}}$ whenever $@_i[\mathsf{F}\,\varphi] \in Q^+$, for some $i \in Id$ and $\varphi \in \mathcal{L}_i$, and every path from Q to a node Q' with $@_i[\varphi] \in Q'^+$ has a node in $Cl_{\mathcal{T}}$.

$$\langle @_i[\mathsf{F}\,\varphi] \in Q^+, Q^-, Q^\bowtie, Q^\#, Q^\times \rangle$$

(F_i^+)

$$\langle (Q^+ \setminus \{@_i[\mathsf{F}\,\varphi]\}) \cup \{@_i[\varphi]\}, Q^-, Q^\bowtie, Q^\#, Q^\times \rangle \quad \langle (Q^+ \setminus \{@_i[\mathsf{F}\,\varphi]\}) \cup \{@_i[\mathsf{X}\,\mathsf{F}\,\varphi]\}, Q^-, Q^\bowtie, Q^\#, Q^\times \rangle$$

$$\langle Q^+, @_i[\mathsf{F}\,\varphi] \in Q^-, Q^\bowtie, Q^\#, Q^\times \rangle$$

(F_i^-)

$$\langle Q^+, (Q^- \setminus \{@_i[\mathsf{F}\,\varphi]\}) \cup \{@_i[\varphi], @_i[\mathsf{X}\,\mathsf{F}\,\varphi]\}, Q^\bowtie, Q^\#, Q^\times \rangle$$

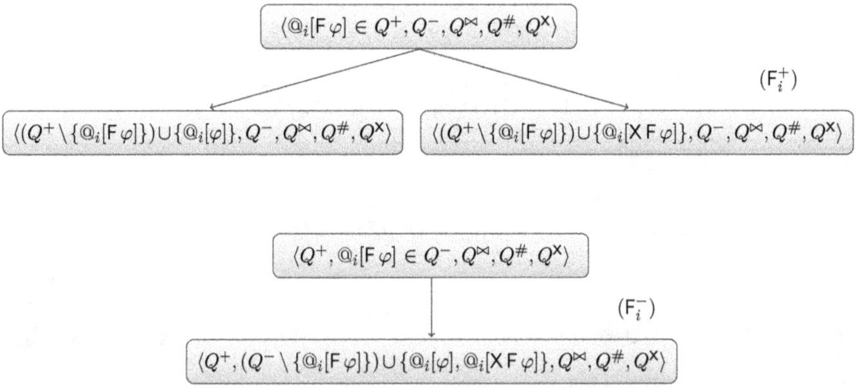

Figure 4.13: Local rules for F.

Example 4.17 *Recall the DTL signature in Example 4.2. Consider the state tuple Q such that $Q^+ = \{@_i[\mathsf{F}\,\textcircled{c}_j[p]], @_j[\mathsf{G}\,\neg p]\}$ and $Q^- = Q^\bowtie = Q^\# = Q^\times = \emptyset$. A tableau for Q is given in Figure 4.14.*

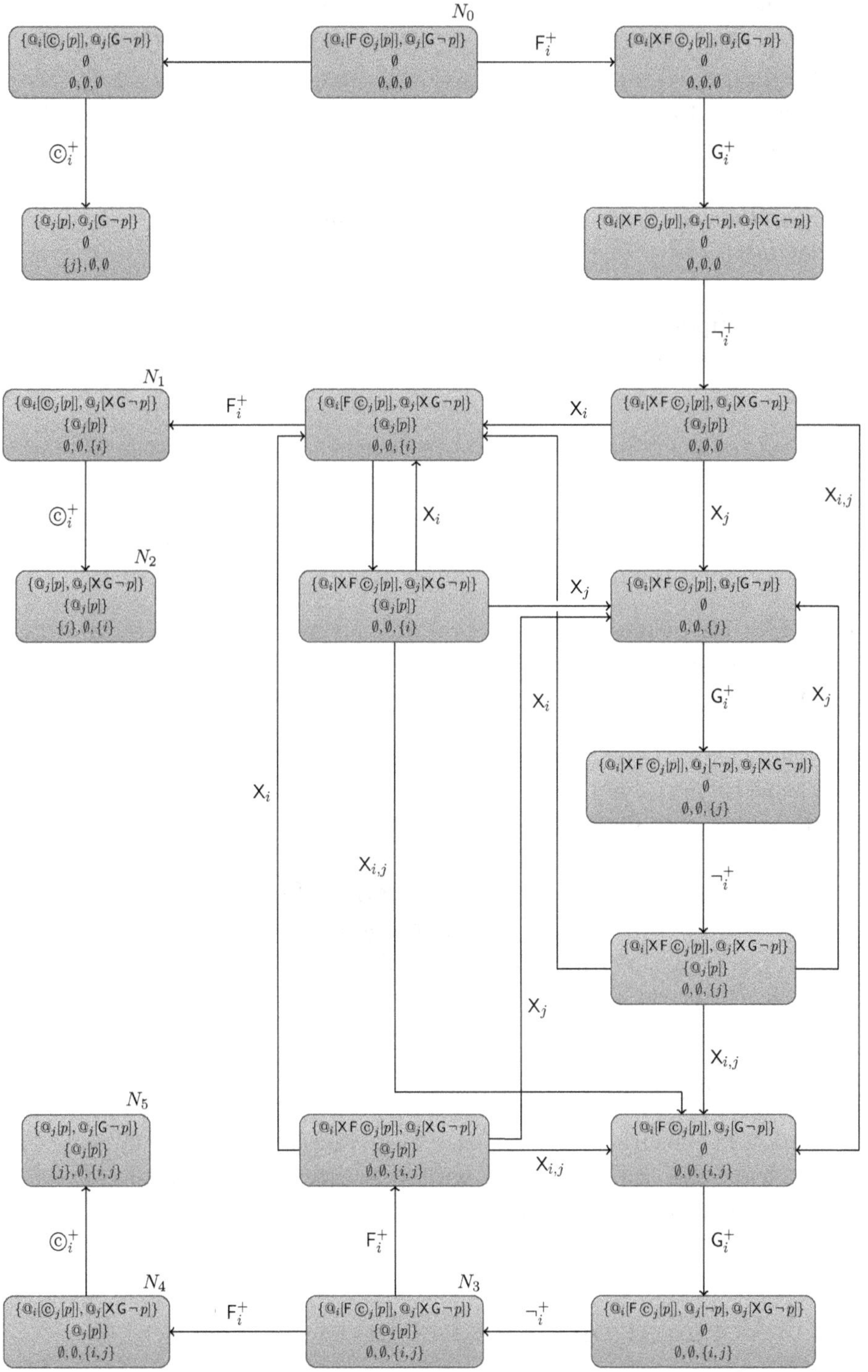

Figure 4.14: Tableau for $\langle\{\{@_i[\mathsf{F}\,©_j[p]], @_j[\mathsf{G}\,\neg p]\}, \emptyset, \emptyset, \emptyset, \emptyset\rangle$.

Nodes N_2 and N_5 close due to $(C1)$ since $(\perp 2)$ holds. Nodes N_1 and N_4 close due $(C2)$. Node N_3 closes due to $(C5)$. Observe that $@_i[\mathsf{F}\,©_j[p]] \in N_3^+$ and the only nodes N such that $@_i[©_j[p]] \in N^+$ are N_1 and N_4. Any path starting in N_3 and ending in either N_1 or N_4 has a closed node since N_1 and N_4 are closed. All the remaining nodes are also closed but we omit the details. Hence, the tableau is unsuccessful. By Corollary 4.10, we conclude that the formula $@_i[\mathsf{F}\,©_j[p]] \wedge @_j[\mathsf{G}\,\neg p]$ is not satisfiable. □

The temporal operator until U cannot be defined by abbreviation using the operators considered herein. We extend the local satisfaction relation with:

- $\mu_i, \xi_i \Vdash_i \varphi \,\mathsf{U}\, \psi$ if there is $\xi_i' \in \Xi_i$ such that $\xi_i \subseteq \xi_i'$ and $\mu_i, \xi_i' \Vdash_i \psi$, and $\mu_i, \xi_i'' \Vdash_i \varphi$ for every $\xi_i'' \in \Xi_i$ such that $\xi_i \subseteq \xi_i'' \subset \xi_i'$.

The rules for this temporal operator are given in Figure 4.15. We also need to extend the definition of the set of closed nodes $Cl_{\mathcal{T}}$ with

(C6) $Q \in Cl_{\mathcal{T}}$ whenever $@_i[\varphi \,\mathsf{U}\, \psi] \in Q^+$, for some $i \in Id$ and $\varphi, \psi \in \mathcal{L}_i$, and every path from Q to a node Q' with $@_i[\psi] \in Q'^+$ has a node in $Cl_{\mathcal{T}}$.

The rules for U are sound and it is not difficult to prove a result similar to Proposition 4.7. Furthermore, the result stated in Proposition 4.9 still holds in this new setting. Hence, the soundness of the tableaux system is preserved.

With respect to completeness, the notion of complete path needs to be adjusted to encompass the temporal operator U. Hence, we say that a path $Q_0, Q_1, \ldots$ is complete if, in addition to the conditions stated in Section 4.5, it also holds that

- if $@_i[\varphi \,\mathsf{U}\, \psi] \in Q_n^+$ for some $i \in Id$ and $\varphi, \psi \in \mathcal{L}_i$, then there is some $m \geq n$ such that $@_i[\psi] \in Q_m^+$.

Lemma 4.11, ensuring the existence of a complete path in any successful tableau, can still be proved with this extended notion of complete path. Finally, the completeness of the tableaux system stated in Proposition 4.14 also holds in this setting.

4.7 Verification of global invariants

We now focus on the proof of global invariants. Recall that we are considering an anchored version of DTL, but, since DTL does not have global

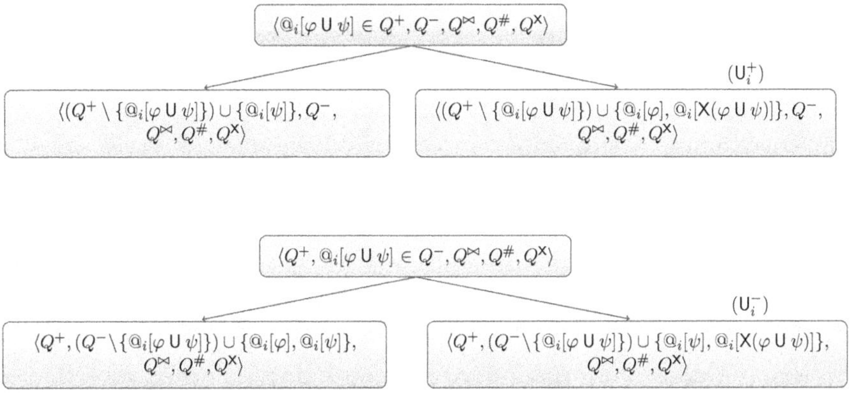

Figure 4.15: Local rules for U.

temporal operators, we cannot take advantage of the usual correspondence between anchored and floating entailment. Given a finite set of global formulas Δ and a global formula δ without temporal operators, our goal is to check if $\Delta \vDash \delta$, i.e., if, for every interpretation structure μ such that $\mu \Vdash \Delta$, it holds that $\mu, \xi \Vdash \delta$ for every global state ξ.

Since our tableaux system does not capture this kind of reasoning we present in Figure 4.18 the procedure ENT for checking whether $\Gamma \vDash \delta$. This algorithm starts by building a tableau $\mathcal{T}$ for $Q_0 = \langle \Gamma, \emptyset, \emptyset, \emptyset, \emptyset \rangle$. If the tableau is unsuccessful then the result follows immediately, by soundness. Otherwise, $\mathcal{T}$ is successful. In this case, the algorithm analyzes the tableau trying to find an interpretation structure μ such that μ satisfies Γ but δ does not hold in some state of μ. If it succeeds, then the algorithm outputs that $\Gamma \nvDash \delta$. Otherwise, it outputs that $\Gamma \vDash \delta$. Before describing the procedure in full detail, we need to introduce some notation.

Let Q be a non closed state of $\mathcal{T}$, π be a path from Q_0 to Q containing only distinct non closed nodes, and

$$\widehat{\pi}_Q = Q_1, \ldots, Q_n$$

be the sequence of states obtained from π omitting all the nodes that are not states. Observe that $Q_n = Q$ and that Q_1 is not necessarily the node Q_0. Moreover, $Q_1^{\mathsf{X}} = \emptyset$. For each $i \in Id$, let

$$\theta_i(\widehat{\pi}_Q) = \begin{cases} \max\{l \in \{2, \ldots n\} \mid i \in Q_l^{\mathsf{X}}\} & \text{if } i \in \bigcup_{l=2}^{n} Q_l^{\mathsf{X}} \\ 1 & \text{otherwise.} \end{cases}$$

The *cut induced by* $\widehat{\pi}_Q$ is the set of tuples

$$C_{\widehat{\pi}_Q} = \{\langle Q_{\theta_i(\widehat{\pi}_Q)}, \Theta_i(\widehat{\pi}_Q)\rangle \mid i \in Id\},$$

where $\Theta_i(\widehat{\pi}_Q) = \{j \in Id \mid \theta_j(\widehat{\pi}_Q) = \theta_i(\widehat{\pi}_Q)\}$. Observe that $\{\Theta_i(\widehat{\pi}_Q) \mid i \in Id\}$ constitutes a partition of Id. Given a state tuple Q, we denote by $\mathcal{C}_Q$ the set of all cuts induced by Q, i.e., the set of all $C_{\widehat{\pi}_Q}$ for every $\widehat{\pi}_Q$ defined as above. Note that the construction of $\mathcal{C}_Q$ can be done in a finite number of steps. An *extension* of the cut $C_{\widehat{\pi}_Q}$ is a set $D = \{\langle Q'_{\theta_i(\widehat{\pi}_Q)}, \Theta_i(\widehat{\pi}_Q)\rangle \mid i \in Id\}$, where each $Q'_{\theta_i(\widehat{\pi}_Q)}$ is a state tuple that it is either $Q_{\theta_i(\widehat{\pi}_Q)}$ or it results from $Q_{\theta_i(\widehat{\pi}_Q)}$ by adding some global formulas. It is an *atomic extension* if we only add atomic formulas. Note that a particular case of extension of a cut is the cut itself. Given $i \in Id$, in the sequel we will often write just D_i for the state tuple $Q'_{\theta_i(\widehat{\pi}_Q)}$.

We say that D is *contradictory* whenever there is $i \in Id$ such that D_i is a contradictory state tuple. Given an interpretation structure μ and $\xi \in \Xi$, we say that $\mu, \xi \Vdash D$ if $\mu, last(\xi|_i){\downarrow} \Vert\!\!\vdash D_i$ for all $i \in Id$. For the sake of simplicity, we assume that $last(\xi|_i){\downarrow} = \emptyset$ whenever $\xi|_i = \emptyset$.

Noting that the synchronization requirements of $Q'_{\theta_i(\widehat{\pi}_Q)}$ and $Q_{\theta_i(\widehat{\pi}_Q)}$ are the same, if D_i is both contradictory and satisfiable, then $Q_{\theta_i(\widehat{\pi}_Q)}$ is also contradictory and satisfiable, contradicting Proposition 4.5. Hence, if D is contradictory, there is no interpretation structure μ and $\xi \in \Xi$ such that $\mu, \xi \Vdash D$.

Let $\mathcal{T}$ be a successful tableau for Q_0, Q a non closed state and $C \in \mathcal{C}_Q$. Let D be a non contradictory atomic extension of C, and let $\mathcal{T}[C \leftarrow D]$ denote the directed graph that results from $\mathcal{T}$ by replacing C_i for D_i for each $i \in Id$. Note that $\mathcal{T}[C \leftarrow D]$ is not necessarily a tableau. Let π_1 be the finite path in $\mathcal{T}$ from Q_0 to Q that was used to obtain the cut C, and assume that Q is the n-th state in π_1. Reasoning as in Lemma 4.11, it is possible to extract from $\mathcal{T}[C \leftarrow D]$ a path $Q, Q'_1, Q'_2, \ldots$ that fulfills all the requirements of a complete path, but that starts in Q and, additionally, satisfies (C3) in Figure 4.5 with respect to the formulas in π_1. Let $\pi_2 = Q'_1, Q'_2, \ldots$ and π_D be the path $\pi_1[C \leftarrow D] \cdot \pi_2$ in $\mathcal{T}[C \leftarrow D]$. Finally, let μ^D be a global interpretation structure obtained from π_D as in the proof of Proposition 4.14, and let ξ^{n-1} be the n-th global state of μ^D again as in Proposition 4.14. Henceforth, we denote this state by ξ^{n_D}. It is not very difficult to conclude that $\mu^D, \xi^{n_D} \Vdash D$, and therefore $\mu^D, \xi^{n_D} \Vdash C$.

We are now ready to present the procedure Ent, which uses two auxiliary procedures, SetGlobal and LocSat, that we informally describe in the sequel.

The procedure SetGlobal, presented in Figure 4.16, receives as input a cut C of $\mathcal{T}$ and a global formula δ. It builds several extensions of C with suitable global subformulas of δ, corresponding to possible ways of falsifying δ. The procedure outputs a (possibly empty) stack S con-

---

SETGLOBAL(C, δ)

Input: a cut C and a global formula δ.
Output: a (possibly empty) stack S such that each of its elements is a non-contradictory extension of C.

R:=**empty**;
S:=**empty**;
push$(\langle \emptyset, \{\delta\} \rangle, R)$;
repeat
    $P :=$ **top**(R);
    pop(R);
    if *P contains a non-atomic global formula* **then**
        let δ' be a a non-atomic global formula in P;
        case δ'
            $\delta'_1 \Rightarrow \delta'_2 \in P^+$:
                push$(\langle P^+ \setminus \{\delta'\}, P^- \cup \{\delta'_1\} \rangle, R)$;
                push$(\langle (P^+ \setminus \{\delta'\}) \cup \{\delta'_2\}, P^- \rangle, R)$;
            $\delta'_1 \Rightarrow \delta'_2 \in P^-$:
                push$(\langle P^+ \cup \{\delta'_1\}, (P^- \setminus \{\delta'\}) \cup \{\delta'_2\} \rangle, R)$;
    else
        if $\perp \notin P^+$ **then**
            $D := C$;
            for each $@_i[\varphi] \in P^+$
                $D_i^+ := D_i^+ \cup \{@_i[\varphi]\}$;
            for each $@_i[\varphi] \in P^-$
                $D_i^- := D_i^- \cup \{@_i[\varphi]\}$;
            if $\neg$ **contradictory**(D) **then push**(D, S) **end**;
        end
    end
until R=**empty**

Figure 4.16: Procedure SETGLOBAL.

taining the extensions that are not contradictory. In this procedure, we use pairs P of sets of global formulas and we denote the first component by P^+ and the second component by P^-.

The procedure LOCSAT, presented in Figure 4.17, receives as input a cut C of $\mathcal{T}$ and a global formula δ. It starts by evaluating SETGLOBAL(C, δ). Then, it transforms the extensions of C in stack S, as returned by SETGLOBAL, into atomic extensions, keeping only the non-contradictory ones. Upon termination, if the stack S is non-empty then it is possible to provide a counterexample for falsifying δ based on C. Otherwise, there is no such counterexample based on C. In this case, the second and third components of the tuple returned by the procedure are not relevant. We use _ to denote an arbitrary value.

Finally, the procedure ENT receives as input a set of global formulas Γ and a global formula δ. It outputs *yes* if $\Gamma \models \delta$, and it outputs *no* to-

LocSat(C, δ)

Input: a cut C and a global formula δ.
Output: a triple $\langle b, \mu, \xi \rangle$.

```
b := false;
SetGlobal(C, δ);
while S ≠empty ∧ ¬b do
    D := top(S);
    if some D_k with k ∈ Id contains a non-state formula then
        pop(S);
        let δ' be a non-state formula in D_k;
        case δ'
            @_i[ψ1 ⇒ ψ2] ∈ D_k^+:
                Q := ⟨D_k^+ \ {δ'}, D_k^- ∪ {@_i[ψ1]}, D_k^⋈, D_k^#, D_k^×⟩;
                if ¬contradictory(Q) then push(D[D_k := Q], S) end
                Q := ⟨(D_k^+ \ {δ'}) ∪ {@_i[ψ2]}, D_k^-, D_k^⋈, D_k^#, D_k^×⟩;
                if ¬contradictory(Q) then push(D[D_k := Q], S) end
            @_i[ψ1 ⇒ ψ2] ∈ D_k^-:
                Q := ⟨D_k^+ ∪ {@_i[ψ1]}, (D_k^- \ {δ'}) ∪ {@_i[ψ2]}, D_k^⋈, D_k^#, D_k^×}⟩;
                if ¬contradictory(Q) then push(D[D_k := Q], S) end
            @_i[©_j[ψ]] ∈ D_k^+:
                if j ∈ D_k^× then
                    Q := ⟨D_k^+ \ {δ'} ∪ {@_j[ψ]}, D_k^-, D_k^⋈, D_k^#, D_k^×⟩;
                    if ¬contradictory(Q) then push(D[D_k := Q], S) end
                end
            @_i[©_j[ψ]] ∈ D_k^-:
                if j ∈ D_k^× then
                    Q := ⟨D_k^+, D_k^- \ {δ'} ∪ {@_j[ψ]}, D_k^⋈, D_k^#, D_k^×⟩;
                    if ¬contradictory(Q) then push(D[D_k := Q], S) end
                else
                    Q := ⟨D_k^+, D_k^- \ {δ'}, D_k^⋈, D_k^#, D_k^×⟩;
                    push(D[D_k := Q], S);
                end
    else
        b := true
    end
end;
if b then
    return ⟨b, μ^top(S), ξ^n_top(S)⟩
else
    return ⟨b, _, _⟩
end
```

Figure 4.17: Procedure LocSat.

gether with a counterexample, otherwise. After building a tableau $\mathcal{T}$ for $\langle \Gamma, \emptyset, \emptyset, \emptyset, \emptyset \rangle$, if $\mathcal{T}$ is successful then, for each cut C induced by each non-closed state Q, it uses procedure LocSat to search for a counterexample for falsifying δ based on C.

$\text{Ent}(\Gamma, \delta)$

Input: a set Γ of global formulas and a global formula δ.
Output: *yes* if $\Gamma \models \delta$, *no* and a counterexample, otherwise.

```
Q0 := ⟨Γ, ∅, ∅, ∅, ∅⟩;
construct a tableau T for Q0;
if T is unsuccessful then
    return yes;
else
    O := {Q in T | Q is not closed};
    b := false;
    while O ≠ ∅ ∧ ¬b do
        let Q ∈ O;
        O := O \ {Q};
        T_Q := C_Q;
        while T_Q ≠ ∅ ∧ ¬b do
            let C ∈ T_Q;
            T_Q := T_Q \ {C};
            ⟨b1, μ, ξ⟩ := LocSat(C, δ);
            b := b ∨ b1;
        end
    end
    if ¬b then
        return yes;
    else
        return ⟨no, μ, ξ⟩;
    end
end
```

Figure 4.18: Procedure Ent.

In order to prove the correctness of procedure Ent, we first establish a number of auxiliary results. In the sequel, we will assume given a cut C induced by a sequence of states $\widehat{\pi}_Q$ as defined above, and a global formula δ without temporal operators. Note that along the execution of $\text{SetGlobal}(C, \delta)$ each P in the stack R only includes subformulas of δ. Observe also that upon termination of $\text{SetGlobal}(C, \delta)$, every D in the stack S is a non-contradictory extension of C and, for every $k \in Id$, the formulas in $D_k^+ \cup D_k^-$ are state formula or atomic global formulas without temporal operators. Moreover, the formulas in $(D_k^+ \setminus C_k^+) \cup (D_k^- \setminus C_k^-)$ are subformulas of δ. These assertions are preserved by each step of the loop of $\text{LocSat}(C, \delta)$.

It is also worthwhile to observe that upon termination of the procedure $\text{SetGlobal}(C, \delta)$, the following holds: for every D in S and $i, k \in Id$, if $@_i[\psi] \in D_k^+ \cup D_k^-$ and $@_i[\psi]$ is not a state formula, then $i \in D_k^{\mathsf{X}}$ whenever $D_k^{\mathsf{X}} \neq \emptyset$. This assertion is again preserved by each step of the loop of LocSat(C, δ).

Lemma 4.18 *Let μ be a global interpretation structure and ξ a global state such that $\mu, \xi \Vdash C$. Upon termination of* SetGlobal(C, δ) *the following holds: $\mu, \xi \nVdash \delta$ if and only if $\mu, \xi \Vdash B$ for some B in S.* □

Lemma 4.19 *Upon termination of* LocSat(C, δ)*, the following hold: for every μ and ξ such that $\mu, \xi \Vdash C$,*

(i) b is true if and only if S is not empty;

(ii) if $\mu, \xi \Vdash \delta$, then $\mu, \xi \nVdash B$ for every B in S;

(iii) if $\mu, \xi \nVdash \delta$, then $\mu, \xi \Vdash B$ for some B in S;

(iv) if S is not empty, then $\mu^{\mathtt{top}(S)}, \xi^{n_{\mathtt{top}(S)}} \nVdash \delta$. □

It is not difficult to conclude that both procedures SetGlobal(C, δ) and LocSat(C, δ) end after a finite number of steps. Proposition 4.20 establishes the correctness of the procedure Ent.

Proposition 4.20 *Let Γ be a set of global formulas and δ a global formula without temporal operators. Then*

(i) if $\Gamma \vDash \delta$, then, upon termination, Ent(Γ, δ) *outputs yes;*

(ii) if $\Gamma \nvDash \delta$, then, upon termination, Ent(Γ, δ) *outputs $\langle no, \mu, \xi \rangle$ such that $\mu \Vdash \Gamma$ and $\mu, \xi \nVdash \delta$.* □

It is also not difficult to conclude that the procedure Ent ends after a finite number of steps. We now present an example illustrating the procedure Ent at work.

Example 4.21 *Consider the usual* mutual exclusion problem, *where two processes A and B try to access a critical section, controlled by a semaphore Z.*

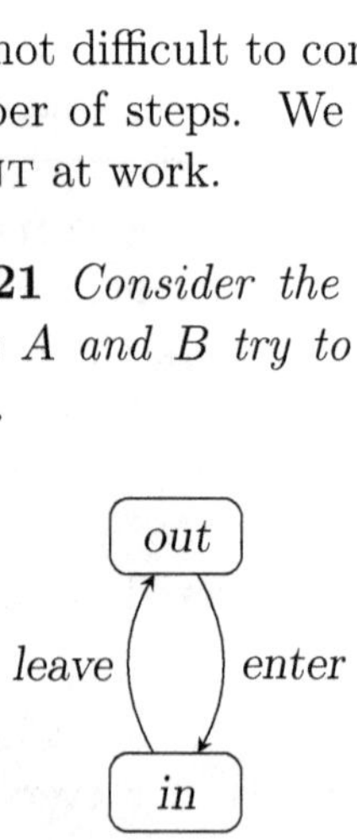

Processes A and B

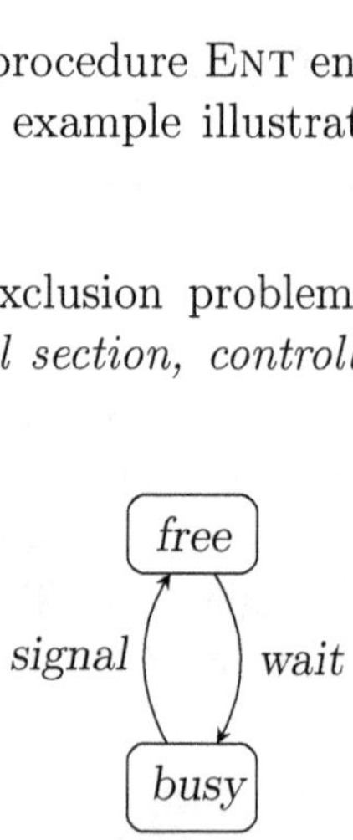

Semaphore Z

For specifying processes A and B, we will use a propositional symbol i to denote that a process is in the critical section and two propositional symbols E and L for the actions of entering and leaving the critical section, respectively. Hence, we have $Prop_A = Prop_B = \{i, E, L\}$. *For the semaphore, we will use a propositional symbol f to denote that the semaphore is free and two propositional symbols for the actions W and S for the actions of wait and signal, respectively. Hence, we have* $Prop_Z = \{f, W, S\}$. *The axioms for A and B are*

(A1) $@_A[\neg i]$

(A2) $@_A[\mathsf{G}(\mathsf{X}\,E \Leftrightarrow (\neg i \wedge \mathsf{X}\,i))]$

(A3) $@_A[\mathsf{G}(\mathsf{X}\,L \Leftrightarrow (i \wedge \neg \mathsf{X}\,i))]$

(A4) $@_A[\mathsf{G}(\mathsf{X}\,L \Leftrightarrow \neg \mathsf{X}\,E)]$

(B1) $@_B[\neg i]$

(B2) $@_B[\mathsf{G}(\mathsf{X}\,E \Leftrightarrow (\neg i \wedge \mathsf{X}\,i))]$

(B3) $@_B[\mathsf{G}(\mathsf{X}\,L \Leftrightarrow (i \wedge \neg \mathsf{X}\,i))]$

(B4) $@_B[\mathsf{G}(\mathsf{X}\,L \Leftrightarrow \neg \mathsf{X}\,E)]$

The axioms for the semaphore are

(Z1) $@_Z[f]$

(Z2) $@_Z[\mathsf{G}(\mathsf{X}\,W \Leftrightarrow (f \wedge \neg \mathsf{X}\,f))]$

(Z3) $@_Z[\mathsf{G}(\mathsf{X}\,S \Leftrightarrow (\neg f \wedge \mathsf{X}\,f))]$

(Z4) $@_Z[\mathsf{G}(\mathsf{X}\,W \Leftrightarrow \neg \mathsf{X}\,S)]$

The interaction axioms are:

(I1) $@_A[\mathsf{G}(E \Rightarrow ©_Z[W])]$

(I2) $@_A[\mathsf{G}(L \Rightarrow ©_Z[S])]$

(I3) $@_B[\mathsf{G}(E \Rightarrow ©_Z[W])]$

(I4) $@_B[\mathsf{G}(L \Rightarrow ©_Z[S])]$

(I5) $@_Z[\mathsf{G}(W \Rightarrow (©_A[E] \vee ©_B[E]))]$

(I6) $@_Z[\mathsf{G}(S \Rightarrow (©_A[L] \vee ©_B[L]))]$

(I7) $@_Z[\mathsf{G}(\neg(©_A[\top] \wedge ©_B[\top]))]$

All the axioms are straightforward. We just point out that axiom (I7) is included to prevent the semaphore from synchronizing with both processes at the same time.

Let Γ *be set of all previous axioms. Our goal is to prove that processes A and B are not simultaneously in the critical section, i.e., we want to establish*

$$\Gamma \models \neg(@_A[i] \wedge @_B[i]).$$

To this end, we apply the procedure ENT *to the set* Γ *and the formula* $\neg(@_A[i] \wedge @_B[i])$. *The first step is to build the tableau for* Γ. *In Figure 4.19, we depict the reachable, non-closed part of a tableau for* Γ. *Furthermore, for the sake of simplicity, we only depict the relevant states.*

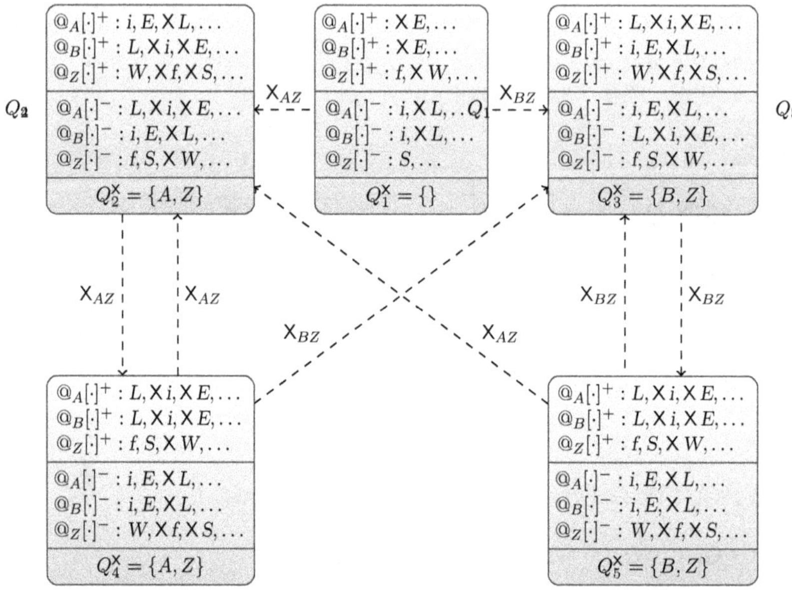

Figure 4.19: Relevant fragment of a tableau for the mutual exclusion problem.

For the tableau in Figure 4.19 we have the following sets of cuts, for each state tuple:

- $\mathcal{C}_{Q_1} = \{\{\langle Q_1, \{A, B, Z\}\rangle\}\}$
- $\mathcal{C}_{Q_2} = \{\{\langle Q_2, \{A, Z\}\rangle, \langle Q_1, \{B\}\rangle\}, \{\langle Q_2, \{A, Z\}\rangle, \langle Q_5, \{B\}\rangle\}\}$
- $\mathcal{C}_{Q_3} = \{\{\langle Q_3, \{B, Z\}\rangle, \langle Q_1, \{A\}\rangle\}, \{\langle Q_3, \{B, Z\}\rangle, \langle Q_4, \{A\}\rangle\}\}$
- $\mathcal{C}_{Q_4} = \{\{\langle Q_4, \{A, Z\}\rangle, \langle Q_1, \{B\}\rangle\}, \{\langle Q_4, \{A, Z\}\rangle, \langle Q_5, \{B\}\rangle\}\}$
- $\mathcal{C}_{Q_5} = \{\{\langle Q_5, \{B, Z\}\rangle, \langle Q_1, \{A\}\rangle\}, \{\langle Q_5, \{B, Z\}\rangle, \langle Q_4, \{A\}\rangle\}\}$.

Procedure ENT *will apply* LOCSAT *to each of these cuts. We briefly sketch the outcome of* LOCSAT *when applied to one of such cuts. Consider for instance* $C = \{\langle Q_2, \{A, Z\}\rangle, \langle Q_5, \{B\}\rangle\} \in \mathcal{C}_{Q_2}$. *Procedure* LOCSAT, *when applied to* C *and* δ *starts by assigning* b *to* `false` *and then calls procedure* SETGLOBAL. *This procedure will process the formula*

$\neg(@_A[i] \wedge @_B[i])$ until only global atomic formulas remain in the top of stack R. After two executions of the loop body, the stack R will only contain the pair $\langle\{@_A[i], @_B[i]\}, \{\perp\}\rangle$. An extension of C is then built during the next execution of the loop body. The formula $@_B[i]$ is added to Q_5^+ leading to a contradictory state tuple that, consequently, is not pushed on top of S. Hence, upon termination of SetGlobal, *the stack S will be empty. This implies that procedure* LocSat *will terminate immediately returning $\langle$`false`$, \_, \_\rangle$. The outcome for the remaining cuts will be the same and, consequently, the output of* Ent *will be yes. Hence, by Proposition 4.20, we can conclude that*

$$\Gamma \models \neg(@_A[i] \wedge @_B[i]).$$

and so, mutual exclusion is guaranteed.

4.8 Concluding remarks

We have proposed a decidable Fisher-Ladner-style tableaux system for an anchored version of the distributed Temporal Logic DTL. Our tableaux system is built on top of a tableaux system for LTL as presented in [8] and integrates in a smooth way both the usual rules for the temporal operators and rules for tackling the specific communication features of DTL. The usual fixpoint properties of temporal operators (from a local perspective) are used in order to explore the global state space.

As we remarked above, the anchored version of DTL is less expressive in terms of global reasoning, but a suitable labeling discipline can be used to control the distribution dimension. However, the tableaux system for such an anchored version can only be used to decide validity of DTL formulas. We have thus endowed the tableaux system with a specific decision procedure for dealing with entailment, which is enough to capture the usual notion of invariant. We have also showed how the tableaux system proposed here can be coupled with a model-checking-like feature for deciding global DTL properties.

As future work, we believe that it would be interesting to extend our verification procedure to the case of entailment $\Delta \models \delta$, where δ contains temporal operations, and to the case of plain, non-anchored, DTL. Moreover, it would be useful to implement our system in order to provide support for automated reasoning. A first step in this direction was already taken [12].

Acknowledgments

This work was partially done under the scope of UID/EEA/50008/2013, financed by the applicable financial framework (FCT/MEC through national funds and when applicable co-funded by FEDER and PT2020). The authors further acknowledge the support of EU FP7 Marie Curie PIRSES-GA-2012-318986 project GeTFun.

Bibliography

[1] D. Basin, C. Caleiro, J. Ramos, and L. Viganò. A labeled tableaux for the distributed temporal logic DTL. In *Proceedings of the 15th Int. Symp. on Temporal Representation and Reasoning (TIME 2008)*, pages 101–109. IEEE Computer Society Press, 2008.

[2] D. Basin, C. Caleiro, J. Ramos, and L. Viganò. Labelled tableaux for distributed temporal logic. *Journal of Logic and Computation*, 19:1245–1279, 2009.

[3] D. Basin, C. Caleiro, J. Ramos, and L. Viganò. Distributed temporal logic for the analysis of security protocol models. *Theoretical Computer Science*, 412(31):4007–4043, 2011.

[4] E. Best and C. Fernández. *Nonsequential processes – A Petri net view.* Springer-Verlag, 1988.

[5] C. Caleiro, L. Viganò, and D. Basin. Metareasoning about security protocols using distributed temporal logic. *Electronic Notes in Theoretical Computer Science*, 125(1):67–89, 2005. Preliminary version presented at IJCAR'04 ARSPA Workshop.

[6] C. Caleiro, L. Viganò, and D. Basin. Relating strand spaces and distributed temporal logic for security protocol analysis. *Logic Journal of the IGPL*, 13(6):637–664, 2005.

[7] H.-D. Ehrich and C. Caleiro. Specifying communication in distributed information systems. *Acta Informatica*, 36:591–616, 2000.

[8] F. Kröger and S. Merz. *Temporal logic and state systems.* Springer, 2008.

[9] O. Lichtenstein and A. Pnueli. Propositional Temporal Logic: Decidability and Completeness. *Logic Journal of the IGPL*, 8(1):55–85, 2000.

[10] K. Lodaya, R. Ramanujam, and P. Thiagarajan. Temporal logics for communicating sequential agents: I. *International Journal of Foundations of Computer Science*, 3(1):117–159, 1992.

[11] K. Lodaya and P. Thiagarajan. A modal logic for a subclass of event structures. In *Proceedings of 14th ICALP*, LNCS 267, pages 290–303. Springer, 1987.

[12] C. Neves. Deciding distributed temporal logic. Master's thesis, IST, Universidade de Lisboa, 2015. Supervised by P. Gouveia.

[13] R. Ramanujam. Locally linear time temporal logic. In *In Proceeding of 11th LICS*, pages 118–127. IEEE Computer Society Press, 1996.

[14] G. Winskel. Event structures. In W. Brauer, W. Reisig, and G. Rozenberg, editors, *Petri Nets: Applications and Relationships to Other Models of Concurrency*, LNCS 255, pages 325–392. Springer-Verlag, 1987.

Appendix: Proofs

Section 4.4

Lemma 4.4

Proof. We provide a sketch of the proof. By definition of state, the only rule applicable at Q_{s_0} is (X). Furthermore, since Q_{s_0} is the only state in this path, we have that $Q^{\mathsf{X}}_{s_1} = Q^{\mathsf{X}}_{s_l}$, for $l = 2, \ldots, n$. Hence, we focus on the case $r = 1$. If $i \notin Q^{\mathsf{X}}_{s_1}$, then $Q^{+}_{s_1}{\downarrow_i} = Q^{+}_{s_0}{\downarrow_i}$ and $Q^{-}_{s_1}{\downarrow_i} = Q^{-}_{s_0}{\downarrow_i}$ and so all i-formulas in $Q^{+}_{1} \cup Q^{-}_{1}$ are state formulas. Hence, no rule triggered by i can be applied at Q_{s_1}.

The only way that a rule triggered by $i \notin Q^{\mathsf{X}}_{s_1}$ can be applied at Q_{s_n} is if some communication rule (for some other agent) has been applied at a previous node. Assume that $@_k[©_i[\varphi]] \in Q^{+}_{s_{n-1}}$ and rule $(©^{+}_{k})$ was applied at $Q_{s_{n-1}}$. Then, $@_i[\varphi] \in Q^{+}_{s_n}$ but also $i \in Q^{\bowtie}_{s_n}$. So, Q_{s_n} would be contradictory due to condition $(\perp 3)$, thus contradicting the assumption.

Similarly, assume that $@_k[©_i[\varphi]] \in Q^{-}_{s_{n-1}}$ and rule $(©^{-}_{k})$ was applied at $Q_{s_{n-1}}$. Then, either $@_i[\varphi] \in Q^{-}_{s_n}$ and $i \in Q^{\bowtie}_{s_n}$, or $i \in Q^{\#}_{s_n}$. In the first case, we can reason as above. The last case is trivial as no non-state i-formulas are introduced and consequently if rules triggered by i were not applicable before, then they remain non-applicable.

The cases where $@_k[©_i[\varphi]] \in Q^{+}_{s_l}$ or $@_k[©_i[\varphi]] \in Q^{-}_{s_l}$, with $1 \leq l < n-1$, are similar. □

Proposition 4.5

Proof. Let μ be an interpretation structure and ξ one of its global states. If Q fulfills either $(\perp 1)$ or $(\perp 2)$, then $\mu, \xi \not\Vdash \widehat{Q}$ and, consequently, $\mu, \xi \not\Vvdash Q$.

If Q fulfills $(\perp 3)$, then $Q^{\bowtie} \not\subseteq Q^{\mathsf{X}}$. Hence, μ and ξ do not meet the synchronization requirements in Q, as $(S3)$ fails, and, consequently, $\mu, \xi \not\Vvdash Q$.

Similarly, if Q fulfills $(\perp 4)$, then $Q^{\#} \cap Q^{\mathsf{X}} \neq \emptyset$. Hence, μ and ξ do not meet the synchronization requirements in Q, as $(S4)$ fails. Once again, $\mu, \xi \not\Vvdash Q$. □

Proposition 4.9

Proof. We complete the proof given in Section 4.4 by proving that if a node is in $Cl_{\mathcal{T}}$ due to $(C2)$, $(C3)$ or $(C4)$, then it does not occur in $\pi^{\mu,\ell}_{Q}$.

(C2) Assume that $Q' \in Cl_{\mathcal{T}}$ because all its successors are in $Cl_{\mathcal{T}}$. If Q' occurs is $\pi_Q^{\mu,\ell}$ then necessarily one of the successors of Q' would also occur in $\pi_Q^{\mu,\ell}$, because this path is infinite. But as all the successors of Q' are closed, by the induction hypothesis, none of them can occur in $\pi_Q^{\mu,\ell}$.

(C3) Assume that $Q' \in Cl_{\mathcal{T}}$ is closed because a formula $@_i[\mathsf{G}\,\varphi] \in Q'^-$ and all paths in $\mathcal{T}$ from Q' to a node Q'' such that $@_i[\varphi] \in Q''^-$ have a node $\widetilde{Q} \in Cl_{\mathcal{T}}$. By the induction hypothesis, no such $\widetilde{Q}$ occurs in $\pi_Q^{\mu,\ell}$.

Assume that Q' occurs in $\pi_Q^{\mu,\ell}$. Then $Q' = Q_{k,m}$, for some $k \in \mathbb{N}_0$ and $0 \le m \le n_k$.

We prove that either $@_i[\mathsf{G}\,\varphi] \in Q^-_{k',m'}$ or $@_i[\mathsf{X}\,\mathsf{G}\,\varphi] \in Q^-_{k',m'}$ for every node $Q_{k',m'}$ in $\pi_Q^{\mu,\ell}$, with $k' \ge k$, and $m \le m' \le n_k$ if $k' = k$ and $0 \le m' \le n_{k'}$ otherwise. To this end it is enough to observe that the assertion holds for $Q_{k,m}$ and to prove that every rule preserves the assertion. The preservation of the assertion is trivial for every rule other than rule (X) whenever $i \in K$ and rule (G_i^-).

- Let us assume that rule (X) with $i \in K$ was applied at the node $Q_{k',m'}$. In this case, $Q_{k',m'}$ is a state, and therefore its successor is $Q_{k'+1,0}$ and $@_i[\mathsf{G}\,\varphi] \notin Q^-_{k',m'}$. Hence, $@_i[\mathsf{X}\,\mathsf{G}\,\varphi] \in Q^-_{k',m'}$ holds and, as a consequence, $@_i[\mathsf{G}\,\varphi] \in Q^-_{k'+1,0}$, which means that the assertion is preserved in this case.
- Assume now that rule (G_i^-) was applied at node $Q_{k',m'}$ with respect to $@_i[\mathsf{G}\,\varphi]$. In this case, $Q_{k',m'}$ is not a state, and therefore its successor in $\pi_Q^{\mu,\ell}$ is $Q_{k',m'+1}$ and either $@_i[\varphi] \in Q^-_{k',m'+1}$ or $@_i[\mathsf{X}\,\mathsf{G}\,\varphi] \in Q^-_{k',m'+1}$. If $@_i[\varphi] \in Q^-_{k',m'+1}$ then the subpath of $\pi_Q^{\mu,\ell}$ starting in $Q_{k,m}$ (that is, node Q') and ending in $Q_{k',m'+1}$ must include a closed node. Observe that this closed node is one of nodes $\widetilde{Q}$ referred above and therefore it could not occur in $\pi_Q^{\mu,\ell}$. Thus, $@_i[\mathsf{X}\,\mathsf{G}\,\varphi] \in Q^-_{k',m'+1}$ necessarily holds and the assertion is also preserved in this case.

Recall that $\mu, \xi^k \Vdash\!\!\vdash Q_{k,m}$ and that $@_i[\mathsf{G}\,\varphi] \in Q^-_{k,m}$. Consequently $\mu, \xi^k \nVdash @_i[\mathsf{G}\,\varphi]$ that is $\mu_i, \xi^k|_i \nVdash_i \mathsf{G}\,\varphi$. Then, there is a local state $\xi' \in \Xi_i$ such that $\xi^k|_i \subseteq \xi'$ and $\mu_i, \xi' \nVdash_i \varphi$. Let us first assume that $\xi^k|_i \subset \xi'$, and therefore $\xi' \neq \emptyset$. Let $k' \in \mathbb{N}$ be such that $\ell(k') =$

$last(\xi')$. Then, $\xi^k \subset \xi^{k'}$ and therefore $k' > k$. Moreover, $\xi^{k'}|_i = \xi'$ and $last(\xi^{k'}) \in Ev_i$. Hence, $\mu, \xi^{k'} \not\Vdash @_i[\varphi]$ holds. Furthermore, as $last(\xi^{k'}|_i) \in Ev_i$, then $i \in Ids(\ell(k')) = Q^{\mathsf{X}}_{k',0}$, by construction of $\pi^{\mu,\ell}_Q$. As proved above, either $@_i[\mathsf{G}\,\varphi] \in Q^{-}_{k'-1,n_{k'-1}}$ or $@_i[\mathsf{X}\,\mathsf{G}\,\varphi] \in Q^{-}_{k'-1,n_{k'-1}}$, but since $Q_{k'-1,n_{k'-1}}$ is a state only the latter holds. Furthermore, rule (X) was applied at $Q_{k'-1,n_{k'-1}}$ with $i \in K$, and therefore $@_i[\mathsf{G}\,\varphi] \in Q^{-}_{k',0}$. In this case, rule (G^{-}_i) must have been applied at node $Q_{k',l}$ with respect to $@_i[\mathsf{G}\,\varphi]$, for some $0 \leq l < n_{k'}$. As $\mu, \xi^{k'} \Vdash\!\vdash Q_{k',l}$ and $\mu, \xi^{k'} \not\Vdash @_i[\varphi]$ then μ and $\xi^{k'}$ satisfy the left premiss. Hence, by construction of $\pi^{\mu,\ell}_Q$, node $Q_{k',l+1}$ is this left premiss. Thus, $@_i[\varphi] \in Q^{-}_{k',l+1}$ holds and the subpath of $\pi^{\mu,\ell}_Q$ starting in $Q_{k,m}$ and ending in $Q_{k',l+1}$ must include a closed node. This closed node is one of the nodes $\widetilde{Q}$ referred at the beginning that, by the induction hypothesis, cannot occur in $\pi^{\mu,\ell}_Q$. We then have a contradiction and therefore Q' cannot occur in $\pi^{\mu,\ell}_Q$. Let us now assume that $\xi^k|_i = \xi'$. Then, $\mu, \xi^k \not\Vdash @_i[\varphi]$ holds. Since $@_i[\mathsf{G}\,\varphi] \in Q^{-}_{k,m}$, rule (G^{-}_i) must have been applied at node $Q_{k,l}$ with respect to $@_i[\mathsf{G}\,\varphi]$, for some $m \leq l < n_k$, and the proof follows similarly.

(C4) Assume that $Q' \in Cl_{\mathcal{T}}$ is closed because there is $i \in Id$ and all paths in $\mathcal{T}$ from Q' to a node Q'' such that $i \in Q''^{\mathsf{X}}$ have a node $\widetilde{Q} \in Cl_{\mathcal{T}}$. By the induction hypothesis, no such $\widetilde{Q}$ occurs in $\pi^{\mu,\ell}_Q$.

Assume that Q' occurs in $\pi^{\mu,\ell}_Q$. Then $Q' = Q_{k,m}$, for some $k \in \mathbb{N}_0$ and $0 \leq m \leq n_k$.

Let $k' = \min\{k'' \in \mathbb{N} \mid k'' > k \text{ and } \ell(k'') \in Ev_i\}$. Then, $i \in Ids(\ell(k'))$ and therefore $i \in Q^{\mathsf{X}}_{k',0}$. Then, the subpath of $\pi^{\mu,\ell}_Q$ starting in $Q_{k,m}$ and ending in $Q_{k',0}$ must include a closed node. This closed node is one the nodes $\widetilde{Q}$ above that, by the induction hypothesis, cannot occur in $\pi^{\mu,\ell}_Q$. Hence, we have a contradiction and therefore Q' cannot occur in $\pi^{\mu,\ell}_Q$.

□

Section 4.5

Lemma 4.13

Proof. Consider the segment $Q_{\restriction (n)}, Q_{k_1}, \ldots, Q_{k_q}, Q_{\restriction (n+1)}$ of π. As $Q_{\restriction (n)}$ is a state then rule (X) was applied from that state to Q_{k_1}. And as $Q_{\restriction (n+1)}$ is the next state then rule (X) was not applied at any other node in this segment. Hence, $Q_k^{\mathsf{X}} = Q^{\mathsf{X}}_{\restriction (n+1)}$, for every $k = k_1, \ldots, k_q$.

Next, we observe that all the formulas in $Q^+_{\restriction (n)} \cup Q^-_{\restriction (n)}$ are atomic or of the form $@_i[\mathsf{X}\,\varphi]$, for some $\varphi \in \mathcal{L}_i$. Furthermore, given that $i \notin K$, where K the set of agent identifiers that was used in rule (X) at $Q_{\restriction (n)}$ to obtain Q_{k_1}, then $Q^+_{\restriction (n)}{\downarrow_i} = Q^+_{k_1}{\downarrow_i}$ and $Q^-_{\restriction (n)}{\downarrow_i} = Q^-_{k_1}{\downarrow_i}$. We can then conclude that no rule triggered by i was applied in this segment. If a new formula for i was to appear in $Q_k^+ \cup Q_k^-$, for some $k \in \{k_1, \ldots, k_q\}$, it would have to be due to some interaction with another agent. That is, either rule ($©_j^+$) or ($©_j^-$) would have been applied to a formula $@_j[©_i[\varphi]]$. In the case of rule ($©_j^-$) the left node would have to be considered. In either of these cases, i would also have to be added to $Q_k^{\bowtie}$. But, given that $i \notin Q_k^{\mathsf{X}}$ then the node would be closed by ($C1$), which is impossible because the path is complete. □

Proposition 4.14

Proof. Herein we complete the proof presented in Section 4.5, by proving that $\mu, \xi^{cnt(n)} \Vdash \widehat{Q}_n$ and μ and $\xi^{cnt(n)}$ meet the requirements in Q_n.

Throughout the proof note that, for every $p \in \mathbb{N}$ and $i \in Id$, it holds that $e_p \in \xi^p$ and if $e_p \in Ev_i$ then $last(\xi^p|_i) = e_p$. Moreover, $i \in Q_p^{\mathsf{X}}$ if and only if $i \in Q^{\mathsf{X}}_{\restriction (cnt(p))}$ and $i \in Q^{\mathsf{X}}_{\restriction (cnt(p))}$ if and only if $e_{cnt(p)} \in Ev_i$.

We first prove that $\mu, \xi^{cnt(n)} \Vdash \widehat{Q}_n$ and μ. The proof follows by induction on the structure of the formula φ, simultaneously for all agents.

1. φ is $p \in Prop_i$: if $@_i[p] \in Q_n^+$ then, by construction of the tableau, $@_i[p] \in Q^+_{\restriction (cnt(n))}$ (recall that $Q_{\restriction (cnt(n))}$ is the first state after Q_n, if Q_n is not a state, otherwise it is Q_n itself). If $cnt(n) = 0$ then $@_i[p] \in Q^+_{\restriction (0)}$ which implies, by model construction, that $p \in \vartheta_i(\emptyset)$. Consequently, $\mu_i, \emptyset \Vdash_i p$ and, thus, $\mu, \xi^{cnt(n)} \Vdash @_i[p]$. If $cnt(n) > 0$ then we have two possibilities: either $e_{cnt(n)} \in Ev_i$ or $e_{cnt(n)} \notin Ev_i$. In the first case, we have that $p \in \vartheta(\xi^{cnt(n)}|_i)$, which implies that $\mu_i, \xi^{cnt(n)}|_i \Vdash_i p$. If $e_{cnt(n)} \notin Ev_i$ then $\xi^{cnt(n)}|_i = \{e_{i_1}, \ldots, e_{i_k}\}$ with $i_k < cnt(n)$. Assume, without loss of generality,

that $cnt(n) = i_k + 1$, that is, e_{i_k} and $e_{cnt(n)}$ correspond to consecutive events in E. As $e_{cnt(n)} \notin Ev_i$ then $i \notin Q^{\mathsf{X}}_{\mid(cnt(n))}$. By Lemma 4.13, we know that if $@_i[p] \in Q^{+}_{\mid(cnt(n))}$ then $@_i[p] \in Q^{+}_{\mid(i_k)}$. By the construction of the model, we have that $p \in \vartheta(\xi^{cnt(n)}|_i)$ and, so, $\mu_i, \xi^{cnt(n)}|_i \Vdash_i p$. In either case, we are able to conclude that $\mu, \xi^{cnt(n)} \Vdash @_i[p]$.

If $@_i[p] \in Q_n^-$ then the proof is similar.

2. φ is $\bot$: as the path is complete then Q_n is not closed and, so, it must be the case that $@_i[\bot] \in Q_n^-$. Then, if follows by definition, that $\mu_i, \xi^{cnt(n)}|_i \nVdash_i \bot$ and, consequently, that $\mu, \xi^{cnt(n)} \nVdash @_i[\bot]$.

3. φ is $\varphi_1 \Rightarrow \varphi_2$: if $@_i[\varphi_1 \Rightarrow \varphi_2] \in Q_n^+$ then, by the tableau construction, there must be some k such that $n \leq k < \mid(cnt(n))$ and rule $(\Rightarrow_i^+)$ was applied at Q_k. Hence, either $@_i[\varphi_1] \in Q_{k+1}^-$ or $@_i[\varphi_2] \in Q_{k+1}^+$. Observe that $cnt(n) = cnt(k) = cnt(k+1)$. Hence, using the induction hypothesis, we have that $\mu_i, \xi^{cnt(k+1)}|_i \nVdash_i \varphi_1$ or $\mu_i, \xi^{cnt(k+1)}|_i \Vdash_i \varphi_2$, which implies that $\mu_i, \xi^{cnt(k+1)}|_i \Vdash_i \varphi_1 \Rightarrow \varphi_2$ from which it follows that $\mu_i, \xi^{cnt(n)}|_i \Vdash_i \varphi_1 \Rightarrow \varphi_2$. Hence, $\mu, \xi^{cnt(n)} \Vdash @_i[\varphi_1 \Rightarrow \varphi_2]$. The proof for the case $@_i[\varphi_1 \Rightarrow \varphi_2] \in Q_n^-$ is similar.

4. φ is $(\mathsf{X}\,\varphi_1)$: if $@_i[\mathsf{X}\,\varphi_1] \in Q_n^+$ then, by tableau construction, $@_i[\mathsf{X}\,\varphi_1] \in Q^{+}_{\mid(cnt(n))}$ and the rule for X was applied at $Q_{\mid(cnt(n))}$ with some set K. There are two situations to consider: either $i \in K$ or $i \notin K$. In the first case $@_i[\varphi_1] \in Q^{+}_{\mid(cnt(n))+1}$. By induction hypothesis, it follows that $\mu, \xi^{cnt(\mid(cnt(n))+1)} \Vdash @_i[\varphi_1]$. As $cnt(\mid(k)+1) = cnt(\mid(k)) + 1$, then $\mu_i, \xi^{cnt(\mid(cnt(n)))+1}|_i \Vdash_i \varphi_1$. Now, let $l = cnt(\mid(cnt(n)))$. Observe that $\xi^{l+1} = \xi^l \cup \{e_{l+1}\}$ and that $e_{l+1} \in Ev_i$. Hence, $\xi^{l+1}|_i = \xi^l|_i \cup \{e_{l+1}\}$ and, so, $\mu_i, \xi^l|_i \cup \{e_{l+1}\} \Vdash_i \varphi_1$ which implies that $\mu_i, \xi^l|_i \Vdash_i \mathsf{X}\,\varphi_1$. Now, observing that $cnt(\mid(n)) = n$, we can conclude that $\mu, \xi^{cnt(n)} \Vdash @_i[\mathsf{X}\,\varphi_1]$. If $i \notin K$, then, given that the path is complete, there must be some state Q_k with $k > \mid(cnt(n))$ where rule (X) was applied with some set K' such that $i \in K'$. Let us consider the first of those states. By Lemma 4.13, we know that $@_i[\mathsf{X}\,\varphi_1] \in Q_k^+$. Furthermore, by model construction, $\xi^{cnt(\mid(cnt(n)))}|_i = \xi^{cnt(k)}|_i$. The rest of the proof follows as in the previous case. The proof for the case where $@_i[\mathsf{X}\,\varphi_1] \in Q_n^-$ is similar.

5. φ is $\mathsf{G}\,\varphi_1$: if $@_i[\mathsf{G}\,\varphi_1] \in Q_n^+$ then there must be some k such that $n \leq k < \mid(cnt(n))$ and rule (G_i^+) was applied at Q_k. Is follows

that $@_i[\varphi_1] \in Q^+_{k+1}$ and $@_i[\mathsf{X}\,\mathsf{G}\,\varphi_1] \in Q^+_{k+1}$. Using the induction hypothesis on the former, we get $\mu_i, \xi^{cnt(n)}|_i \Vdash_i \varphi_1$ as in previous cases. From the latter, we conclude that $@_i[\mathsf{X}\,\mathsf{G}\,\varphi_1] \in Q^+_{\mid\,(cnt(n))}$. Hence, using an inductive argument and reasoning as in the previous case, we can establish that $\mu_i, \xi^k|_i \Vdash_i \varphi_1$ for every $k \geq cnt(n)$ and, thus, conclude that $\mu_i, \xi^{cnt(n)}|_i \Vdash_i \mathsf{G}\,\varphi_1$. Hence, $\mu, \xi^{cnt(n)} \Vdash @_i[\mathsf{G}\,\varphi_1]$.

Assume now that $@_i[\mathsf{G}\,\varphi_1] \in Q^-_n$. As the path is complete, there is some $k \geq n$ such that $@_i[\varphi_1] \in Q^-_k$. By induction hypothesis, $\mu_i, \xi^{cnt(k)}|_i \not\Vdash_i \varphi_1$. As cnt is monotonic, $cnt(k) \geq cnt(n)$ implying that $\mu_i, \xi^{cnt(n)}|_i \not\Vdash_i \mathsf{G}\,\varphi_1$. Hence, $\mu, \xi^{cnt(n)} \not\Vdash @_i[\mathsf{G}\,\varphi_1]$.

6. φ is $©_j[\varphi_1]$: if $@_i[©_j[\varphi_1]] \in Q^+_n$ then there must be some k such that $n \leq k < \mid(cnt(n))$ and the rule $(©^+_i)$ was applied at Q_k. It follows that $@_j[\varphi_1] \in Q^+_{k+1}$ and $j \in Q^{\bowtie}_{k+1}$. The former, using the induction hypothesis, implies that $\mu_j, \xi^{cnt(k+1)}|_j \Vdash_j \varphi_1$. The latter implies that $j \in Q^{\bowtie}_{\mid\,(cnt(n))}$. Consequently, as $Q_{\mid\,(cnt(n))}$ is not closed, it must be the case that $j \in Q^{\mathsf{X}}_{\mid\,(cnt(n))}$, and therefore $Q^{\mathsf{X}}_{\mid\,(cnt(n))} = Q^{\mathsf{X}}_k \neq \emptyset$. Hence, we can use Lemma 4.4 to conclude that $i \in Q^{\mathsf{X}}_k = Q^{\mathsf{X}}_{\mid\,(cnt(n))}$. So, $e_{cnt(n)} = last(\xi^{cnt(n)}|_i) = last(\xi^{cnt(n)}|_j) \in Ev_i \cap Ev_j$. Hence, $\xi^{cnt(n)}|_i \neq \emptyset$ and $last(\xi^{cnt(n)}|_i) \in Ev_j$. In addition, given that $cnt(k+1) = cnt(n)$ it follows that $\xi^{cnt(k+1)}|_j = \xi^{cnt(n)}|_j = e_{cnt(n)}\!\downarrow_j$. Hence, $\mu_i, \xi^{cnt(n)}|_i \Vdash_i ©_j[\varphi_1]$ and, consequently, $\mu, \xi^{cnt(n)} \Vdash @_i[©_j[\varphi_1]]$.

   Assume now that $@_i[©_j[\varphi]] \in Q^-_n$. Again, there is some k such that $n \leq k < \mid(cnt(n))$ and rule $(©^-_i)$ was applied at Q_k. Then, either $@_j[\varphi] \in Q^-_{k+1}$ and $j \in Q^{\bowtie}_{k+1}$ or $j \in Q^{\#}_{k+1}$. In the first case, the proof follows as above. So assume that $j \in Q^{\#}_{k+1}$, which implies that $j \in Q^{\#}_{\mid\,(cnt(n))}$. If $\xi^{cnt(n)}|_i = \emptyset$, then $\mu_i, \xi^{cnt(n)}|_i \not\Vdash_i ©_j[\varphi_1]$ and, consequently, $\mu, \xi^{cnt(n)} \not\Vdash @_i[©_j[\varphi_1]]$. Otherwise, $\xi^{cnt(n)}|_i \neq \emptyset$ and therefore $cnt(n) > 0$. Let $k' = \mid(cnt(n)-1)$. Then $k' < n \leq k$ and, reasoning as above, by Lemma 4.4, we can conclude that $i \in Q^{\mathsf{X}}_{k'+1}$. As $\mid(cnt(k'+1)) = \mid(cnt(n))$, then $i \in Q^{\mathsf{X}}_{\mid\,(cnt(n))}$. Given that $Q_{\mid\,(cnt(n))}$ is not closed and $j \in Q^{\#}_{\mid\,(cnt(n))}$, then $j \notin Q^{\mathsf{X}}_{\mid\,(cnt(n))}$. Hence, $e_{cnt(n)} = last(\xi^{cnt(n)}|_i) \notin Ev_j$, which implies that $\mu_i, \xi^{cnt(n)}|_i \not\Vdash_i ©_j[\varphi_1]$, and we can also conclude that $\mu, \xi^{cnt(n)} \not\Vdash @_i[©_j[\varphi_1]]$.

We now prove that μ and $\xi^{cnt(n)}$ meet the requirements in Q_n. As we observe below, this is straightforward by the construction of the model.

Since there is no closed state tuple in π, no state tuple is contradictory and, therefore, $Q_n^{\bowtie} \subseteq Q_n^{\times}$ and $Q_n^{\#} \cap Q_n^{\times} = \emptyset$. Hence, conditions $(S3)$ and $(S4)$ hold. Observe now that $Q_n^{\times} = \emptyset$ if $n \leq \; |\,(0)$, and $Q_n^{\times} \neq \emptyset$ otherwise. When $n > \; |\,(0)$, condition $(S1)$ holds trivially. If $n \leq \; |\,(0)$ then $\xi^{cnt(n)} = \xi^0 = \emptyset$, and therefore condition $(S1)$ also holds. Finally, we address condition $(S2)$. If $n \leq \; |\,(0)$, condition $(S2)$ holds trivially. Otherwise, we have that $|\,(m-1) < n \leq \; |\,(m)$ for some $m \in \mathbb{N}$. Then $cnt(n) = m$ and $Q_n^{\times} = Q^{\times}_{|\,(cnt(n))}$. Since $i \in Q^{\times}_{|\,(cnt(n))}$ if and only if $e_{cnt(n)} \in Ev_i$, we have $\{i \in Id : e_{cnt(n)} \in Ev_i\} = Q^{\times}_{|\,(cnt(n))} = Q_n^{\times}$. Clearly, $e_{cnt(n)} \in \xi^{cnt(n)}$ and $e_{cnt(n)}$ is maximal with respect to $\xi^{cnt(n)}$ by construction of $\xi^{cnt(n)}$ and the definition of global causality. Therefore, $(S2)$ holds. □

Section 4.7

Lemma 4.18

Proof. $(\rightarrow)$ We start by proving that the assertion

$$\mu, \xi \Vdash \widehat{P} \text{ for some } P \text{ in } R \text{ or } \mu, \xi \Vdash B \text{ for some } B \text{ in } S \tag{4.1}$$

is an invariant of the loop, where $\widehat{P} = \bigwedge_{\alpha \in P^+} \alpha \wedge \bigwedge_{\beta \in P^-} \neg \beta$. Let us prove that (4.1) is preserved by each step of the loop. Assume that it holds at the beginning of the step. We only consider the case where $\mu, \xi \Vdash \widehat{P}$ and P is the top of stack R, since in all the other cases (4.1) is trivially preserved.

If $\delta_1' \Rightarrow \delta_2' \in P^-$, then $\mu, \xi \Vdash \delta_1'$ and $\mu, \xi \nVdash \delta_2'$. Hence, $\mu, \xi \Vdash \widehat{P'}$ where $P' = \langle P^+ \cup \{\delta_1'\}, (P^- \setminus \{\delta_1' \Rightarrow \delta_2'\}) \cup \{\delta_2'\}\rangle$. Since, P' is pushed onto stack R, the assertion follows. The proof when $\delta_1' \Rightarrow \delta_2' \in P^+$ is similar. Assume now that P only contains atomic global formulas. Since $\mu, \xi \Vdash \widehat{P}$, it cannot be the case that $\perp \in P^+$. Furthermore, for every $@_i[\varphi] \in P^+$, we have $\mu, \xi \Vdash @_i[\varphi]$, that is, $\mu_i, \xi|_i \Vdash_i \varphi$ which implies that $\mu, last(\xi|_i){\downarrow} \Vdash @_i[\varphi]$. Moreover, for every $@_i[\varphi] \in P^-$, it holds $\mu, \xi \nVdash @_i[\varphi]$ and therefore $\mu, last(\xi|_i){\downarrow} \nVdash @_i[\varphi]$. We also know that $\mu, last(\xi|_i){\downarrow} |\!\!\Vdash C_i$ for every $i \in Id$. Hence, $\mu, last(\xi|_i){\downarrow} |\!\!\Vdash D_i$ for every $i \in Id$, where D is the extension of C built by the procedure. Then $\mu, \xi \Vdash D$ and therefore D is not contradictory. Hence, D is pushed onto S, and (4.1) follows once again. Assertion (4.1) is thus preserved by each step of the loop.

Assume now that $\mu, \xi \not\Vdash \delta$. Then, $\mu, \xi \Vdash \widehat{\langle \emptyset, \{\delta\} \rangle}$ and therefore $\mu, \xi \Vdash \widehat{P}$ for some P in the stack $\texttt{push}(\langle \emptyset, \{\delta\} \rangle, \texttt{empty})$. As a consequence, (4.1) holds at the beginning of the loop, and, since it is an invariant of the loop, it also holds upon its termination. As R is then empty, it must be the case that $\mu, \xi \Vdash B$ for some B in S.

($\leftarrow$) We prove that if $\mu, \xi \Vdash \delta$ then $\mu, \xi \not\Vdash B$ for every B in S. To begin with, we prove that

$$\mu, \xi \not\Vdash \widehat{P} \text{ for every } P \text{ in } R \text{ and } \mu, \xi \not\Vdash B \text{ for every } B \text{ in } S \qquad (4.2)$$

is invariant of the loop. Assume that (4.2) holds at the beginning of the step. Let P be the top of stack R. Then $\mu, \xi \not\Vdash \widehat{P}$.

Assume that $\delta_1' \Rightarrow \delta_2' \in P^-$ and let $P' = \langle P^+ \cup \{\delta_1'\}, (P^- \setminus \{\delta_1' \Rightarrow \delta_2'\}) \cup \{\delta_2'\} \rangle$. Then either $\mu, \xi \Vdash \delta_1' \Rightarrow \delta_2'$, or $\mu, \xi \not\Vdash \alpha$ for some $\alpha \in P^+$, or $\mu, \xi \Vdash \alpha$ for some $\alpha \in P^- \setminus \{\delta_1' \Rightarrow \delta_2'\}$. In the second case, since $\alpha \in P'^+$ we conclude that $\mu, \xi \not\Vdash \widehat{P'}$. The third case is similar. In the first case, $\mu, \xi \not\Vdash \delta_1'$ or $\mu, \xi \Vdash \delta_2'$ and therefore $\mu, \xi \not\Vdash \widehat{P'}$. Hence, assertion (4.2) follows. The proof when $\delta_1' \Rightarrow \delta_2' \in P^+$ is similar. Assume now that P only contains atomic global formulas. If $\perp \in P^+$, then S is not updated, and therefore the step preserves (4.2). Otherwise, let D be obtained from C and P as described in the procedure. If D is contradictory then (4.2) is preserved for the same reason. Assume that D is not contradictory. and therefore it is pushed onto S. Since $\mu, \xi \not\Vdash \widehat{P}$ and $\perp \notin P^+$, then $\mu, \xi \not\Vdash @_i[\varphi]$ for some $@_i[\varphi] \in P^+$ or $\mu, \xi \Vdash @_i[\varphi]$ for some $@_i[\varphi] \in P^-$. In the first case, $\mu_i, \xi|_i \not\Vdash_i \varphi$ and therefore $\mu, \mathit{last}(\xi|_i){\downarrow} \not\Vdash @_i[\varphi]$. Thus, $\mu, \mathit{last}(\xi|_i){\downarrow} \not\Vdash D_i$ and $\mu, \xi \not\Vdash D$. Again we conclude that (4.2) holds. The other case is similar. Assertion (4.2) is thus preserved by each step of the loop.

Assume that $\mu, \xi \Vdash \delta$. Then, $\mu, \xi \not\Vdash \widehat{\langle \emptyset, \{\delta\} \rangle}$ and therefore $\mu, \xi \not\Vdash \widehat{P}$ for all P in the stack $\texttt{push}(\langle \emptyset, \{\delta\} \rangle, \texttt{empty})$. Since S is empty, (4.2) holds at the beginning of the loop, and, as it is an invariant of the loop, it also holds upon its termination. Hence, in particular, $\mu, \xi \not\Vdash B$ for every B in S. □

Lemma 4.19

Proof. Recall that upon termination of SetGlobal(C, δ), every D in the stack S is an extension of C. Moreover, LocSat(C, δ) only pushes onto the stack extensions of C. Hence, as $\mu, \xi \Vdash C$, along its execution μ and $\mathit{last}(\xi|_k){\downarrow}$ meet all the synchronization requirements in D_k for every D in S and $k \in \mathit{Id}$.

Assertion (i) is straightforward. In order to establish (ii) we start by proving that

$$\mu, \xi \nVdash B \text{ for every } B \text{ in } S \tag{4.3}$$

is an invariant of the loop. Assume that (4.3) holds at the beginning of the step. Let D be the top of stack S. Then $\mu, \xi \nVdash D$. If D_k only contains state formulas for every $k \in Id$, then (4.3) is clearly preserved since S is not updated. Otherwise, let $k \in Id$ be such that there is a non sate formula δ' in D_k. The relevant case occurs when $\mu, \mathit{last}(\xi|_k)\!\downarrow\, \nVdash D_k$. Let $\xi' = \mathit{last}(\xi|_k)\!\downarrow$. If δ' is $@_i[\psi_1 \Rightarrow \psi_2]$ the proof is similar to the one presented for Lemma 4.18. If δ' is $@_i[©_j[\psi]] \in D_k^+$ and $j \notin D_k^{\times}$, we just remove D from S and therefore (4.3) is preserved. Assume then $j \in D_k^{\times}$ and let $Q = \langle D_k^+ \setminus \{\delta'\} \cup \{@_j[\psi]\}, D_k^-, D_k^{\bowtie}, D_k^{\#}, D_k^{\times}\rangle$. The important case is when $\mu, \xi' \nVdash \delta'$. Observe that $i, j \in D_k^{\times}$. Since μ and ξ' meet all the synchronization requirements in D_k, reasoning as in the proof of Proposition 4.7, we conclude that the local sates $\xi'|_i$ and $\xi'|_j$ are non-empty, and $\mathit{last}(\xi'|_i) = \mathit{last}(\xi'|_j) \in Ev_j$. Therefore, $\mu_j, \mathit{last}(\xi'|_i)\!\downarrow_j \nVdash_j \psi$. Since, $\mathit{last}(\xi'|_j)\!\downarrow_j = \xi'|_j$ then $\mathit{last}(\xi'|_i)\!\downarrow_j = \xi'|_j$. Thus, $\mu_j, \xi'|_j \nVdash_j \psi$ and so $\mu, \xi' \nVdash @_j[\psi]$. As a consequence, $\mu, \mathit{last}(\xi|_k)\!\downarrow\, \nVdash Q$ and $\mu, \xi \nVdash D[D_k := Q]$. Assertion (4.3) is then preserved. Let δ' be $@_i[©_j[\psi]] \in D_k^-$. If $j \in D_k^{\times}$, the proof is similar to the one above. Assume that $j \notin D_k^{\times}$ and let $Q = \langle D_k^+, D_k^- \setminus \{\delta'\}, D_k^{\bowtie}, D_k^{\#}, D_k^{\times}\rangle$. If $D_k^{\times} = \emptyset$ then $\xi' = \emptyset$ since μ and ξ' meet the synchronization requirements in D_k. Hence, $\xi'|_i = \emptyset$ and therefore $\mu, \xi' \nVdash \delta'$. If $D_k^{\times} \neq \emptyset$ then $i \in D_k^{\times}$. Thus, in particular, $\mathit{last}(\xi'|_i) \notin Ev_j$. Consequently, $\mu, \xi' \nVdash \delta'$ again holds. Given that we are assuming $\mu, \xi' \nVdash D_k$ and $\delta' \in D_k^-$, then $\mu, \xi' \nVdash \alpha$ for some $\alpha \in D_k^+$ or $\mu, \xi' \Vdash \alpha$ for some $\alpha \in D_k^- \setminus \{\delta'\}$. As a consequence, $\mu, \mathit{last}(\xi|_k)\!\downarrow\, \nVdash Q$. Then $\mu, \xi \nVdash D[D_k := Q]$, and (4.3) is again preserved. We can then conclude that (4.3) holds at the end of the step.

Assume now that $\mu, \xi \Vdash \delta$. Then, by Lemma 4.18, (4.3) holds at the beginning of the loop. Since (4.3) is also invariant of the loop, it holds upon its termination. Hence, upon termination of LocSat(C, δ), assertion (ii) holds.

In order to prove (iii) we now show that

$$\mu, \xi \Vdash B \text{ for some } B \text{ in } S \tag{4.4}$$

is an invariant of the loop. Assume that (4.4) holds at the beginning of the step. Let D be the top of stack S. Assertion (4.4) is preserved if D_k only contains state formulas for every $k \in Id$ as S is not updated. Otherwise, let $k \in Id$ be such that there is a non state formula δ' in D_k. The relevant case occurs when $\mu, \xi \Vdash D$. If δ' is $@_i[\psi_1 \Rightarrow \psi_2]$ the proof is analogous to the one for Lemma 4.18. Let $\xi' = \mathit{last}(\xi|_k)\!\downarrow$. Assume that δ' is $@_i[\copyright_j[\psi]] \in D_k^+$. Since $\mu, \xi' \Vdash\!\!\vdash D_k$, then $\mu, \xi' \Vdash @_i[\copyright_j[\psi]]$, and therefore $\xi'|_i \neq \emptyset$ and $last(\xi'|_i) \in Ev_j$. Recall that μ and ξ' meet all the synchronization requirements in D_k. Thus, if $D_k^{\mathsf{X}} = \emptyset$ then $\xi' = \emptyset$ and thus $\xi'|_i = \emptyset$. Hence, $D_k^{\mathsf{X}} \neq \emptyset$ and, as a consequence, $i \in D_k^{\mathsf{X}}$. Moreover, if $j \notin D_k^{\mathsf{X}}$ then $last(\xi'|_i) \notin Ev_j$. Hence, $j \in D_k^{\mathsf{X}}$. Let $Q = \langle D_k^+ \setminus \{\delta'\} \cup \{@_j[\psi_1]\}, D_k^-, D_k^{\bowtie}, D_k^{\#}, D_k^{\mathsf{X}}\rangle$. Since $\mu, \xi' \Vdash @_i[\copyright_j[\psi]]$, then $\mu_j, last(\xi'|_i) \downarrow_j \Vdash_j \psi$. Reasoning as above, $last(\xi'|_i) \downarrow_j = \xi'|_j$ holds and therefore $\mu_j, \xi'|_j \Vdash_j \psi$. As a consequence, $\mu, \xi' \Vdash @_j[\psi]$ and thus $\mu, \xi' \Vdash\!\!\vdash Q$. Noting that Q and C_k have the same synchronization requirements, we can conclude that Q cannot be contradictory. The assertion (4.4) is then preserved because $\mu, \xi \Vdash D[D_k := Q]$. Assume now that δ' is $@_i[\copyright_j[\psi]] \in D_k^-$. Since $\mu, \xi' \Vdash\!\!\vdash D_k$, if $j \notin D_k^{\mathsf{X}}$ then $\mu, \xi \Vdash D[D_k := Q]$ trivially holds because we only remove δ' from D_k^- to get Q. Assume $j \in D_k^{\mathsf{X}}$ and let $Q = \langle D_k^+, D_k^- \setminus \{\delta'\} \cup \{@_j[\psi]\}, D_k^{\bowtie}, D_k^{\#}, D_k^{\mathsf{X}}\rangle$. In particular, $\mu, \xi' \nVdash @_i[\copyright_j[\psi]]$. Reasoning as above we conclude that $\mu, \xi' \nVdash @_j[\psi]$, that $\mu, \xi' \Vdash\!\!\vdash Q$, and that Q is not contradictory. Hence, $\mu, \xi \Vdash D[D_k := Q]$ and (4.4) is again preserved. We then conclude that (4.4) holds at the end of the step.

Assume that $\mu, \xi \nVdash \delta$. Then, by Lemma 4.18, (4.4) holds at the beginning of the loop. Since (4.4) is invariant of the loop, it holds upon its termination. Hence, upon termination of LocSat(C, δ), assertion (iii) holds.

(iv) First of all note that upon termination of SetGlobal(C, δ), every D in stack S is a non contradictory extension of C. Moreover, it is also easy to conclude that this assertion also holds upon termination of the loop in LocSat(C, δ). Assume that upon termination of LocSat(C, δ) the stack S is not empty. Then, by (i), b is *true*. It is easy to conclude that then $\mathtt{top}(S)$ is, in particular, a non contradictory atomic extension of C. Thus, $\mu^{\mathtt{top}(S)}, \xi^{n_{\mathtt{top}(S)}} \Vdash \mathtt{top}(S)$ and therefore $\mu^{\mathtt{top}(S)}, \xi^{n_{\mathtt{top}(S)}} \Vdash C$. Then, the result follows by (ii). □

Proposition 4.20

Proof. (i) Assume that $\Gamma \vDash \delta$. If no μ satisfies Γ then, by Proposition 4.14, any tableau for Γ is unsuccessful. Consequently, Ent(Γ, δ) outputs *yes*. Otherwise, by Corollary 4.10, there is a successful tableau $\mathcal{T}$ for $\langle \Gamma, \emptyset, \emptyset, \emptyset, \emptyset\rangle$. Assume that Ent$(\Gamma, \delta)$ outputs $\langle no, \mu, \xi\rangle$. Then b is *true* upon termination of LocSat(C, δ) for some cut C. By Lemma 4.19, the stack S is not empty. Moreover, μ is $\mu^{\mathtt{top}(S)}$, ξ is $\xi^{n_{\mathtt{top}(S)}}$, and $\mu^{\mathtt{top}(S)}, \xi^{n_{\mathtt{top}(S)}} \nVdash \delta$. Taking into account, the definition of $\mu^{\mathtt{top}(S)}$ and the proof of Proposition 4.14, we know that $\mu^{\mathtt{top}(S)} \Vdash \Gamma$. We reach a contradiction since we are assuming that $\Gamma \vDash \delta$. Hence, $\textsc{Ent}(\Gamma, \delta)$ outputs *yes*.

(ii) Assume that $\Gamma \nvDash \delta$. Then, in particular, there is μ such that $\mu \Vdash \Gamma$. Again by Corollary 4.10, we know that there is a successful tableau $\mathcal{T}$ for $Q_0 = \langle \Gamma, \emptyset, \emptyset, \emptyset, \emptyset \rangle$.

Consider a linearization ℓ of μ as in the proof of Proposition 4.9, and let k be such that $\xi^k = \xi$. Moreover, recall $\pi^{\mu,\ell}_{Q_0}$ and consider the k-th state, $Q_{\mid (k)}$, of $\pi^{\mu,\ell}_{Q_0}$. Let C be the cut induced by $(\widehat{\pi}^{\mu,\ell}_{Q_0})_{Q_{\mid (k)}}$. By construction, we know that $\mu, \xi^k \Vdash C$. Hence, by (iii) of Lemma 4.19, S is not empty. Furthermore, by (i) and (iv) of the same lemma, b is *true* and $\mu^{\mathtt{top}(S)}, \xi^{n_{\mathtt{top}(S)}} \nVdash \delta$. Consequently, the result follows. □

5

THAT'S ENOUGH: ASYNCHRONY WITH STANDARD CHOREOGRAPHY PRIMITIVES

Luís Cruz-Filipe and Fabrizio Montesi
Dept. Mathematics and Computer Science, University of Southern Denmark
{lcf,fmontesi}@imada.sdu.dk

Abstract

Choreographies are widely used for the specification of concurrent and distributed software architectures. Since asynchronous communications are ubiquitous in real-world systems, previous works have proposed different approaches for the formal modelling of asynchrony in choreographies. Such approaches typically rely on ad-hoc syntactic terms or semantics for capturing the concept of messages in transit, yielding different formalisms that have to be studied separately.

In this work, we take a different approach, and show that such extensions are not needed to reason about asynchronous communications in choreographies. Rather, we demonstrate how a standard choreography calculus already has all the needed expressive power to encode messages in transit (and thus asynchronous communications) through the primitives of process spawning and name mobility. The practical consequence of our results is that we can reason about real-world systems within a choreography formalism that is simpler than those hitherto proposed.

5.1 Introduction

Today, concurrent and distributed systems are widespread. Multi-core hardware and large-scale networks represent the norm rather than the exception. However, programming such systems is challenging, because it is difficult to program correctly the intended interactions among components executed concurrently (e.g., services). Empirical investigations of bugs in concurrent and distributed software [18, 19] reveal that most errors are due to: deadlocks (e.g., a component that was supposed to be ready for interaction at a given time is actually not); violations of atomicity intentions (e.g., a component is performing some action when not intended to); or, violations of ordering intentions (some components perform the right actions, but not when intended). If the design and implementation of a concurrent system are initially difficult, they get even harder as the system evolves and has to be maintained. Without proper tool support, introducing new actions at components may have unexpected effects due to side-effects. To mitigate this problem, *choreographies* can be used as high-level formal specifications of the intended interactions among components [1, 2, 3, 4, 16, 17, 25, 27].

Example 5.1 *We use a choreography to define a scenario where a buyer, Alice (*a*), purchases a product from a seller (*s*) through her bank (*b*).*

1. a.*title* -> s;
2. s.*price* -> a;
3. s.*price* -> b;
4. if b $\overset{\leftarrow}{=}$ a then
5.   b -> s[*ok*]; b -> a[*ok*];
6.   s.*book* -> a;
7. else b -> s[*ko*]; b -> a[*ko*]

In Line 1, the term a.*title* -> s *denotes an interaction whereby* a *communicates the title of the book that Alice wishes to buy to* s. *The seller then sends the price of the book to both* a *and* b. *In Line 4,* a *sends the price she expects to pay to* b, *which confirms that it is the same amount requested by* s *(stored internally at* b*). If so,* b *notifies both* s *and* a *of the successful transaction (Line 5) and* s *sends the book to* a *(Line 6). Otherwise,* b *notifies* s *and* a *of the failure (Line 7) and the choreography terminates.*

Choreographies are the foundations of an emerging development paradigm, called Choreographic Programming [21, 22], where an automatic projection procedure is used to synthesise a set of compliant local implementations (the implementations of the single components) from a choreography [4, 17, 25]. This procedure is formally proven to be correct, preventing deadlocks, ordering errors, and atomicity violations. This ensures, critically, that updates to either the choreography or the local implementations do not introduce bugs and that developers always know what communications their systems will enact (by looking at the choreography). In the previous example, the implementation inferred for, e.g., Alice (a), would be: send the book title to s; receive the price from s; send the price to b for confirmation; await the success/failure notification from b; in case of success, receive the book from s.

Choreography languages come in all sizes and flavours, with different sets of primitives inspired by practical applications, such as adaptation [10, 11], channel mobility [5, 6], or web services [2, 4, 27]. However, this multiplicity makes it increasingly difficult to reuse available theory and tools, because of the differences and redundancies among these models. For this reason, we previously introduced the model of Core Choreographies (CC) [8], a minimal and representative theoretical model of Choreographic Programming. In CC, components are modelled as concurrent processes that run independently and possess own memory, inspired by process calculi [26]. Example 5.1 is written in the syntax of CC described in § 5.2.

In this paper, we are interested in studying asynchronous communications in choreographies. As a motivation, consider the two communications in Lines 2 and 3 of Example 5.1: typically, in a realistic system, we would expect s to send the price to a and then immediately proceed to sending it also to b, without waiting for a to receive its message. Typically, asynchronous communications are formalised in choreography models by defining ad-hoc extensions to their syntax and semantics [5, 12, 16, 17, 23, 24], causing a substantial amount of duplication in their technical developments (many of which are even incompatible with each other).

Unfortunately, there are still no foundational studies that provide an elegant and general understanding of asynchrony in choreographies. Here, we pursue such a study in the context of CC. We depict our overall development in Figure 5.1, and describe it in the following.

We first present our development for the computational fragment of CC, called Minimal Choreographies (MC) [8]. We take inspiration from how asynchrony is modelled in foundational process models, specifically the π-calculus [20]. The key idea there is to use processes to repre-

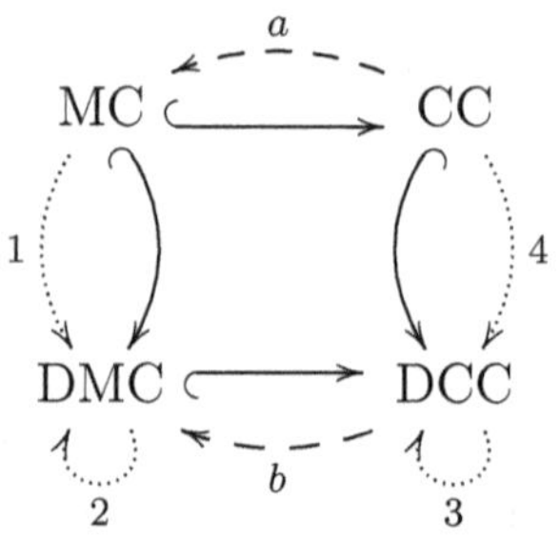

Figure 5.1: Choreography calculi and encodings.

sent messages in transit, allowing the sender to proceed immediately after having sent a message without having to synchronise with the receiver [26]. In an asynchronous system, there is no bound to the number of messages that could be transiting in the network; this means that MC is not powerful enough for our purposes, because it can only capture a finite number of processes (the same holds for CC). For this reason, we extend MC with two standard notions, borrowed from process calculi and previous choreography models: process spawning – the ability to create new processes at runtime – and name mobility – the ability to send process references, or names. We call this new language Dynamic Minimal Choreographies (DMC). MC is a strict sub-language of DMC, denoted by the arrow $\hookrightarrow$ on the left-hand side of Figure 5.1. In general, all arrows of shape $\hookrightarrow$ in that figure denote (strict) language inclusion.

The dotted arrow (1) in Figure 5.1 is the cornerstone of our development: every choreography in MC can be encoded in an asynchronous implementation in DMC, by using auxiliary processes to represent messages in transit. Since DMC extends MC with new primitives, it makes sense to extend this encoding to the whole language of DMC (2). This syntactic interpretation of asynchrony in choreographies is our main contribution. Specifically, our results show that asynchronous communications can be modelled in choreographies using well-known notions, i.e., process spawning and name mobility (studied, e.g., in [5, 7]), without the need for ad-hoc constructions. Coming back to the title: we already have enough.

The fact that our encoding can be extended from MC to DMC is evidence that our approach is robust, and the simplicity of DMC makes it a convenient foundational calculus to use in future developments of choreographies. However, one of the expected advantages of using a foundational theory such as DMC for capturing asynchrony is indeed that we can reuse existing formal techniques based on standard primi-

$$C ::= \mathbf{0} \mid \eta; C \mid \text{if } \mathsf{p} \overset{\leftarrow}{=} \mathsf{q} \text{ then } C_1 \text{ else } C_2 \mid \text{def } X = C_2 \text{ in } C_1 \mid X$$
$$\eta ::= \mathsf{p}.e \texttt{ -> } \mathsf{q} \mid \mathsf{p} \texttt{ -> } \mathsf{q}[l] \qquad e ::= v \mid * \mid \ldots$$

Figure 5.2: Core Choreographies, Syntax.

tives for choreographies. (This is a common scenario in π-calculus, where many techniques apply to its sub-languages [26].) We show an example of such reuses. Core Choreographies (CC) [8] is MC with the addition of a primitive for communicating choices explicitly as messages, called selection [4, 14, 16, 28] (the terms in Lines 5 and 7 in Example 5.1 are selections). An important property of CC is that selections can be encoded in the simpler language MC – the dashed arrow (a) in Figure 5.1. What happens if we add selections to DMC? Ideally, the resulting calculus (called Dynamic Core Choreographies, or DCC) should *both* have an asynchronous interpretation through the techniques introduced in this paper and still possess the property that selections are encodable using the simpler language DMC. This is indeed the case. We extend our encoding to yield an interpretation of asynchronous selections, yielding (3) and (4). The second property (encodability of selections in DCC) follows immediately from language inclusion, giving us (b) for free.

5.2 Background

We briefly introduce CC and MC, from [8], and summarise their key properties.

The syntax of CC is given in Figure 5.2, where C ranges over choreographies. Processes $(\mathsf{p}, \mathsf{q}, \ldots)$ run in parallel, and each process stores a value in a local memory cell.[1] Each process can access its own value using the syntactic construct $*$, but it cannot read the contents of another process. Term $\eta; C$ is an interaction between two processes, read "the system may execute η and proceed as C". In a value communication $\mathsf{p}.e$ -> q, p sends its local evaluation of expression e to q, which stores the received value. In a label selection p -> $\mathsf{q}[l]$, p communicates label l to q. The set of labels is immaterial, as long as it contains at least two elements. In a conditional $\text{if } \mathsf{p} \overset{\leftarrow}{=} \mathsf{q} \text{ then } C_1 \text{ else } C_2$, q sends its value to p, which checks if the received value is equal to its own; the choreography proceeds as C_1, if that is the case, or as C_2, otherwise. In all these ac-

[1] In the original presentation, values were restricted to natural numbers; we drop this restriction here since it is orthogonal to our development.

$$\frac{e[\sigma(\mathsf{p})/*] \downarrow v}{\mathsf{p}.e \text{ -> } \mathsf{q}; C, \sigma \to C, \sigma[\mathsf{q} \mapsto v]} \;\lfloor \mathrm{C|Com} \rceil$$

$$\frac{}{\mathsf{p} \text{ -> } \mathsf{q}[l]; C, \sigma \to C, \sigma} \;\lfloor \mathrm{C|Sel} \rceil$$

$$\frac{i = 1 \text{ if } \sigma(\mathsf{p}) = \sigma(\mathsf{q}),\ i = 2 \text{ o.w.}}{\mathsf{if}\ \mathsf{p} \overset{\leftarrow}{=} \mathsf{q}\ \mathsf{then}\ C_1\ \mathsf{else}\ C_2, \sigma \to C_i, \sigma} \;\lfloor \mathrm{C|Cond} \rceil$$

$$\frac{C_1, \sigma \;\to\; C_1', \sigma'}{\mathsf{def}\ X = C_2\ \mathsf{in}\ C_1, \sigma \;\to\; \mathsf{def}\ X = C_2\ \mathsf{in}\ C_1', \sigma'} \;\lfloor \mathrm{C|Ctx} \rceil$$

$$\frac{C_1 \preceq C_2 \quad C_2, \sigma \to C_2', \sigma' \quad C_2' \preceq C_1'}{C_1, \sigma \to C_1', \sigma'} \;\lfloor \mathrm{C|Struct} \rceil$$

Figure 5.3: Core Choreographies, Semantics.

tions, the two interacting processes must be different. Definitions and invocations of recursive procedures (X) are standard. The term **0** is the terminated choreography.

The semantics of CC uses reductions of the form $C, \sigma \to C', \sigma'$, where the total state function σ maps each process name to its value. We use v, w, ... to range over values. The reduction relation $\to$ is defined by the rules given in Figure 5.3.

These rules formalise the intuition presented earlier. In the premise of $\lfloor$C|Com$\rceil$, we write $e[\sigma(\mathsf{p})/*]$ for the result of replacing $*$ with $\sigma(\mathsf{p})$ in e. In the reductum, $\sigma[\mathsf{q} \mapsto v]$ denotes the updated state function σ where q now maps to v.

Rule $\lfloor$C|Struct$\rceil$ uses the structural precongruence relation $\preceq$, which gives a concurrent interpretation to choreographies by allowing non-interfering actions to be executed in any order. The key rule defining $\preceq$ is

$$\frac{\mathsf{pn}(\eta) \cap \mathsf{pn}(\eta') = \emptyset}{\eta; \eta' \;\equiv\; \eta'; \eta} \;\lfloor \mathrm{C|Eta\text{-}Eta} \rceil$$

where $C \equiv C'$ stands for $C \preceq C'$ and $C' \preceq C$ and $\mathsf{pn}(C)$ returns the set of all process names occurring in C. The other rules for $\preceq$ are standard, and support recursion unfolding and garbage collection of unused definitions.

CC was designed as a core choreography language, in which in particular it is possible to implement any computable function. Furthermore, CC choreographies can always progress until they terminate.

Theorem 5.2 *If C is a choreography, then either $C \preceq \mathbf{0}$ (C has terminated) or, for all σ, $C, \sigma \to C', \sigma'$ for some C' and σ' (C can reduce).*

Label selections are not required for Turing completeness, and thus the simpler fragment MC obtained from CC by omitting them is interesting as an intermediate language for compilers and, also, for theoretical analysis. One of the reasons for having label selection is to make choice propagation explicit in choreographies; in a system implementation, this allows, e.g., to monitor distributed choices without having to inspect the message payload. Another reason is *projectability*: the possibility of automatically generating processes implementations that satisfy the choreographic specification. In Example 5.1, the label selections in Lines 5 and 7 are important in order for b to let a and s know whether or not they should communicate.

Choices communicated by label selections can also be encoded as data in value communications, by sending a boolean value to determine which one of two branches was selected. This is the key idea behind the encoding presented in [8] – arrow (a) in Figure 5.1 – which transforms a choreography in CC to one in MC by encoding selections as value communications and nested conditionals.

We do not need to concern ourselves with projectability in this work, and we will thus omit its details. This is because CC and MC enjoy a projectability property that is not altered by our development. Formally, there exists a procedure $\mathsf{Amend}(\cdot)$ that, given any choreography, returns a choreography in CC that is projectable. Then, given a projectable CC choreography, the encoding $(\![\cdot]\!)$ transforms it into a choreography in MC, by encoding selections as value communications and conditionals. These transformations preserve the computational meaning of choreographies, as formally stated in the following theorem ($\cdot^+$ extends a state function to the auxiliary processes introduced by the transformations in a systematic way).

Theorem 5.3 *Let C, C' be MC choreographies and σ, σ' be states. If $C, \sigma \to^* C', \sigma'$, then $(\![\mathsf{Amend}(C)]\!)^+, \sigma^+ \to^* (\![\mathsf{Amend}(C')]\!), \sigma'^+$.*

The main limitation of CC is that its semantics is synchronous. Indeed, in a real-world scenario implementation of Example 5.1, we would expect s to proceed immediately to sending its message in Line 3 after having sent the one in Line 2, without waiting for a to receive the latter. Capturing this kind of asynchronous behaviour is the main objective of our development in the remainder of this paper.

$$C ::= \cdots \mid \mathsf{def}\, X(\tilde{\mathsf{p}}) = C_2 \,\mathsf{in}\, C_1 \mid X\langle\tilde{\mathsf{p}}\rangle$$
$$\eta ::= \cdots \mid \mathsf{p}\,\mathsf{starts}\,\mathsf{q} \qquad e ::= \mathsf{p} \mid \cdots$$

Figure 5.4: Dynamic Core Choreographies, Syntax.

5.3 Asynchrony in MC

In this section, we extend CC with primitives to implement asynchronous communication, obtaining a calculus of Dynamic Core Choreographies (DCC). We focus on MC and first show that any MC choreography can be encoded in DMC – the fragment of DCC that does not use label selection – in such a way that communication becomes asynchronous.

More precisely, we provide a mapping $\{\!\!\{\cdot\}\!\!\} : \text{MC} \to \text{DMC}$ such that every communication action $\mathsf{p}.e$ -> $\mathsf{q} \in C \in \text{MC}$ becomes split into a send/receive pair in $\{\!\!\{C\}\!\!\} \in \text{DMC}$, with the properties that: p can continue executing without waiting for q to receive its message (and even send further messages to q); and messages from p to q are delivered in the same order as they were originally sent.

The system DCC. We briefly motivate DCC. In CC, there is a bound on the number of values that can be stored at any given time by the system: since each process can hold a single value, the maximum number of values the system can know is equal to the number of processes in the choreography, which is fixed. However, in an asynchronous setting, the number of values that need to be stored is unbounded: a process p may loop forever sending values to q, and q may wait an arbitrary long time before receiving any of them. Therefore, we need to extend CC with the capability to generate new processes. As discussed in [7], this requires enriching the language with two additional abilities: parameters to recursive procedures (in order to be able to use a potentially unbounded number of processes at the same time) and action to communicate process names.

Formally, the differences between the syntax of CC and that of DCC are highlighted in Figure 5.4: procedure definitions and calls now have parameters; there is a new term for generating processes; and, the expressions sent by processes can also be process names. The possibility of communicating a process name (p.q -> r) ensures name mobility. We will use the abbreviation p : r <-> q as shorthand for p.q -> r; p.r -> q.

The semantics for DCC includes an additional ingredient, borrowed from [7]: a graph of connections G, keeping track of which pairs of

$$\frac{\mathsf{p} \overset{G}{\longleftrightarrow} \mathsf{q} \quad e[\sigma(\mathsf{p})/*] \downarrow v}{G, \mathsf{p}.e \texttt{ -> } \mathsf{q}; C, \sigma \ \rightarrow \ G, C, \sigma[\mathsf{q} \mapsto v]} \ \lfloor \mathsf{C}|\mathrm{Com} \rceil$$

$$\frac{\mathsf{p} \overset{G}{\longleftrightarrow} \mathsf{q}}{G, \mathsf{p} \texttt{ -> } \mathsf{q}[l]; C, \sigma \ \rightarrow \ G, C, \sigma} \ \lfloor \mathsf{C}|\mathrm{Sel} \rceil$$

$$\frac{}{G, \mathsf{p}\ \mathsf{starts}\ \mathsf{q}; C, \sigma \ \rightarrow \ G \cup \{\mathsf{p} \leftrightarrow \mathsf{q}\}, C, \sigma[\mathsf{q} \mapsto \bot]} \ \lfloor \mathsf{C}|\mathrm{Start} \rceil$$

$$\frac{\mathsf{p} \overset{G}{\longleftrightarrow} \mathsf{q} \quad \mathsf{p} \overset{G}{\rightarrow} \mathsf{r}}{G, \mathsf{p}.\mathsf{r} \texttt{ -> } \mathsf{q}; C, \sigma \ \rightarrow \ G \cup \{\mathsf{q} \rightarrow \mathsf{r}\}, C, \sigma} \ \lfloor \mathsf{C}|\mathrm{Intro} \rceil$$

$$\frac{i = 1 \text{ if } \sigma(\mathsf{p}) = \sigma(\mathsf{q}), \quad i = 2 \text{ otherwise}}{G, \mathsf{if}\ \mathsf{p} \overset{\texttt{<-}}{=} \mathsf{q}\ \mathsf{then}\ C_1\ \mathsf{else}\ C_2, \sigma \ \rightarrow \ G, C_i, \sigma} \ \lfloor \mathsf{C}|\mathrm{Cond} \rceil$$

$$\frac{C_1 \preceq C_2 \quad G, C_2, \sigma \ \rightarrow \ G', C_2', \sigma' \quad C_2' \preceq C_1'}{G, C_1, \sigma \ \rightarrow \ G', C_1', \sigma'} \ \lfloor \mathsf{C}|\mathrm{Struct} \rceil$$

Figure 5.5: Dynamic Core Choreographies, Semantics.

processes are allowed to communicate. This graph is directed, and an edge from p to q in G (written $\mathsf{p} \overset{G}{\rightarrow} \mathsf{q}$) means that p knows the name of q. In order for an actual message to flow between p and q, both processes need to know each other, which we write as $\mathsf{p} \overset{G}{\longleftrightarrow} \mathsf{q}$.[2] The reduction relation now has the form $G, C, \sigma \rightarrow G', C', \sigma'$, where G and G' are the connection graphs before and after executing C, respectively. The complete rules are given in Figure 5.5, with $\preceq$ defined similarly to CC. In rule $\lfloor \mathsf{C}|\mathrm{Start} \rceil$, the fresh process q is assigned a default value $\bot$.

The proof of Theorem 5.2 can be generalised to DCC, but this requires an extra ingredient: a simple type system (which we do not detail, as it is a subsystem of that presented in [7]). This type system checks that all processes that attempt at communicating are connected in the communication graph, e.g., by being properly introduced using name mobility. Furthermore, we can define a target process calculus for DCC and an EndPoint Projection that will automatically synthetise correct-by-construction deadlock-free implementations of (projectable) choreographies, using techniques from [8] and [7]. Although these constructions are not technically challenging, we omit them for brevity, since

[2] In some process calculi, the weaker condition $\mathsf{p} \overset{G}{\rightarrow} \mathsf{q}$ is typically sufficient for p to send a message to q. Our condition is equivalent to that found in the standard model of Multiparty Session Types [16]. This choice is orthogonal to our development.

they are immaterial for our results.

The fragment of DCC that does not contain label selections is called Dynamic Minimal Choreographies (DMC). Amendment and label selection elimination hold for DMC and DCC just as for MC and CC, so that, for any DMC choreography C, $(\!|\mathsf{Amend}(C)|\!)$ is always projectable, and Theorem 5.3 also holds for these calculi (arrow (b) in Figure 5.1).

The encoding. We focus now on the mapping (1) from Figure 5.1, as this is the key ingredient to establish the remaining connections in that figure. Let C be a choreography in MC. In order to encode C in DMC, we use a function $M_C : \mathcal{P}^2 \to \mathbb{N}$, where $\mathcal{P} = \mathsf{pn}(C)$ is the set of process names in C. Intuitively, $\{\!\{C\}\!\}$ will use a countable set of auxiliary processes $\{\mathsf{pq}^i \mid \mathsf{p},\mathsf{q} \in \mathcal{P}, i \in \mathbb{N}\}$, where pq^i will hold the ith message from p to q.

First, we setup initial channels for communications between all processes occurring in C.

$$\{\!\{C\}\!\} = \{\mathsf{p\,start\,pq}^0;\ \mathsf{p : q \mathrel{<\!\!-\!\!>} pq}^0\}_{p,q\in\mathcal{P},\mathsf{p}\neq\mathsf{q}}\ ;\ \{\!\{C\}\!\}_{M_0}$$

Here, $M_0(\mathsf{p},\mathsf{q}) = 0$ for all p and q. For simplicity, we write pq^M for $\mathsf{pq}^{M(\mathsf{p},\mathsf{q})}$ and pq^{M+} for $\mathsf{pq}^{M(\mathsf{p},\mathsf{q})+1}$. The definition of $\{\!\{C\}\!\}_M$ is given in Figure 5.6.

$$\begin{aligned}
\{\!\{\mathsf{p}.e\ \mathsf{\text{->}}\ \mathsf{q}; C\}\!\}_M &= \mathsf{p}.e\ \mathsf{\text{->}}\ \mathsf{pq}^M; \mathsf{p\,start\,pq}^{M+}; \mathsf{p} : \mathsf{pq}^M \mathsf{\text{<->}} \mathsf{pq}^{M+};\\
&\quad \mathsf{pq}^M.\mathsf{q}\text{->}\mathsf{pq}^{M+}; \mathsf{pq}^M.\mathsf{pq}^{M+}\text{->}\mathsf{q}; \mathsf{pq}^M.*\ \text{->}\ \mathsf{q};\\
&\quad \{\!\{C\}\!\}_{M[(\mathsf{p},\mathsf{q})\mapsto M(\mathsf{p},\mathsf{q})+1]}\\
\{\!\{\mathsf{if}\ \mathsf{p} \overset{\leftarrow}{=} \mathsf{q}\ \mathsf{then}\ C_1\ \mathsf{else}\ C_2\}\!\}_M &= \mathsf{q}.*\ \text{->}\ \mathsf{qp}^M; \mathsf{q\,start\,qp}^{M+}; \mathsf{q} : \mathsf{qp}^M \text{<->} \mathsf{qp}^{M+};\\
&\quad \mathsf{qp}^M.\mathsf{p}\ \text{->}\ \mathsf{qp}^{M+}; \mathsf{qp}^M.\mathsf{qp}^{M+}\ \text{->}\ \mathsf{p};\\
&\quad \mathsf{if}\ \mathsf{p} \overset{\leftarrow}{=} \mathsf{qp}^M\ \mathsf{then}\ \{\!\{C_1\}\!\}_{M[(\mathsf{p},\mathsf{q})\mapsto M(\mathsf{p},\mathsf{q})+1]}\\
&\quad\qquad \mathsf{else}\ \{\!\{C_2\}\!\}_{M[(\mathsf{p},\mathsf{q})\mapsto M(\mathsf{p},\mathsf{q})+1]}\\
\{\!\{\mathbf{0}\}\!\}_M &= \mathbf{0}\\
\{\!\{X\}\!\}_M &= X\langle \tilde{M} \rangle\\
\{\!\{\mathsf{def}\ X = C_2\ \mathsf{in}\ C_1\}\!\}_M &= \mathsf{def}\ X(\overline{M_0}) = \{\!\{C_2\}\!\}_{M_0}\ \mathsf{in}\ \{\!\{C_1\}\!\}_M
\end{aligned}$$

Figure 5.6: Encoding MC in DMC

We write $\tilde{M}$ for $\{\mathsf{pq}^M \mid \mathsf{p},\mathsf{q} \in \mathcal{P}, \mathsf{p} \neq \mathsf{q}\}$, where we assume that the order of the values of M is fixed. In recursive definitions, we reset M to M_0; note that the parameter declarations act as binders, so these process names are still fresh. We can use α-renaming on $\{\!\!\{C\}\!\!\}$ to make all bound names distinct.

In order to encode $\mathsf{p}.e$ -> q, p uses the auxiliary process pq^M to store the value it wants to send to q. Then, p creates a fresh process (to use in the next communication) and sends its name to pq^M. Afterwards, p is free to proceed with execution. In turn, pq^M communicates q's name to the new process, which now is ready to receive the next message from p. Finally, pq^M waits for q to be ready to receive both the value being communicated and the name of the process that will store the next value.

The behaviours of the choreographies C and $\{\!\!\{C\}\!\!\}$ are closely related, as formalised in the following theorems.

Theorem 5.4 *Let* $\mathsf{p} \in \mathsf{pn}(C)$ *and* $\mathsf{pq} \in \mathsf{pn}(\{\!\!\{C\}\!\!\}) \setminus \mathsf{pn}(C)$. *If* $G, \{\!\!\{C\}\!\!\}, \sigma \to^* G', C_1, \sigma_1 \to G', C_2, \sigma_2$ *where in the last transition a value* v *is sent from* p *to* pq, *then there exist* G'', C_3, σ_3, C_4 *and* σ_4 *such that* $G', C_2, \sigma_2 \to^* G'', C_3, \sigma_3 \to G'', C_4, \sigma_4$ *and in the last transition the same value* v *is sent from* pq *to some process* $\mathsf{q} \in \mathsf{pn}(C)$.

Theorem 5.4 states that messages sent from p to q are eventually received by q.

Theorem 5.5 *If* $G, \{\!\!\{C\}\!\!\}_M, \sigma \to^* G_1, C_1, \sigma_1$, *then there exist* G'', C', σ' *and* σ'' *such that* $C, \sigma \to^* C', \sigma'$, *and* $G_1, C_1, \sigma_1 \to^* G'', \{\!\!\{C'\}\!\!\}_M, \sigma''$, *and* σ' *and* σ'' *coincide on the values stored at* $\mathsf{pn}(C)$.

Theorem 5.5 states that the encoding does not add any additional behaviour to the original choreography, aside from expanding communications into several actions.

Example 5.6 *We partially show the result of applying this transformation to Lines 1–3 of Example 5.1. We only include the initializations of the channels that are used in this fragment; the numbers indicated refer to the line numbers in the original example.*

The first three lines initialize three channels: from a *to* s; *from* s *to* a; *and from* s *to* b. *Then one message is passed in each of these channels, as dictated by the encoding. All communications are asynchronous in the sense explained above, as in each case the main sender process sends its message to a dedicated intermediary* (as^0, sa^0 *or* sb^0, *respectively), who will eventually deliver it to the recipient. Moreover, causal dependencies are kept: in Step 2,* s *can only send its message to* sa^0 *after receiving the message sent by* a *in Step 1. However, in Step 3* s *can send its message to* sb^0 *without waiting for* a *to receive the previous message, as the action* $\mathsf{s}.\mathit{price}$ -> sa^0 *can swap with the three actions immediately preceding it.*

$\quad$ a start as^0; $\mathsf{a} : \mathsf{as}^0$ <-> s;

$\quad$ s start sa^0; $\mathsf{s} : \mathsf{sa}^0$ <-> a;

$\quad$ s start sb^0; $\mathsf{s} : \mathsf{sb}^0$ <-> b;

1. $\mathsf{a}.\mathit{title}$ -> as^0; a start as^1; $\mathsf{a} : \mathsf{as}^1$ <-> as^0;
   $\mathsf{as}^0.\mathsf{as}^1$ -> s; $\mathsf{as}^0.\mathsf{s}$ -> as^1; $\mathsf{as}^0.*$ -> s;
2. $\mathsf{s}.\mathit{price}$ -> sa^0; s start sa^1; $\mathsf{s} : \mathsf{sa}^1$ <-> sa^0;
   $\mathsf{sa}^0.\mathsf{sa}^1$ -> a; $\mathsf{sa}^0.\mathsf{a}$ -> sa^1; $\mathsf{sa}^0.*$ -> a;
3. $\mathsf{s}.\mathit{price}$ -> sb^0; s start sb^1; $\mathsf{s} : \mathsf{sb}^1$ <-> sb^0;
   $\mathsf{sb}^0.\mathsf{sb}^1$ -> b; $\mathsf{sb}^0.\mathsf{b}$ -> sb^1; $\mathsf{sb}^0.*$ -> b;

$\quad$ ...

We briefly illustrate Theorems 5.4 and 5.5 in this setting. Theorem 5.4 states that, e.g., the action $\mathsf{a}.\mathit{title}$ -> as^0 *is eventually followed by a communication of title from* as^0 *to some other process in the original choreography (in this case,* s*). Theorem 5.5 implies that if, e.g.,* $\mathsf{a}.\mathit{title}$ -> as^0 *is executed, then it must be "part" of an action in the original choreography (in this case,* $\mathsf{a}.\mathit{title}$ -> s*), and furthermore it is possible to find an execution path that will execute the remaining actions generated from that one (the remaining five actions in Step 1).*

5.4 The General Case

The calculus DMC is in itself synchronous, just like MC. We now show that we can extend $\{\!\!\{\cdot\}\!\!\}$ to the full language of DMC – arrow (2) in Figure 5.1 – thereby obtaining a systematic way to write asynchronous communications in DMC. By further marking which communications we want to treat as synchronous (so that they are untouched by $\{\!\!\{\cdot\}\!\!\}$) we obtain a calculus in which we can have both synchronous and asynchronous communication, compiled in itself. This is similar (albeit dual) to the situation in asynchronous π-calculus, where we can also encode synchronous communication without extending the language.

The main challenge is dealing with M, as the source choreography can now include process spawning. This means that the domain of M can be dynamically extended throughout the computation of $\{\!\!\{C\}\!\!\}_M$, which renders our parameter-passing in recursive calls invalid (since the number of parameters in the procedures generated by our encoding is fixed). However, since each procedure $X(\mathsf{p}_1, \ldots, \mathsf{p}_n)$ in DMC can only use (by convention) the processes $\mathsf{p}_1, \ldots, \mathsf{p}_n$ in its body, we can restrict the additional parameters introduced by the encoding to the $n(n-1)$ auxiliary processes currently assigned by M to communications between the p_is. For example, $\{\!\!\{\mathsf{def}\, X(\mathsf{p},\mathsf{q}) = C_2 \,\mathsf{in}\, C_1\}\!\!\}_M$ would be $\mathsf{def}\, X(\mathsf{p},\mathsf{q},\mathsf{pq}^0,\mathsf{qp}^0) = \{\!\!\{C_2\}\!\!\}_{M_0} \,\mathsf{in}\, \{\!\!\{C_1\}\!\!\}_M$. We will not write this definition formally.

With this in mind, we can easily define the new cases for $\{\!\!\{C\}\!\!\}_M$.

$$\begin{aligned}
\{\!\!\{\mathsf{p\ start\ q}; C\}\!\!\}_M &= \mathsf{p\ start\ q};\ \mathsf{p\ start\ pq}^0;\ \mathsf{q\ start\ qp}^0;\\
&\quad\ \mathsf{p : q <\!\!-\!\!> pq}^0;\ \mathsf{q : p <\!\!-\!\!> qp}^0;\\
&\quad\ \{\!\!\{C\}\!\!\}_{M[(\mathsf{p},\mathsf{q})\mapsto 0,(\mathsf{q},\mathsf{p})\mapsto 0]}\\
\{\!\!\{\mathsf{p.q \mathbin{-\!\!>} r}; C\}\!\!\}_M &= \mathsf{p\ start\ qr}^0;\ \mathsf{p.qr}^0 \mathbin{-\!\!>} \mathsf{pq}^M;\ \mathsf{p.qr}^0 \mathbin{-\!\!>} \mathsf{pr}^M;\\
&\quad\ \mathsf{p\ start\ pq}^{M+};\ \mathsf{p : pq}^M <\!\!-\!\!> \mathsf{pq}^{M+};\\
&\quad\ \mathsf{p\ start\ pr}^{M+};\ \mathsf{p : pr}^M <\!\!-\!\!> \mathsf{pr}^{M+};\\
&\quad\ \mathsf{pq}^M\mathsf{.q} \mathbin{-\!\!>} \mathsf{pq}^{M+};\ \mathsf{pr}^M\mathsf{.r} \mathbin{-\!\!>} \mathsf{pr}^{M+};\\
&\quad\ \mathsf{pq}^M\mathsf{.pq}^{M+} \mathbin{-\!\!>} \mathsf{q};\ \mathsf{pq}^M\mathsf{.qr}^0 \mathbin{-\!\!>} \mathsf{q};\\
&\quad\ \mathsf{pr}^M\mathsf{.pr}^{M+} \mathbin{-\!\!>} \mathsf{r};\ \mathsf{pr}^M\mathsf{.qr}^0 \mathbin{-\!\!>} \mathsf{r};\\
&\quad\ \{\!\!\{C\}\!\!\}_{M[(\mathsf{p},\mathsf{q})\mapsto M(\mathsf{p},\mathsf{q})+1,(\mathsf{p},\mathsf{r})\mapsto M(\mathsf{p},\mathsf{r})+1,(\mathsf{q},\mathsf{r})\mapsto 0]}
\end{aligned}$$

In $\{\!\!\{\mathsf{p\ start\ q}; C\}\!\!\}_M$, we simply create the asynchronous communication channels between p and q – the only step where these process will need to synchronize – and extend M in the continuation. The encoding of $\mathsf{p.q \mathbin{-\!\!>} r}$ is better understood by reading it as a composition: first, p creates the new asynchronous communication channel from q to r, then uses its own channels to send this name to these processes. Note that the auxiliary channels do not communicate, so this encoding will introduce asymmetries in the graph of communications.

Theorems 5.4 and 5.5 still hold for this extended encoding.

Projections. Finally, we extend this encoding to the whole language of DCC – arrow (4) in Figure 5.1 – by adding the clause

$$\begin{aligned}
\{\!\!\{\mathsf{p} \mathbin{-\!\!>} \mathsf{q}[l]; C\}\!\!\}_M &= \mathsf{p} \mathbin{-\!\!>} \mathsf{pq}^M[l];\ \mathsf{p\ start\ pq}^{M+};\\
&\quad\ \mathsf{p : pq}^M <\!\!-\!\!> \mathsf{pq}^{M+};\ \mathsf{pq}^M\mathsf{.q} \mathbin{-\!\!>} \mathsf{pq}^{M+};\\
&\quad\ \mathsf{pq}^M\mathsf{.pq}^{M+} \mathbin{-\!\!>} \mathsf{q};\ \mathsf{pq}^M \mathbin{-\!\!>} \mathsf{q}[l];\\
&\quad\ \{\!\!\{C\}\!\!\}_{M[(\mathsf{p},\mathsf{q})\mapsto M(\mathsf{p},\mathsf{q})+1]}
\end{aligned}$$

to the definition of $\{\!\!\{C\}\!\!\}_M$. Restricting this encoding to the language of CC yields arrow (3) in Figure 5.1.

We finish this section with a brief informal note on projectability. As we discussed in § 5.2, a formal presentation of projection for DMC and

DCC is beyond the scope of this paper. However, we point out that our encoding for asynchronous communications preserves projectability, i.e., if C is projectable, then so is $\{\!\!\{C\}\!\!\}$.

5.5 Related Work

To the best of our knowledge, this is the first work presenting an interpretation of asynchronous communications in choreographies based solely on the expressive power of primitives for the creation of processes and their connections, via name mobility.

Our work recalls the development of the asynchronous π-calculus [13] (Aπ for short, using the terminology from [26]). Aπ has a synchronous semantics, in the sense that two processes can communicate when they are both ready to, respectively, perform compatible input and output actions. However, an output action can have no sequential continuation, but can instead only be composed in parallel with other behaviour. Thus, the interpretation of communications in Aπ is asynchronous, since outputs can be seen as messages in transit over a network. The synchronisation between (the process holding) a message in transit and the intended receiver models then the extraction of the message from the medium by the receiver. Differently from our work, Aπ is obtained from the standard π-calculus by *restricting* the syntax of processes such that all communications necessarily conform to this asynchronous interpretation. It is then shown that Aπ is expressive enough to encode the synchronous communications from standard π-calculus, by using acknowledgement messages. DMC and DCC exhibit the dual behaviour: communications are naturally synchronous, but we can always encode them to be asynchronous by passing them through intermediary processes.

Other studies have investigated asynchronous communications in choreographies. The distinctive feature of our work is that it does not rely on any ad-hoc syntax or semantics for capturing asynchrony. In [15], choreographies are used as types for communication protocols and are related to asynchronous communications by encoding choreographies in types for terms in a variant of the π-calculus. However, asynchrony can only be observed in the semantics of processes, not at the level of choreographies, and the syntax of processes is equipped with ad-hoc runtime terms[3] that represent messages in transit. The first work defining an asynchronous semantics for choreographies is [5], by defining an

[3]Runtime terms are assumed never to be used by the programmer, but only produced as the result of execution.

ad-hoc rule in the semantics of choreographies that allows nested (not appearing at the top level) communications to be executed if, among other conditions: the sender is the same as the one in the communication at the top level of the choreography, the receiver is not involved in the nested communication. This technique has been later adopted also in [23] – for defining the composition of asynchronous choreographies with legacy process code – and in [16] (the journal version of [15]) – to formulate a semantics for communication protocols represented as choreographies. In [12], choreographies (not processes, for example as in [15]) are equipped with runtime terms to represent messages in transit.

Process spawning and name mobility are the key additions to DCC and DMC, from CC and MC, that yield the expressive power to represent asynchronous communications. Process spawning in choreographies has been studied also in the works [4, 5, 23], but in a different form where processes have to synchronise over a shared channel to proceed. Name mobility in choreographies was introduced in [5], but for channel rather than process names. Our process spawning and name mobility primitives are simplifications of those presented in [7], which makes all results from that work applicable to DCC (and thus DMC).

5.6 Conclusions

Choreographies are widely used in the context of concurrent and distributed software architectures, in order to specify precisely how the different components of a system should interact [2, 27]. Previous formalisations of asynchronous communications in choreographies exchange expressivity for simplicity, yielding ad-hoc models with unclear connections. In this work, we showed that a choreography calculus with process spawning and process name mobility can capture asynchronous communications. Therefore, all such calculi with similar primitives have the same power. Our development is conservative wrt previous work, allowing us to import existing techniques developed for previous calculi. For example, the techniques shown in [7, 8] could be reapplied to DCC to synthesise deadlock-free process implementations. Here, we showed how to import the result of selection elimination from [8]. In conclusion, we now have a setting where we can reason about asynchronous communications in choreographies by considering a simple synchronous semantics, just like it can be done in the seminal model of π-calculus for mobile processes.

This work extends results previously published in [9].

Acknowledgements

Montesi was supported by CRC (Choreographies for Reliable and efficient Communication software), grant no. DFF–4005-00304 from the Danish Council for Independent Research.

Bibliography

[1] S. Basu and T. Bultan. Choreography conformance via synchronizability. In *WWW*, pages 795–804, 2011.

[2] Business Process Model and Notation. `http://www.omg.org/spec/BPMN/2.0/`.

[3] N. Busi, R. Gorrieri, C. Guidi, R. Lucchi, and G. Zavattaro. Choreography and orchestration conformance for system design. In *COORDINATION*, volume 4038 of *LNCS*, pages 63–81. Springer, 2006.

[4] M. Carbone, K. Honda, and N. Yoshida. Structured communication-centered programming for web services. *ACM Trans. Program. Lang. Syst.*, 34(2):8, 2012.

[5] M. Carbone and F. Montesi. Deadlock-freedom-by-design: multiparty asynchronous global programming. In R. Giacobazzi and R. Cousot, editors, *POPL*, pages 263–274. ACM, 2013.

[6] Chor. Programming Language. `http://www.chor-lang.org/`.

[7] L. Cruz-Filipe and F. Montesi. A language for the declarative composition of concurrent protocols. *CoRR*, abs/1602.03729, 2016. Submitted for publication.

[8] L. Cruz-Filipe and F. Montesi. A core model for choreographic programming. In O. Kouchnarenko and R. Khosravi, editors, *FACS*, volume 10231 of *LNCS*, pages 17–35. Springer, 2017.

[9] L. Cruz-Filipe and F. Montesi. Encoding asynchrony in choreographies. In *SAC 2017*, pages 1175–1177, 2017.

[10] M. Dalla Preda, M. Gabbrielli, S. Giallorenzo, I. Lanese, and J. Mauro. Dynamic choreographies - safe runtime updates of distributed applications. In T. Holvoet and M. Viroli, editors, *COORDINATION*, volume 9037 of *LNCS*, pages 67–82. Springer, 2015.

[11] M. Dalla Preda, S. Giallorenzo, I. Lanese, J. Mauro, and M. Gabbrielli. AIOCJ: A choreographic framework for safe adaptive distributed applications. In B. Combemale, D. Pearce, O. Barais, and J. Vinju, editors, *SLE*, volume 8706 of *LNCS*, pages 161–170. Springer, 2014.

[12] P.-M. Deniélou and N. Yoshida. Multiparty compatibility in communicating automata: Characterisation and synthesis of global session types. In F. Fomin, R. Freivalds, M. Kwiatkowska, and D. Peleg, editors, *ICALP (II)*, volume 7966 of *LNCS*, pages 174–186. Springer, 2013.

[13] K. Honda and M. Tokoro. An object calculus for asynchronous communication. In *ECOOP*, volume 512 of *LNCS*, pages 133–147, Berlin, Heidelberg, New York, Tokyo, 1991. Springer.

[14] K. Honda, V. Vasconcelos, and M. Kubo. Language primitives and type disciplines for structured communication-based programming. In C. Hankin, editor, *ESOP*, volume 1381 of *LNCS*, pages 122–138. Springer, 1998.

[15] K. Honda, N. Yoshida, and M. Carbone. Multiparty asynchronous session types. In G. C. Necula and P. Wadler, editors, *POPL*, pages 273–284. ACM, 2008.

[16] K. Honda, N. Yoshida, and M. Carbone. Multiparty asynchronous session types. *J. ACM*, 63(1):9, 2016.

[17] I. Lanese, C. Guidi, F. Montesi, and G. Zavattaro. Bridging the gap between interaction- and process-oriented choreographies. In A. Cerone and S. Gruner, editors, *SEFM*, pages 323–332. IEEE, 2008.

[18] T. Leesatapornwongsa, J. Lukman, S. Lu, and H. Gunawi. TaxDC: A taxonomy of non-deterministic concurrency bugs in datacenter distributed systems. In *ASPLOS*, pages 517–530. ACM, 2016.

[19] S. Lu, S. Park, E. Seo, and Y. Zhou. Learning from mistakes: a comprehensive study on real world concurrency bug characteristics. *ACM SIGARCH Computer Architecture News*, 36(1):329–339, 2008.

[20] R. Milner, J. Parrow, and D. Walker. A calculus of mobile processes, I and II. *Information and Computation*, 100(1):1–40,41–77, Sept. 1992.

[21] F. Montesi. *Choreographic Programming.* Ph.D. thesis, IT University of Copenhagen, 2013. `http://fabriziomontesi.com/files/choreographic_programming.pdf`.

[22] F. Montesi. Kickstarting choreographic programming. In T. Hildebrandt, A. Ravara, J. van der Werf, and M. Weidlich, editors, *WS-FM*, volume 9421 of *LNCS*, pages 3–10. Springer, 2016.

[23] F. Montesi and N. Yoshida. Compositional choreographies. In P. D'Argenio and H. Melgratti, editors, *CONCUR*, volume 8052 of *LNCS*, pages 425–439. Springer, 2013.

[24] D. Mostrous, N. Yoshida, and K. Honda. Global principal typing in partially commutative asynchronous sessions. In *ESOP*, pages 316–332, 2009.

[25] Z. Qiu, X. Zhao, C. Cai, and H. Yang. Towards the theoretical foundation of choreography. In C. Williamson, M. Zurko, P. Patel-Schneider, and P. Shenoy, editors, *WWW*, pages 973–982. ACM, 2007.

[26] D. Sangiorgi and D. Walker. *The π-calculus: a Theory of Mobile Processes.* Cambridge University Press, 2001.

[27] W3C WS-CDL Working Group. Web services choreography description language version 1.0. http://www.w3.org/TR/2004/WD-ws-cdl-10-20040427/, 2004.

[28] N. Yoshida, R. Hu, R. Neykova, and N. Ng. The Scribble protocol language. In *TGC*, pages 22–41, 2013.

6

MONOSEQUENT PROOF SYSTEMS

Jean-Yves Béziau
Federal University of Rio de Janeiro - UFRJ
Brazilian Research Council - CNPq
jyb@ufrj.br

Abstract

In this paper we present a sequent sequent with monosequents (sequents with only one formula on each side) for the logic of conjunction and disjunction. We first start by a general discussion about sequent systems. We then introduce the system $\mathbf{S11}_{\wedge,\vee}$ and prove some basic results, in particular the fact that distributivity does not hold. In a following part we show that the logic generated by this system is indeed the fibring of conjunction with disjunction. We furthermore discuss the relation between this system and lattice theory. We end up with the story of the development of this work and some personal recollections.

Dedicated to Amílcar Sernadas for his x-th birthday ($0 \leq x \leq \omega$)

Navegar é preciso, viver não é preciso.

Fernando Pessoa

6.1 Historico-philosophical remarks on systems of sequents

Systems of sequents are proof systems due to Gerhard Gentzen (1935). The expression "sequent calculus" (plural "sequent calculi") is nowadays often used, but was originally not used by Gentzen himself. In his seminal work *Investigation into logical deduction*, published in two parts ([28] and [29] (translated in French in 1955 with useful comments [30] and in 1969 in English [31]), Gentzen introduced a particular use of the word "sequent" (in German: *Sequenz*), but he used neither the expression "systems of sequents", nor "sequent calculus".

In this 1935 work, where sequent systems are introduced, he presents two kinds of systems both with a classical version and an intuitionistic version. He uses the expression "Calculus of natural deduction" (*Kalkül des natürlichen Schließens*) for the first kind of systems, hence the acronyms NJ and NK (J stands for *intuitionistischer* and K stands for *klassischer*, the transformation of "j" into "i" has been attributed to a misreading of the typographer); and he uses the expression "Logistic calculus" (*logistischer Kalkül*) for the second type, hence the acronyms LJ and LK.

The expression "systems of propositions" (*Satzsysteme*) was used by Paul Hertz for systems from which Gentzen's sequent systems are directly inspired. These systems were presented in a series of papers all written in German: [34], [35], [36], [37]. The first one has been written September 15, 1921 and published in *Mathematische Annalen* in September 1922, Volume 87, Issue 3, pp.246–269. It is entitled *Über Axiomensysteme für beliebige Satzsysteme. I. Teil.* It has been translated for the first time in English in the book *Universal Logic: an Anthology*, published in 2012 with the title *Axiomatic Systems for Arbitrary Systems of Sentences. Part I* with an introductory paper by Javier Legris entitled "Paul Hertz and the origins of structural reasoning" (Legris wrote several other papers about Hertz, see references in the bibliography of this latter paper, [45]).

The subtitle of this universal logic anthology is *From Paul Hertz to Dov Gabbay.* The reason to choose Hertz as the starting point of this anthology is the revolutionary position he undertook having a prophetic vision of logic where objects of reasoning are not restricted to sentences or propositions. Hertz uses the word "Satz" at the meta-level [1].

Hertz's position is well summarized in Gentzen's first paper entitled

[1]Note that the word "Satz" in German is neutral relatively to the divide "sentence" / "proposition", it can also mean *principle*, as in the expression *Der Satz vom zureichenden Grund* which is equivalent to *The principle of sufficient reason.*

Über die Existenz unabhängiger Axiomensysteme zu unendlichen Satzsystemen [27] ("On the existence of independent axiom systems for infinite sentence systems"):

> A *proposition* has the form
>
> $$u_1, ..., u_n \to v.$$
>
> The u's and v's are called *elements.* We might think of them as events and the 'proposition' then reads: The happening of the events $u_1, ..., u_n$ causes the happening of v.
>
> The 'proposition' may also be understood this: A domain of elements containing the elements $u_1, ..., u_n$ also contains the element v.
>
> The elements may furthermore be thought of as properties and then 'proposition' can then be interpreted thus: An object with the properties $u_1, ..., u_n$ also has the property v.
>
> Or we imagine the elements to stand for 'propositions', in the sense of the propositional calculus, and the 'proposition' then reads: If the propositions $u_1, ..., u_n$ are true, then the proposition v is also true..
>
> Our considerations do not depend on any particular kind of informal interpretations of the 'propositions', since we are concerned only with their formal structure.

This paper is about Hertz's systems, Gentzen does not introduce yet his own systems here but already coins a proper terminology, in particular the word "cut" that has become so famous.

Paul Hertz (1881-1940) Gerhard Gentzen (1909-1945)

There are two reasons to qualify Hertz's work as *structural.* The rules introduced by Hertz are not about logical operators such as connectives or quantifiers. They have been called by Gentzen in 1935 "structural rules" (*Struktur-schlußfiguren*). The second reason, connected with the first, is that Hertz anticipated the idea of a logic as a mathematical structure of the type $\mathcal{L} = \langle \mathbb{F}; \vdash \rangle$, where:

• $\mathbb{F}$ is a set of objects, called *formulas* (sets of formulas are called *theories*);

• $\vdash$ is a binary relation between theories and formulas, i.e. $\vdash \subseteq \mathcal{P}(\mathbb{F}) \mathcal{X} \mathbb{F}$, called *consequence relation.* Let us call such kind of structure, an *abstract logic.* It is a structure of the same kind (but of different type) as an abstract algebra $\mathcal{A} = \langle \mathbb{A}; f_{(i \in I)} \rangle$ where:

• $\mathbb{A}$ is a set of objects;

• $f_{(i \in I)}$ is a family of functions.

Garrett Birkhoff promoted *Universal Algebra* as the study of the class of all abstract algebras ([16] for more details and bibliographical references). By analogy we have promoted *Universal Logic* as the study of abstract structures of the type $\mathcal{L} = \langle \mathbb{F}; \vdash \rangle$ (see [4] and [5]).[2]

According to this approach logical operators have to be understood from the point of view of a more fundamental notion (cf. [43] and [44]). Whatever name is given to her, what is important is the distinction between two strata. In Poland, Alfred Tarski (see [55] and [58]), not so much later, independently promoted a similar distinction, introducing the notion of consequence operator. What is similar in the work of Hertz and Tarski is that they have both developed a work at this pure level of abstraction. Tarski (jointly with Łukasiewicz, [46]) went on applying this abstract approach to logical operators, Hertz didn't, but Gentzen did it for him.

The notion of deduction (or inference) appears in a sequent system at three levels:

(1) as the connective "$\supset$" of implication,

(2) as the symbol "$\rightarrow$" at the middle of the sequent linking the antecedent formulas to the consequent formulas,

(3) as a horizontal line "————————" linking the premisses to the conclusion.

In a Hertz's system, in a Hilbert's system and in a Gentzen's natural deduction system it appears only at two levels. (1) does not appear in a Hertz's system. In the two other cases (1) generally appears but not (2).

---

[2] The reason to use "$\mathbb{F}$" and not "$\mathbb{L}$" as the name of the domain is to avoid to make a connection with *language*. The elements of the domain of a logic are objects that can be events, etc. (in the spirit of Hertz's approach). Formulas are at best names for these objects.

Tarski's original theory of consequence operator has only one level (which corresponds to none of the three levels above). When applied to specific logical operators it has one more level, the level (1). It is important to emphasize that Tarski's theory is not a deductive system. Transitivity of the consequence relation/operator in Tarski's theory should not be confused with the cut rule of Gentzen. Such a confusion prevents to understand the cut-elimination theorem (for more on this topic see [7]).

6.2 The architecture of sequent systems

Let us first explain here why we prefer to use the expression "sequent system" than "sequent calculus". There are three reasons. Firstly, as rightly pointed out by Paul Halmos ([33]), the word "calculus" is highly ambiguous, having different heterogeneous meanings. The semantical field of "system" is wide but less heterogeneous. Secondly, one of the dominating meanings of "calculus" is connected with "computing". The relation between proof theory and computability makes sense, but considering that mathematical proofs are totally reducible to some computable processes is highly controversial. The work of Gentzen has certainly contributed to show than in many ways mathematical proofs are computable, in particular one of the applications of the cut-elimination theorem is, in some cases, decidability. But on the one hand sequent systems are not necessarily leading to a reduction to computability and on the other hand they have virtues not limited to computability. The third reason, related to the second one, is to keep a connection with the perspective of Paul Hertz, who used the word "System".

For Hertz, a *Satz* is of the form:

$$a_1, ..., a_n \to b$$

Originally he was using a comma, which later on was withdrawn. Hertz is considering in fact that the antecedent is a set of objects, that he calls "complex" (*Komplex*) and that he represents by a upper case letter, e.g. 'K', the order having no importance despite the comma and natural numbers as subscripts. A rule for Hertz is therefore something of the form:

$$\frac{K \to a}{L \to b}$$

Gentzen transformed Hertz's *Satz* in somtehing he called a "sequent" (*Sequenz*). He kept the arrow at the middle, writing a sequent as follows ([28], par 1, section 2.3):

$$\mathfrak{A}_1, ..., \mathfrak{A}_u \rightarrow \mathfrak{B}_1, ..., \mathfrak{B}_v$$

Gentzen was using Fraktur Gothic letters instead of lower case letters of the Latin alphabet. Fraktur Gothic letters are used to denote what he was calling "formulas", which are not anymore as for Hertz any kind of objects, but formulas of what is nowadays called "first-order logic". Lower case Latin letters were used, as it is nowadays used, as names for constants and variables of objects (by contrast to predicates / relations).

Something like $\mathfrak{A}_1, ..., \mathfrak{A}_u$ is a *sequence* of formulas in the usual sense of the word "sequence" in mathematics (there is an informal notion of sequence and a formal one developed in set theory, but the writing is the same). It means that the order is important and that $\mathfrak{A}, \mathfrak{A}$ is a sequence different from $\mathfrak{A}, \mathfrak{A}, \mathfrak{A}$. However Gentzen uses the word "sequence" (*Sequenz*) specifically for two sequences of formulas with at the middle Hertz's arrow of which he says: "the $\rightarrow$, like commas is an auxiliary symbol and not a logical symbol" ([28], part 1, section 2.3). In English this idiosyncratic use of *Sequenz* has been translated by the word "sequent", a good choice to avoid ambiguity. Nevertheless it is good to remember that for Gentzen a sequent is a pair of sequences of formulas, that we could write:

$$< \Sigma_1; \Sigma_2 >$$

The semicolon ";" is a symbol which is used in mathematics for representing pairs. Dummett in his textbook on intuitionistic logic [24] chose the nearby symbol ":" and does not used "<" and ">". This is not so nice but avoid the confusion with implication which is nowadays generally written using "$\rightarrow$" rather than "$\supset$" as it was done by Whitehead-Russell and Hilbert. It is anyway a better option than to use "$\vdash$" as it is frequently done generating a confusion between a sequent system and the logic it generates. Maybe it would be good to introduce an additional symbol for sequent like "$\succ$" (this what we have done in [7], see also [39], in particular section 1.21). But here we will use the original Hertz/Gentzen's writing, emphasizing the relation with Hertz's original work and also because we are using Sam Buss's LaTeX style file where this option has been chosen.

Gentzen considered the case of an empty sequence and he also famously considered a sequent system where the length of the right sequence is at most 1: the sytsem *LJ* for intuitionistic logic. And he has this surprising result according to which the difference between sequent systems for intuitionistic logic (LJ) and classical logic (LK) is structural in the sense that the rules for the logical operators are exactly the same, the difference being in the structure of the sequents. However in *LJ* the structural rules are also *mutatis mudandis* the same as in *LK*. At this point it is important to make a distinction between internal and structural determinations, structural principles being either on the internal side (structural rules) or on the external side (configuration of the sequents). This is described in the following table (from [8]).

	INTERNAL DETERMINATIONS	
	STRUCTURAL RULES	LOGICAL RULES
EXTERNAL DETERMINATIONS		
	e.g.	
SEQUENT	Identity	(Dealing
-monosequent/multisequent		with the
-finite sets	Cut	morphology
-sequences		of the logic)
-structures, e.g.		
idempotent abelian semigroup		
	Weakening	
		e.g.
		Connectives
RULE		
-order on premises		Quantifiers
-cardinality of premises	Permutation	
		Modalities
	Contraction	
PROOF		
-sequence/tree		
-length/depth	Associativity	
(finite or not)		
-order type		
INDUCED LOGIC		
e.g.		
$T \vdash a \Leftrightarrow \exists\ To$ finite, subset of T $To \rightarrow a$ is derivable		
$T \vdash a \Leftrightarrow T \rightarrow a$ is derivable		
STRUCTURAL PRINCIPLES		

Table 1: THE ARCHITECTURE OF SEQUENT SYSTEMS

It is also possible to reduce the size of the sequence to at most 1 on the left in a sequent, leaving the right sequence having a variable length. This has been proposed by Igor Urbas [57] in a system he called LDJ, the "D" standing for "Dual". The logic generated by LDJ is a paraconsistent logic dual of intuitionistic logic (Note that not every paraconsistent logic is a dual of intuitionistic logic and that there are different dualizations of intuitionistic logic).[3]

[3]An interesting variation of sequents to deal with other non-classical logics are *hypersquents* introduced by G.Pottinger [51] and A.Avron[2].

What we are proposing in the present paper is to consider sequents where we have one and only one formula both on the left and on the right. Hertz already considered such kind of things, he called them "first-degree propositions". Gentzen renamed them *linear* propositions [27], but when introducing sequent systems he did not consider this configuration. We will call them here *monosequents*. The expression *first-degree propositions* would be ambiguous in particular because we are not calling here sequents propositions and the expression *linear sequent* would be ambiguous due to the use of the expression *linear logic* promoted by Jean-Yves Girard [32]. A *monosequent system* is a system of sequents where all sequents are monosequents.

The consequence relation of an abstract logic $\mathcal{L} = \langle \mathbb{F}; \vdash \rangle$ generated by a monosequent systems $S11$ is defined as follows:[4]

$$T \vdash v \text{ iff there is a formula } u \text{ of } T \text{ such that } u \to v \text{ is derivable in } S11$$

Let us consider the following sequent rule:

$$\frac{\mathfrak{A}, \Gamma \to \Theta \quad \mathfrak{B}, \Gamma \to \Theta}{\mathfrak{A} \vee \mathfrak{B}, \Gamma \to \Theta}$$

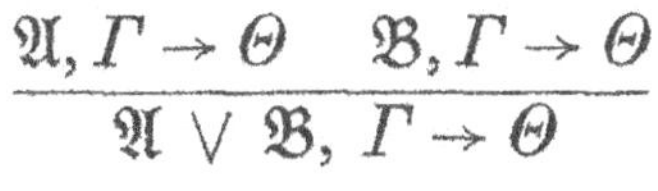

This is a photograph of the original left rule for disjunction presented by Gentzen ([28] p.192). With the modern technology of LaTeX it is possible to produce something fairly similar:

$$\frac{\mathfrak{A}, \Gamma \to \Theta \quad \mathfrak{B}, \Gamma \to \Theta}{\mathfrak{A} \vee \mathfrak{B}, \Gamma \to \Theta} \vee \text{ left}$$

In the present paper we will however replace the Fraktur Gothic letters by lower case Latin alphabet letters. We are therefore coming back to Hertz's original writing. This is not a problem since we are dealing only with propositional logic. Using this writing, the above rule is written as follows:

$$\frac{a, \Gamma \to \Theta \quad b, \Gamma \to \Theta}{a \vee b, \Gamma \to \Theta} \vee \text{ left}$$

And considering only monosequents we have:

$$\frac{a \to c \quad b \to c}{a \vee b \to c} \vee \text{ left}$$

---

[4]Compare monosequents and this definition with what is called FMLA-FMLA framework in [39].

6.3 The monosequent proof systems $\mathbf{S11}_{\wedge,\vee}$ for the logic of conjunction and disjunction

6.3.1 Definition of $\mathbf{S11}_{\wedge,\vee}$

We consider in this paper only the monosequent proof system $\mathbf{S11}_{\wedge,\vee}$ for the logic of conjunction and disjunction.[5] We have on the one hand the axiom of identity and the cut rule, on the other hand logical rules for conjunction and disjunction. Both the axiom and the rules are schemes. a, b, c are any formulas.

AXIOM:

$$a \longrightarrow a$$

CUT RULE:

$$\dfrac{a \longrightarrow c \qquad c \longrightarrow b}{a \longrightarrow b}\ \text{cut}$$

LOGICAL RULES:

$$\dfrac{a \longrightarrow c}{a \wedge b \longrightarrow c}\ \wedge_{l1} \qquad \dfrac{b \longrightarrow c}{a \wedge b \longrightarrow c}\ \wedge_{l2} \qquad \dfrac{c \longrightarrow a \qquad c \longrightarrow b}{c \longrightarrow a \wedge b}\ \wedge_{r}$$

$$\dfrac{c \longrightarrow b}{c \longrightarrow a \vee b}\ \vee_{r1} \qquad \dfrac{c \longrightarrow a}{c \longrightarrow a \vee b}\ \vee_{r2} \qquad \dfrac{a \longrightarrow c \qquad b \longrightarrow c}{a \vee b \longrightarrow c}\ \vee_{l}$$

6.3.2 Proofs in $\mathbf{S11}_{\wedge,\vee}$

Let us see some examples of proofs in $\mathbf{S11}_{\wedge,\vee}$:

$$\dfrac{\dfrac{b \longrightarrow b}{a \wedge b \longrightarrow b}\ \wedge_{l2} \qquad \dfrac{a \longrightarrow a}{a \wedge b \longrightarrow a}\ \wedge_{l1}}{a \wedge b \longrightarrow b \wedge a}\ \wedge_{r}$$

$$\dfrac{\dfrac{\dfrac{a \longrightarrow a}{a \wedge (b \wedge c) \longrightarrow a}\ \wedge_{l1} \qquad \dfrac{\dfrac{b \longrightarrow b}{b \wedge c \longrightarrow b}\ \wedge_{l1}}{a \wedge (b \wedge c) \longrightarrow b}\ \wedge_{l2}}{a \wedge (b \wedge c) \longrightarrow a \wedge b}\ \wedge_{r} \qquad \dfrac{\dfrac{c \longrightarrow c}{b \wedge c \longrightarrow c}\ \wedge_{l2}}{a \wedge (b \wedge c) \longrightarrow c}\ \wedge_{l2}}{a \wedge (b \wedge c) \longrightarrow (a \wedge b) \wedge c}\ \wedge_{r}$$

[5] **S11** has to be pronounced as "S-one-one" not as "S-eleven".

$$\dfrac{a\longrightarrow a \qquad \dfrac{a\longrightarrow a}{a\wedge b\longrightarrow a}\,\wedge_{l1}}{a\vee(a\wedge b)\longrightarrow a}\,\vee_{l}$$

$$\dfrac{\dfrac{\dfrac{a\longrightarrow a}{a\wedge b\longrightarrow a}\,\wedge_{l1} \qquad \dfrac{\dfrac{b\longrightarrow b}{a\wedge b\longrightarrow b}\,\wedge_{l2}}{a\wedge b\longrightarrow b\vee c}\,\vee_{r1}}{a\wedge b\longrightarrow a\wedge(b\vee c)}\,\wedge_{r} \qquad \dfrac{\dfrac{a\longrightarrow a}{a\wedge c\longrightarrow a}\,\wedge_{l1} \qquad \dfrac{\dfrac{c\longrightarrow c}{c\longrightarrow b\vee c}\,\vee_{r2}}{a\wedge c\longrightarrow b\vee c}\,\wedge_{l2}}{a\wedge c\longrightarrow a\wedge(b\vee c)}\,\wedge_{r}}{(a\wedge b)\vee(a\wedge c)\longrightarrow a\wedge(b\vee c)}\,\vee_{l}$$

As an exercise the reader can perform proofs in $\mathbf{S11}_{\wedge,\vee}$ showing that:

$a\wedge a \dashv\vdash a \qquad a\vee a \dashv\vdash a$

$a\wedge b \dashv\vdash b\wedge a \qquad a\vee b \dashv\vdash b\vee a$

$a\wedge(b\wedge c) \dashv\vdash (a\wedge b)\wedge c \qquad a\vee(b\vee c) \dashv\vdash (a\vee b)\vee c$

$a\wedge(a\vee b) \dashv\vdash a \qquad a\vee(a\wedge b) \dashv\vdash a$

6.3.3 Metaproofs about $\mathbf{S11}_{\wedge,\vee}$

ATOMIZATION OF IDENTITY The axiom of identity can be atomized, i.e. it is possible to replace it by the following one where p is an any atomic formula:

$$p\longrightarrow p$$

This is proved by recurrence, showing that it is possible to decrease the degree of the formula subject to identity. There are two cases: conjunction and disjunction.

$$\dfrac{\dfrac{a\longrightarrow a}{a\wedge b\longrightarrow a}\,\wedge_{l1} \qquad \dfrac{b\longrightarrow b}{a\wedge b\longrightarrow b}\,\wedge_{l2}}{a\wedge b\longrightarrow a\wedge b}\,\wedge_{r}$$

$$\dfrac{\dfrac{a\longrightarrow a}{a\longrightarrow a\vee b}\,\vee_{r1} \qquad \dfrac{b\longrightarrow b}{b\longrightarrow a\vee b}\,\vee_{r2}}{a\vee b\longrightarrow a\vee b}\,\vee_{l}$$

CUT-ELIMINATION Cut-elimination has been proved by Gentzen by double recurrence (maybe the first use of it): on the complexity of the formula and the rank of the cut. Let us see how we can in $\mathbf{S11}_{\wedge,\vee}$ lower the complexity of the cut formula and the rank of the cut.

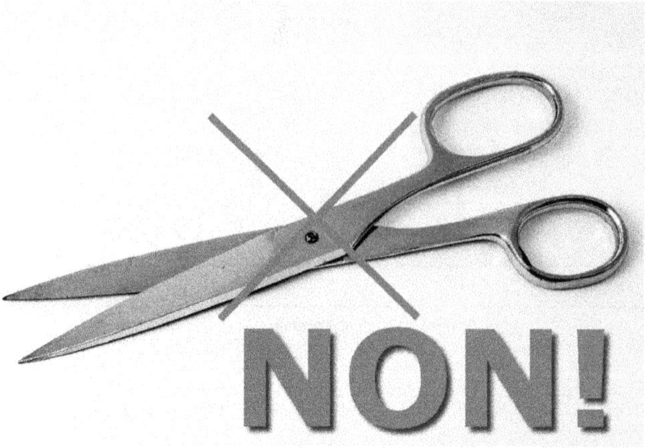

Let us first examine the question of the *complexity* of the cut formula. We have only two cases. The cut formula is of the form $a \wedge b$ or of the form $a \vee b$. We examine the first case, and leave the other one, which is dual, for the reader.
We have the following derivation:

$$\dfrac{\dfrac{u \rightarrow a \qquad u \rightarrow b}{u \rightarrow a \wedge b}\,\wedge_r \qquad \dfrac{a \rightarrow v}{a \wedge b \rightarrow v}\,\wedge_{l1}}{u \rightarrow v}\,\text{cut}$$

that we transform in the following derivation:

$$\dfrac{u \rightarrow a \qquad a \rightarrow v}{u \rightarrow v}\,\text{cut}$$

The situation where the rule $\wedge_{l2}$ is used instead of $\wedge_{l1}$ is similar.

Now let us examine the question of the *rank* of the cut. We have six cases corresponding to the six logical rules we have. We examine only the case when the last rule used before the cut is $\vee_l$ and leave the other cases for the reader.

We have the following derivation:

$$\dfrac{\dfrac{a \rightarrow c \qquad b \rightarrow c}{a \vee b \rightarrow c}\,\vee_l \qquad c \rightarrow u}{a \vee b \rightarrow u}\,\text{cut}$$

that we transform in following derivation:

$$\dfrac{\dfrac{a \rightarrow c \qquad c \rightarrow u}{a \rightarrow u}\,\text{cut} \qquad \dfrac{b \rightarrow c \qquad c \rightarrow u}{b \rightarrow u}\,\text{cut}}{a \vee b \rightarrow u}\,\vee_l$$

DECIDABILITY As a corollary of cut-elimination we have the decidability of the logic generated by $\mathbf{S11}_{\wedge,\vee}$.

NON-DISTRIBUTIVITY
Using cut-elimination we can also perform a negative meta-proof, showing that there is no proof in $\mathbf{S11}_{\wedge,\vee}$ of the sequent (where a, b, c are atomic formulas):

$$a \wedge (b \vee c) \rightarrow (a \wedge b) \vee (a \wedge c)$$

The last used rule is either a $\wedge$ left or a $\vee$ right. Suppose that it is a $\wedge$ left left (i.e. $\wedge_{l1}$). Then we have the two following routes, which both lead nowhere:

$$\dfrac{\dfrac{a \rightarrow a \wedge b}{a \rightarrow (a \wedge b) \vee (a \wedge c)}\ \vee \text{ right}}{a \wedge (b \vee c) \rightarrow (a \wedge b) \vee (a \wedge c)}\ \wedge \text{ left}$$

$$\dfrac{\dfrac{a \rightarrow a \wedge c}{a \rightarrow (a \wedge b) \vee (a \wedge c)}\ \vee \text{ right}}{a \wedge (b \vee c) \rightarrow (a \wedge b) \vee (a \wedge c)}\ \wedge \text{ left}$$

Suppose now that it is a $\wedge$ right left (i.e. $\wedge_{l2}$). Then the next application of rule is either the $\vee$ left or a $\vee$ right. Consider that it is the the $\vee$ left. Then we have:

$$\dfrac{\dfrac{b \rightarrow (a \wedge b) \vee (a \wedge c) \qquad c \rightarrow (a \wedge b) \vee (a \wedge c)}{b \vee c \rightarrow (a \wedge b) \vee (a \wedge c)}\ \vee \text{ left}}{a \wedge (b \vee c) \rightarrow (a \wedge b) \vee (a \wedge c)}\ \wedge \text{ left}$$

It easy to meta-prove that $b \rightarrow (a \wedge b) \vee (a \wedge c)$ cannot be derived in $\mathbf{S11}_{\wedge,\vee}$. This is indeed also the case of $c \rightarrow (a \wedge b) \vee (a \wedge c)$.

Consider now that the last application is a $\vee$ right, then we have the two following routes which both lead nowhere:

$$\dfrac{\dfrac{b \vee c \rightarrow a \wedge b}{b \vee c \rightarrow (a \wedge b) \vee (a \wedge c)}\ \vee \text{ right}}{a \wedge (b \vee c) \rightarrow (a \wedge b) \vee (a \wedge c)}\ \wedge \text{ left}$$

$$\dfrac{\dfrac{b \vee c \rightarrow a \wedge c}{b \vee c \rightarrow (a \wedge b) \vee (a \wedge c)}\ \vee \text{ right}}{a \wedge (b \vee c) \rightarrow (a \wedge b) \vee (a \wedge c)}\ \wedge \text{ left}$$

Coming back to the root of our meta-proof, we have now to examine the situation where the last application is a $\vee$ right. The two cases are symmetric. So we just consider the case of the $\vee$ left right. We have four roads leading nowhere:

$$\dfrac{\dfrac{\dfrac{a \rightarrow a}{a \wedge (b \vee c) \rightarrow a}\,\wedge\text{ left} \qquad \dfrac{b \vee c \rightarrow b}{a \wedge (b \vee c) \rightarrow b}\,\wedge\text{ left}}{a \wedge (b \vee c) \rightarrow a \wedge b}\,\wedge\text{ right}}{a \wedge (b \vee c) \rightarrow (a \wedge b) \vee (a \wedge c)}\,\vee\text{ right}$$

$$\dfrac{\dfrac{\dfrac{a \rightarrow a}{a \wedge (b \vee c) \rightarrow a}\,\wedge\text{ left} \qquad \dfrac{a \rightarrow b}{a \wedge (b \vee c) \rightarrow b}\,\wedge\text{ left}}{a \wedge (b \vee c) \rightarrow a \wedge b}\,\wedge\text{ right}}{a \wedge (b \vee c) \rightarrow (a \wedge b) \vee (a \wedge c)}\,\vee\text{ right}$$

$$\dfrac{\dfrac{a \rightarrow a \wedge b}{a \wedge (b \vee c) \rightarrow a \wedge b}\,\wedge\text{ right}}{a \wedge (b \vee c) \rightarrow (a \wedge b) \vee (a \wedge c)}\,\vee\text{ right}$$

$$\dfrac{\dfrac{b \vee c \rightarrow a \wedge b}{a \wedge (b \vee c) \rightarrow a \wedge b}\,\wedge\text{ right}}{a \wedge (b \vee c) \rightarrow (a \wedge b) \vee (a \wedge c)}\,\vee\text{ right}$$

At the end there is no way out, we cannot prove distributivity. It is also possible to show that we cannot prove the dual form of distributivity.

6.4 Fibring of conjunction with disjunction

In previous works we have been discussing combination of logics in particular the combination of the logic of conjunction with the logic of disjunction (see [9], [11], [12], [40]) . The basic idea of combination of logic crystallized as "fibring" by Dov Gabbay [26] is that nothing more should appear when putting two logics together. There must be no new things, product of an interaction. But it is not just a juxtaposition or a superposition, putting the two logics side by side or one on the top of the other. It is an intertwining, a weaving, ..., a mix, but a sterile mix. It is however not a pointless mix. Fibring is useful to construct or deconstruct step by step a logical structure [20].

The title of one of our papers is "A paradox in the combination of logics" [9]. In this note we point out that if we put together the logic of conjunction with the logic of disjunction combining valuations then we don't get fibring because we have distributivity, the same with standard sequent rules. Carlos Caleiro has however proved a general result in his PhD [18] according to which putting Hilbert-style rules together is not productive, i.e. that it leads to fibring. Here are some Hilbert rules for conjunction and disjunction (cf. [49]):

$$\frac{a\wedge b}{a} \qquad \frac{a\wedge b}{b} \qquad \frac{a \quad b}{a\wedge b} \qquad \frac{a}{a\vee b} \qquad \frac{a}{a\vee a} \qquad \frac{a\vee b}{b\vee a} \qquad \frac{a\vee(b\vee c)}{(a\vee b)\vee c}$$

We can use this set of Hilbert rules and Caleiro's general theorem to prove that $\mathbf{S11}_{\wedge,\vee}$ generates the fibring of conjunction and disjunction. It is enough to prove that the above set of Hilbert's rules $\mathbf{H}_{\wedge,\vee}$ is equivalent to $\mathbf{S11}_{\wedge,\vee}$.

To do that we can transform each rule of $\mathbf{H}_{\wedge,\vee}$ into a rule of $\mathbf{S11}_{\wedge,\vee}$. For example the first one is transformed into:

$$\frac{c\rightarrow a\wedge b}{c\rightarrow a}\ \wedge_{e1}$$

It is easy to show using the cut rule that $\wedge_{e1}$ and $\wedge_{l1}$ are equivalent.

$$\dfrac{c\rightarrow a\wedge b \qquad \dfrac{a\rightarrow a}{a\wedge b\rightarrow a}\ \wedge_{l1}}{c\rightarrow a}\ \text{cut}$$

$$\dfrac{\dfrac{a\wedge b\rightarrow a\wedge b}{a\wedge b\rightarrow a}\ \wedge_{e1} \qquad a\rightarrow c}{a\wedge b\rightarrow c}\ \text{cut}$$

The difficulty is how to derive the rule $\vee_l$ of $\mathbf{S11}_{\wedge,\vee}$ in (the sequent transposition of) $\mathbf{H}_{\wedge,\vee}$. A hint to prove this is to use lattice theory.

6.5 $\mathbf{S11}_{\wedge,\vee}$, lattice theory and logical algebra

The logic generated by $\mathbf{S11}_{\wedge,\vee}$ which is, as we have seen in the previous section, the fibring of the logic of conjunction with the logic of disjunction, has a very close connection to lattice theory. In some sense we can say: it is a lattice. And this is true, up to certain point. But approximate truth is easy, if not trivial (everything is true up to certain point ...), precise truth is another kettle of fish. Let's dive into the kettle.

First let us specify what a lattice is. If we want to compare two things, it is good to know what we have on both sides. There are a thousand ways to define a lattice. This variation of definition is a fascinating aspect of mathematics (Marshall Stone was surprised to see/prove that an idempotent ring is the same as a complemented distributive lattice, i.e. two different ways to define a so-called Boolean algebra). Are all these *different* ways the *same*? And are we able to precisely explain what "the same" means? There are different ways to do it, and they are not necessarily the same! For example we can say that two mathematical structures are "equivalent" iff they have a common expansion by definition up to isomorphism. This can be understood informally, in particular through some examples, but if we want to precisely formalize this, it is not so easy. We can do it in first-order model theory, but this theory is not so obvious, it took several decades until reaching a mature stage and it is only one way to do the job which has its own limitations.

Let us consider the following definition of lattice, which is the one given for example by Haskell Curry in his book *Foundations of mathematical logic* [23]:

A lattice is a structure with a partial order $\leq$ and two binary functions $\wedge$ and $\vee$ obeying the following axioms:[6]

$a \wedge b \leq a$ ($\wedge_{l1@}$)
$a \wedge b \leq b$ ($\wedge_{l2@}$)
if $c \leq a$ and $c \leq b$ then $c \leq a \wedge b$ ($\wedge_{r@}$)

$a \leq a \vee b$ ($\vee_{r1@}$)
$b \leq a \vee b$ ($\vee_{r2@}$)
if $a \leq c$ and $b \leq c$ then $a \vee b \leq c$ ($\vee_{l@}$)

If we write the axiom $\wedge_{r@}$ in a two dimensional figure, erasing the words, putting the two premisses up and the conclusion down, writing "$\rightarrow$" instead of "$\leq$", we have then exactly the rule $\wedge_{r\S}$ (we add here an "§" to the name of the sequent rule to better stress the contrast) of the sequent system $\mathbf{S11}_{\wedge,\vee}$; the same with $\vee_{l@}$.

If we want to operate similar transformations with the other axioms, we have first to replace them with equivalent axioms. Let us consider the axiom $\wedge_{l1@}$, we will leave the other cases as exercises. We replace this axiom by

if $a \leq c$ then $a \wedge b \leq$ c ($\wedge_{l1@\S}$)

Let us prove that $\wedge_{l1@}$ is equivalent to $\wedge_{l1@\S}$.

From right to left. Due to reflexivity of $\leq$, we have $a \leq a$, then by $\wedge_{l1@\S}$ we have the desired result: $a \wedge b \leq a$.

From left to right. If $a \leq c$, by $\wedge_{l1@}$ we have $a \wedge b \leq a$, then by transitivity we have the desired result: $a \wedge b \leq c$.

Let us compare these two proofs with the two following derivations in $\mathbf{S11}_{\wedge,\vee}$:

$$\frac{a \rightarrow a}{a \wedge b \rightarrow a}\,\wedge_{l1\S} \qquad \frac{a \wedge b \rightarrow a \qquad a \rightarrow c}{a \wedge b \rightarrow c}\,\text{cut}$$

There are two differences between proofs in lattice theory and derivations within a sequent system. A first difference is that proofs in lattice theory, like proofs in mathematics in general, are performed in an informal way. It is not necessarily good to use the word "formal" to characterize such a difference, because it is quite ambiguous (see [10]).

We can write our two proofs in lattice theory in a similar way as the two above derivations in $\mathbf{S11}_{\wedge,\vee}$, writing "$\leq$" instead of "$\rightarrow$", "$\wedge_{l1@}$" instead of "$\wedge_{l1\S}$", "transitivity" instead of "cut". But doing that is not a sufficient condition to go from lattice theory to sequent system: although

[6]It would be absurd to write here "$\leq$" because the sign "$\leq$" is not itself a relation of order. If we want to use "$\leq$", we can say: "$\leq$" denote a relation of order. The same for "$\wedge$" and "$\vee$" which are names of binary functions.

useful, it is not a necessary condition, because a sequent system does not reduce to such a way of writing, to a way of writing. Useful, but also dangerous, because we may have the illusion to have really brought about the transformation. A man can make up himself into a monkey, and someone may think he is a monkey. But this is just an appearance. If we better examine the situation, we can see that the real transformation has not been achieved, that we are facing a fake monkey.

When doing proofs in lattice theory we are using a lot of resources. The idea of proof theory is to examine precisely how these resources work, capturing/describing them with a theory, namely proof theory. Proof systems such as sequent systems are part of proof theory but proof theory does not reduce to proof systems, it also includes the reasoning about these systems, which is itself generally carried on at an informal level, although it can also be formalized up to a certain point. In proof theory there is also a *meta* level. That is why it can be ambiguous to call proof theory "metamathematics", as it was done by Hilbert.

To avoid confusion we can use the word "derivation" for proofs within a sequent system, i.e. *formalized proofs*. This is the translation of the word "Herleitung" used by Gentzen, who considered it as a way shorther than "Beweisfigur" to speak about formalized proofs (cf 3.2. in [28]). It is not clear that Gentzen is thinking here of short in the sense of the length of the word, because "Herleitung" has only one letter less than "Beweisfigur". This is rather a thought shortcut than a word shortcut.

When developing a proof system the use of the signs which are used is made quite precise. Gentzen's paper ([28]) starts with a first section of five pages (178–182) entitled "Bezeichnungsfestsetzung." This 23 letter words has been wrongly translated in [31] as "TERMINOLOGY AND NOTATIONS". The French translation "Nomenclature des notations" in [30] is good.

This semiotic precision is often called "syntax". Considering the etymology of this word, "to put together in order', this is not absurd to use it. But syntax is often conceived by opposition to semantics. And some people want/ed to reduce semantics to syntax, considering that the meaning is fully given by "arrangement" of signs. What an "arrangement" is has to be clarified. Some people consider these arrangements as manipulations of signs, reducible to some writing devices. We can call this approach "syntactism". "Syntactism" has to be distinguished from a general position, that we can called "functionalism", according to which the meaning of a sign is given by the rule(s) to use it. A rule is not necessarily a writing device.

Let us now consider the second difference between proof in lattice theory and derivations within a sequent system. This difference is not, like the first one, directly about proof and derivation but about the objects dealt with. "$a \wedge b$" in lattice theory is not the same thing as "$a \wedge b$" in a standard sequent system. This latter difference is not a particularity of sequent system but of propositional languages in general. A propositional language is in general an absolutely free algebra, i.e. an algebra generated by some initial atoms, called "atomic formulas" (The word "formula" being used for any object of the structure, "molecular formulas" for formulas which are not atomic). In such an algebra the object denoted by "$a \wedge a$" is not the same as the one denoted by "a" even if $a \wedge a \dashv\vdash a$.[7]

In a lattice we have $a \wedge a \leq a$ and $a \leq a \wedge a$ and this means that $a \wedge a = a$, i.e. that a and $a \wedge a$ are the same object; in other words: "a" and "$a \wedge a$" denote the same object. In logic we are not obliged to work with a language which is an absolutely free algebra, we can just work with an algebra, this is has been done by Roman Suszko (see [41]). We can work with a sequent system with such a domain of formulas. In this case we are getting closer to lattice theory, but we are losing the possibility to make metaproofs on the complexity of formulas, because formulas don't have anymore an atomic ground. How to prove cut-elimination? Having a more complex structure, an absolutely free algebra, allows us to prove interesting results, having at the end in some sense the same structure. It is not exactly the same but were are able to precisely explain the relation between the two: we have on the one hand a logical structure, the fibring of conjunction with disjunction, on the other hand the factorization of this structure which is a lattice.

Two important remarks. Firstly what we have said is not completely exact. We have to consider a logic structure where there is only one object on the left of the consequence relation $\vdash$. The factorization is then obtained using the corresponding relation $\dashv\vdash$ which has also only one element on both sides. Secondly this procedure does not always work. A way to artificially make it work it to consider as logics, only structures where $\dashv\vdash$ is a congruence relation.

Let's examine more closely the situation where we have a sequent system on an arbitrary abstract algebra (not necessarily an absolutely free one). This situation has up to now not being investigated (however see [48]). Suszko and his collaborators have studied a consequence operator on a whatsoever abstract algebra but this approach is different from the proof-theoretical approach. This approach is interesting for example to go on the opposite direction which generally prevails, algebraization of logic. It can be called logicization of algebra. Sequent systems are

[7] For more comments about that see [1], p.183 and [6].

useful for that, in particular monosequent systems such as the one we have presented. But we can also work with monosequent systems where the intended meaning of the sign "$\rightarrow$" is identity, writing "$=$" instead of "$\rightarrow$". This logicization of algebra is different from *logicism*. It does not mean that we want to reduce algebra to logic, but the idea is to apply logical methodology to algebra. It is dual to *algebra of logic* conceived as application of algebraic methods to logic. Maybe the expression "algebra of logic" is better than "algebraization of logic" because it does not suggest a reductive approach. The expression "algebraic logic" is also not necessarily supporting the reductive stream. We can use the expression "logical algebra" for the kind of methodology we are proposing here. "Logic of algebra" would be too ambiguous.

If we have a sequent systems with sequent with "$=$" as a middle sign and objects of an arbitrary abstract algebra on both sides, we are closer to propositional logic than to first-order logic. This leads to a formalization of lattice theory proofs different to the one using the standard first-order sequent system LK. It is a formalization of the universal part of lattice theory, the one with only universal quantifiers. Putting then more structure on the algebra, it is possible to prove the decidability of this part of lattice theory, as an application of cut-elimination. This is not just a different transcription, it is a different conceptualization, leading to a different methodology with different results.

6.6 Development of this work and personal recollections

I don't remember exactly when I had the idea to work with monosequents, it is sometimes at the beginning of the 1990s. I wrote a short paper at this time with the proof that distributivity does not hold for this kind of systems, but I never developed and published this paper (I also wrote a short paper at the same time about logicization of algebra, that had the same fate). I decided to write the present paper to celebrate the anniversary of Amílcar, because it is connected with one of his favorite topics: combination of logics.

In 2004 Amílcar organized a congress on combination of logics in Lisbon (with the support of W.A.Carnielli). The event **ComBlog'04** – *Workshop on Combination of Logics: Theory and Applications* – took place where Amílcar is working, i.e. at the Department of Mathematics of IST (Instituto Superior Tecnico), Lisbon, Portugal. July 28-30, 2004. with many famous scholars including Dov Gabbay, Joseph Goguen, Jospeh Halpern, Dick de Jongh, Till Mossakowski, Don Pigozzi, Gabriel

Sandu, Ventura Verdú, Frank Wolter. I presented there the talk "A paradox in combination of logic" [9] discussing the fact that when we put conjunction and disjunction together in a natural logical way appears an additional property. This additional property is in fact distributivity and I mentioned at this time, without entering in details, that a way to solve this paradox would be the use of monosequents.

I have been interested since many years to sequent systems. I studied this topic following a master class by Jean-Yves Girard in 1990 at the Department of Mathematics of the University Paris 7 and by myself studying the original work of Gentzen. Girard was quite fascinated by sequent systems and his work on linear logic [32] is much related with these systems. He also told us that the cut-elimination theorem was one of the most important results of modern logic (for him it was in particular in view of the Curry-Howard isomorphism, not from a more philosophical perspective, the transition from Aristotle's syllogistic to sequent systems operated by Hertz calling the cut rule "syllogismus" [38]– about that see [17]).

I then wrote a Master's thesis on paraconsistent logic [3]. This work includes a sequent system for Newton da Costa's logic C1 ([21]) as well a the cut-elimination theorem. Andrés Raggio, a former student of Paul Bernays, had tried to do that in the 60s. Later on, when in Wrocław, Poland, in 1993, I was in touch with Igor Urbas. He was doing research at the University of Konstanz, Germany, in a group directed by André Fuhrmann. I didn't met him, but one of my colleagues of Wrocław University, Tom Skura, was going there frequently. Urbas gave me a Konstanz's pre-print of his paper about *LDJ* (Dual-Intuitionistic Logic) [57], where he studies a system of sequents with at most one formula on the left which generates a paraconsistent logic (Ubas did his PhD in Canberra on paraconsistency, see [56]).

In 1994 I wrote my PhD in mathematical logic (defended in 1995, see [5]) proving in particular a theorem establishing a connection between sequent systems and valuations justifying an intuitive semantical reading of the rules of sequent systems on certain conditions This results justifies what Gentzen wrote on section 2.4. of [28]: "The sequent $\mathfrak{A}_1, ..., \mathfrak{A}_u \rightarrow \mathfrak{B}_1, ..., \mathfrak{B}_v$ has exactly the same informal meaning as the formula $\mathfrak{A}_1, ..., \mathfrak{A}_u \supset \mathfrak{B}_1, ..., \mathfrak{B}_v$." This *informal meaning* in my theorem is nothing else than the meaning given by the standard classical truth-tables for conjunction, disjunction and implication. This result has been inspired by the first paper by Gentzen which is on Hertz's systems [27] that I read at this time. My PhD is entitled *Recherches sur la logique universelle*, with subtitle *Excessivité, Négation, Séquents.* I coined the expression *Universal Logic* when in Poland in 1993 (see [15]) as a name for a general theory of logics inspired by *Universal Algebra* as developed by Birkhoff (see [16]).

I organized the 1st UNILOG (*World Congress and School on Universal Logic*) in Montreux, Switzerland in 2005 and the second in Xi'an, China in 2007. Amílcar liked very much the spirit of universal logic and strongly encouraged me to organize the third edition in Lisbon. We did that in 2010 with the support of his team, in particular Carlos Caleiro, also very much interested in the idea of universal logic. The 3rd World Congress and School on Universal Logic happened Abril 18-25, 2010 in Estoril, nearby Lisbon. In 2008 I started to prepare an anthology on universal logic [13], and Amílcar and Carlos agreed to write for this anthology a paper [19] presenting Dov Gabbay's "Fibred semantics and the weaving of logics" [25].

In 2008 I also moved back to Brazil from Switzerland and since then generally stop in Lisbon when travelling to Europe for a few days. So I had the opportunity to meet Amílcar on a regular basis. In 2012 Carlos launched together with João Marcos (from Natal, Brazil) a Marie Curie Exchange program between Brazil and Europe called **GeTFun** (*Generalizing Truth-Functionality*). I then spent prolonged stays in Lisbon developing contact with more Portuguese colleagues: Sérgio Marcelino, a former student of Carlos, and also Olga Pombo, director of the Center for Philosophy of Sciences of the University of Lisbon (CFCUL). Olga organized a worskhop entitled *The Place of Philosophy of Science at Lisbon University* on February 12-15, 2013, for commemorating the unification of the Techinal University of Lisbon with the University of

Lisbon. I presented there the talk "Philosophy, Logic and Mathematics" [14]) and Amílcar presented a talk entitled O "triunfo do formalismo" ("The triumph of formalism") [54].

I had some discussions at this stage with Amílcar about this unification of the two main Universities of Lisbon and he was strongly supporting it (the institute he was working, the IST, was part of the Technical University). I had also with him other discussions about various subjects, in particular during lunches at *El Corte Inglés* he kindly invited me to take part to with my wife Catherine and his wife Cristina. One of Amílcar qualities is that, beside being a very good logician, he has interest to think and discuss on all topics ... in a logical way! Being therefore a true universal logician.

6.7 Acknowledgments

Thanks to the logic team of IST / University of Lisbon. I presented a talk related to this paper there on September 9, 2016. Thanks to Arnon Avron and Lloyd Humberstone for having carefully read this paper and making useful comments and suggestions. Thanks also to Sara Negri, Hermógenes Oliveira and Anna Zamanski.

Bibliography

[1] A.Anderson and N.D.Belnap, *Entailment - The Logic of Relevance and Necessity - Volume 1*, Princeton University Press, Princeton, 1975.

[2] A. Avron, "A constructive analysis of RM", *Journal of Symbolic Logic*, **52** (1987), 939–951.

[3] J.-Y.Beziau, *La logique paraconsistante C1 de Newton da Costa*, Master Thesis, Department of Mathematics, University Denis Diderot (Paris 7), Paris, 1990.

[4] J.-Y.Beziau, "Universal logic", in Logica94 *Proceedings of the 8th International Symposium*, T.Childers and O.Majer (eds), Prague, 1994, 73-93.

[5] J.-Y.Beziau, *Recherches sur la logique universelle (Excessivité, Négation, Séquents*, PhD Thesis, Department of Mathematics, University Denis Diderot (Paris 7), Paris, 1995.

[6] J.-Y.Beziau, "Logic may be simple", *Logic and Logical Philosophy*, **5** (1997), 129–147.

[7] J.-Y.Beziau, "Rules, derived rules, permissible rules and the various types of systems of deduction", in E.H.Hauesler and L.C.Pereira (eds), *Proof, types and categories*, PUC, Rio de Janeiro, 1999, 159–184.

[8] J.-Y.Beziau, "Sequents and bivaluation", *Logique et Analyse*, **44** (2001), 373–394.

[9] J.-Y.Beziau, "A paradox in the combination of logics", in W.A.Carnielli, F.M.Dionísio and P.Mateus (eds), *Workshop on Combination of Logics: Theory and Applications*, IST, Lisbon, 2004, 75-78.

[10] J.-Y.Beziau, "What is *formal logic*?", in Myung-Hyun-Lee (ed), *Proceedings of the XXII World Congress of Philosophy, vol.13, Korean Philosophical Association*, Seoul, 2008, 9-22.

[11] J.-Y.Beziau,, "The Challenge of Combining Logics", Preface of a special issue of the *Logic Journal of the IGPL* on combination of logic, **19** (2010), 543.

[12] J.-Y.Beziau and M.E.Coniglio, "To distribute or not to distribute?", *Logic Journal of the Interest Group in Pure and Applied Logics*, **19** (2011), 566–583.

[13] J.Y.Beziau (ed), *Universal Logic : an Anthology - From Paul Hertz to Dov Gabbay*, Birkhäuser, Basel, 2012.

[14] J.-Y.Beziau, "Three Sisters : Philosophy, Mathematics and Logic", in N.Nabais and O.Pombo (eds), *O lugar da Filosofia da Ciência na nova Universidade de Lisboa*, University of Lisbon, 2013, 271-291.

[15] J.-Y.Beziau, "Logical Autobiography 50", in A.Koslov and A.Buchsbaum (eds), *The Road to Universal Logic Festschrift for the 50th Birthday of Jean-Yves Béziau, Volume II*, Birkhäuser, Basel, 2015.

[16] J.-Y.Beziau, "The relativity and universality of logic", *Synthese - Special Issue Istvan Németi 70th Birthday*, **192** (2015), 1939–1954.

[17] J.-Y.Beziau, "Is modern logic non-Aristotelian?", in D.Zaitsev (ed), *Nikolai Vasiliev's Logical Legacy and Modern Logic*, Springer, Dordrecht, 2016.

[18] C.Caleiro, *Combining Logics*, PhD, IST, Lisbon, 2000.

[19] C.Caleiro and A.Sernadas, "Fibring logics", in [13], 389–395.

[20] W.Carnielli, M.Coniglio, D.M.Gabbay, P.Gouveia, and C.Sernadas, *Analysis and Synthesis of Logics - How to Cut and Paste Reasoning Systems*, Springer, Heidelberg, 2008.

[21] da Costa, N.C.A., "Calculs propositionnels pour les systèmes formels inconsistants", *Compte Rendus de l'Académie des Science de Paris* , **257** (1963) 790–3792.

[22] H.Curry, *Leçons de logique algébrique*, Gauthiers-Villars, Paris and E.Nauwelaerts, Louvain, 1952. English translation and presentation by J.Seldin in [13], 125–160.

[23] H.Curry, *Foundations of mathematical logic*, McGraw and Hill, New York, 1963.

[24] M.Dummett, *Elements of intuitionism*, Clarendon, Oxford, 1977.

[25] D. Gabbay, "Fibred semantics and the weaving of logics", *The Journal of Symbolic Logic* , **61** (1996) 1057–1120. Re-edited and presented by C.Caleiro and A.Sernadas [19] in [13], 389-396.

[26] D.Gabbay, 1999, *Fibring logics*, Clarendon, Oxford, 1999.

[27] G.Gentzen, "Über die Existenz unabhängiger Axiomensysteme zu unendlichen Satzsystemen", *Mathematische Annalen*, **107** (1933), 329–350 (English translation in [31], 29–52).

[28] G.Gentzen, "Untersuchungen über das logische Schließen. I", *Mathematische Zeitschrift*, **39** (1935), 176–210 (English translation in [31], 68–103).

[29] G.Gentzen, "Untersuchungen über das logische Schließen. II", *Mathematische Zeitschrift*, **39** (1935), 405–431 (English translation in [31], 103–131).

[30] G.Gentzen, *Recherche sur la déduction logique*, Presses Universitaires de France, Paris, 1955. French translation of [28] and [29] with extensive comments by R.Feys and J.Ladrière.

[31] G.Gentzen, *The collected papers of Gerhard Gentzen*, edited by M.E.Szabo, North-Holland, Amsterdam, 1969.

[32] J.-Y.Girard, "Linear logic", *Theoretical Computer Science*, **50** (1987), 1–102.

[33] P.Halmos, "How to write Mathematics", *L'Enseignement Mathématique*, **16** (1970), 123–152.

[34] P.Hertz, "Über Axiomensysteme für beliebige Satzsysteme. I. Teil. Sätze ersten Grades. (Über die Axiomensysteme von der kleinsten Satzzahl und den Begriff des idealen Elementes.)", *Mathematische Annalen*, **87** (1922), 246–269. English translation in [13], 11–30.

[35] P.Hertz, "Über Axiomensysteme für beliebige Satzsysteme. Teil II. Sätze höheren Grades." *Mathematische Annalen*, **89** (1923), 76-102.

[36] P.Hertz, "Reichen die üblichen syllogistischen regeln für das schließen in der positiven logik elementarer sätze aus?", *Annalen der Philosophie und philosophischen Kritik*, **7** (1928), 272–277.

[37] P.Hertz, "Über Axiomensysteme für beliebige Satzsysteme", *Mathematische Annalen*, **101** (1929), 457–514.

[38] P.Hertz, "Von Wesen des Logischen, insbesondere der Bedeutung des modus Barbara", *Erkenntnis*, **2** (1931), 369–392.

[39] L.Humberstone, *The Connectives*, MIT Press, Cambridge, MA, 2011.

[40] L.Humberstone, "Béziau on And and Or", in A.Koslov and A.Buchsbaum (eds), *The Road to Universal Logic Festschrift for the 50th Birthday of Jean-Yves Béziau, Volume I*, Birkhäuser, Basel, 2015. 283–307.

[41] R.Jansana, "Bloom, Brown and Suszkos work on abstract logics", in [13], 251–256.

[42] R.Kahle and M.Rathjen (eds), *Gentzen's Centenary – The Quest for Consistency*, Springer International Publishing Switzerland, Cham, 2015.

[43] A.Koslow, *A structuralist theory of logic*, Cambridge University Press, New York, 1992.

[44] A.Koslow, "Structuralist logic: implications, inferences, and consequences", *Logica Universalis*, **1** (2007), 167–181.

[45] J.Legris, "Paul Hertz and the origins of structural reasoning", in [13], 3-10.

[46] Łukasiewicz and A.Tarski, "Untersuchungen über den Aussagenkalkül", *Comptes Rendus des séances de la Société des Sciences et des Lettres de Varsovie, Classe III*, **23** (1930), 30–50.

[47] S.Negri and J.von Plato, *Structural proof theory*, Cambridge University Press, Cambridge, 2001.

[48] S.Negri and J.von Plato, "Proof systems for lattice theory", *Mathematical Structures in Computer Scienc*, **14** (2004), 507–526.

[49] S.Marcelino and C. Caleiro, "On the characterization of fibred logics, with applications to conservativity and finite-valuedness" *Journal of Logic and Computation*, 2016.

[50] E.Menzler-Trott, *Gentzens Problem*, Birkhäuser, Basel, 2001. English translation: *Logic's lost genius - The life of Gerhard Gentzen*, American Mathematical Society, Providence, 2007.

[51] G. Pottinger, "Uniform, cut-free formulation of T,S4 and S5" (abstract), *Journal of Symbolic Logic*, **48** (1983), 900.

[52] A.Raggio, "Propositional sequence-calculi for inconsistent systems", *Notre Dame Journal of Formal Logic*, **9** (1968), 359–366.

[53] P.Schroeder-Heister, "Resolution and the origins of structural reasoning: Early proof-theoretic ideas of Hertz and Gentzen", *Bulletin of Symbolic Logic*, **8** (2002), 246–265.

[54] A.Sernadas, "O triunfo do formalismo", Unpublished draft, University of Lisbon, 2013. http://filcc-ulutl.fc.ul.pt/docs/ACS.pdf

[55] A.Tarski, "Remarque sur les notions fondamentales de la méthodologie des mathématiques", *Annales de la Société polonaise de mathématique*, **7** (1928), 270–272. English translation and presentation by J.Zygmunt [58] and in [13], 59–70.

[56] I.Urbas, *On Brazilian paraconsistent logics*, PhD, Australian National University, Canberra, 1987.

[57] I.Urbas, "Dual-intutionistic logic", *Notre Dame Journal of Formal Logic*, **37** (1996), 440–451.

[58] J.Zygmunt, "Tarski's first contribution to general metamathematics", in [13], 59–66.

7

MODAL LOGIC **S4** AS A PARACONSISTENT LOGIC WITH A TOPOLOGICAL SEMANTICS

Marcelo E. Coniglio[1] and Leonardo Prieto-Sanabria[2]

[1]Centre for Logic, Epistemology and the History of Science (CLE)
University of Campinas (UNICAMP), Brazil
coniglio@cle.unicamp.br

[2]Pontifical Catholic University of Campinas (PUC-Campinas), Brazil
lprieto@utp.edu.co

Dedicated to Amílcar Sernadas,
a brilliant mind

Abstract

In this paper the propositional logic **LTop** is introduced, as an extension of classical propositional logic by adding a paraconsistent negation. This logic has a very natural interpretation in terms of topological models. The logic **LTop** is nothing more than an alternative presentation of modal logic **S4**, but in the language of a paraconsistent logic. Moreover, **LTop** is a logic of formal inconsistency in which the consistency and inconsistency operators have a nice topological interpretation. This constitutes a new proof of **S4** as being "the logic of topological spaces", but now under the perspective of paraconsistency.

7.1 Topology, Modal Logic and Paraconsistency

The studies on the relationship between modal logic, topology and paraconsistency, have a relatively long history.

By extending the Stone representation theorem for Boolean algebras, McKinsey and Tarski (see [14]) proved in 1944 that it is possible to characterize modal logic **S4** by means of a topological semantics. Within this semantics, the necessity operator $\Box$ and the possibility operator $\Diamond$ are interpreted as the interior and the closure topological operators, respectively, and so this result states that **S4** is, in a certain sense, "the logic of topological spaces". Moreover, they prove that **S4** is semantically characterized by the real line (with the usual topology) or, in general, by any dense-in-itself separable metrizable space. Several variants and generalizations of McKinsey and Tarski's result have been proposed in the literature (see, for instance, [18] and [12]).

Semantics for Paraconsistent logics (that is, logics having a negation which produces some non-trivial contradictory theories) have been defined in topological terms by several authors. For instance, Mortensen studies in [16] some topological properties by means of paraconsistent and paracomplete logics. Goodman already proposes in [10] an "anti-intuitionistic logic" which is paraconsistent and it is endowed with a topological semantics. Along the same lines, Priest analyzes a paraconsistent negation obtained by dualizing the intuitionistic negation, defining a topological semantics for such negation (see [17]). From a broader perspective, Başkent proposes in [1] an interesting study of topological models for paraconsistency and paracompleteness.

By its turn, the relationship between paraconsistency and modal logic is also very close. Already in 1948, Jaśkowski presented in [11] his "discussive logic", which is considered the first formal system for a paraconsistent logic, and it was formalized in terms of modalities. Beziau observes in [2] (see also [3]) that the operator $\neg\alpha \stackrel{\text{def}}{=} {\sim}\Box\alpha$ defines in modal logic **S5** a paraconsistent negation (here, $\sim$ denotes the classical negation). This relation between modalities and paraconsistent negation was already observed by Beziau in 1998 (despite the paper was published only in 2006, see [4]), from the perspective of Kripke semantics, when defining the logic **Z**. However, already in 1987, de Araújo et al. observed in [9] that a Kripke-style semantics can be given for Sette's 3-valued paraconsistent logic **P1**, based on Kripke frames for the modal logic **T**. In that semantics, the formula $\neg\alpha$ (for the paraconsistent negation $\neg$ of **P1**) is interpreted exactly as the modal formula ${\sim}\Box\alpha$. Beziau's approach was generalized by Marcos in [13], showing that there is a close correspondence between non-degenerate modal logics and the paraconsistent logics known as *logics of formal inconsistency* (see Section 7.6).

This paper contributes to this discussion by introducing a propositional logic, called **LTop**, which extends classical propositional logic with a paraconsistent negation. This logic has a very natural interpretation in terms of topological models which associate to each formula a set (not necessarily open or close). The classical connectives are interpreted as usual, and the paraconsistent negation is interpreted as the topological closure of the complement, in a dual form to the usual interpretation of the intuitionistic negation (namely, the interior of the complement). This topological interpretation of a paraconsistent negation is very natural, and it was already proposed in [10], [16] and [1]. Modalities $\Box$ and $\Diamond$ can be defined in the language of **LTop**, which are interpreted as the interior and closure operator, respectively. As expected, the logic **LTop** is nothing more than an alternative presentation of modal logic **S4**, but in a language (and through a Hilbert calculus) corresponding to an extension of classical logic by means of a paraconsistent negation. It is worth noting that the logic **Z** in [4] introduces an axiomatization of **S5** as an extension of classical logic with a paraconsistent negation, and so the present result is a kind of generalization of such result, provided that **S5** can be obtained from **S4** by adding an additional axiom. Additionally, it is proved that **LTop** is a logic of formal inconsistency (see Section 7.6) in which the consistency and inconsistency operators have a nice topological interpretation. This constitutes a new proof of **S4** as being "the logic of topological spaces", but now under the perspective of paraconsistent logics. It is finally shown that intuitionistic propositional logic can be interpreted in **LTop** through a very natural conservative translation.

7.2 The Propositional Logic LTop

In this section, the propositional logic **LTop** will be introduced by means of a Hilbert calculus, with a modal-like notion of derivations.

Definition 7.1 *Let* $\mathcal{V} = \{p_n \ : n \geq 1\}$ *be a denumerable set of* propositional variables. *Given the propositional signature* $\Sigma = \{\sim, \neg, \rightarrow\}$ *let* $\mathbb{L}$ *be the language generated by the set* $\mathcal{V}$ *over the signature* Σ.

As suggested by the notation, $\sim$ and $\neg$ are unary connectives representing two different negations, while $\rightarrow$ is a binary connective which represents an implication (in the logic **LTop** to be defined below). The connective $\sim$ will represent a classical negation, while $\neg$ will represent a paraconsistent negation. The implication will be also classical. The following usual abbreviations can be introduced in the language $\mathbb{L}$:

(conj) $\alpha \wedge \beta \stackrel{\text{def}}{=} {\sim}(\alpha \to {\sim}\beta)$

(disj) $\alpha \vee \beta \stackrel{\text{def}}{=} {\sim}\alpha \to \beta$

Definition 7.2 (Propositional Logic LTop) *The logic* **LTop** *is given by the Hilbert calculus over the language* $\mathbb{L}$ *defined by following the axioms and inference rules:*

Axiom schemas:

$$\alpha \to (\beta \to \alpha) \quad \textbf{(Ax1)}$$

$$(\alpha \to (\beta \to \gamma)) \to ((\alpha \to \beta) \to (\alpha \to \gamma)) \quad \textbf{(Ax2)}$$

$$({\sim}\alpha \to \beta) \to (({\sim}\alpha \to {\sim}\beta) \to \alpha) \quad \textbf{(Ax3)}$$

$$\alpha \to \neg{\sim}\alpha \quad \textbf{(Ax4)}$$

$$\neg{\sim}\neg{\sim}\alpha \to \neg{\sim}\alpha \quad \textbf{(Ax5)}$$

$$\neg(\alpha \wedge \beta) \to \neg\alpha \vee \neg\beta \quad \textbf{(Ax6)}$$

$${\sim}\neg(\alpha \to \alpha) \quad \textbf{(Ax7)}$$

Inference rules:

$$\frac{\alpha \quad \alpha \to \beta}{\beta}\ \textit{(MP)} \qquad \frac{\alpha \to \beta}{\neg\beta \to \neg\alpha}\ \textit{(CR)} \qquad \frac{\alpha \quad \beta}{\alpha \wedge \beta}\ \textit{(DR)}$$

The Logic **LTop** has the following notion of derivation:

Definition 7.3 (Derivations in LTop)
(1) A derivation *of a formula* α *in* **LTop** *is a finite sequence of formulas* $\alpha_1 \ldots \alpha_n$ *such that* α_n *is* α *and every* α_i *is either an instance of an axiom or it is the consequence of some inference rule whose premises appear in the sequence* $\alpha_1 \ldots \alpha_{i-1}$. *We say that* α *is* derivable in **LTop**, *and we write* $\vdash_{\mathbf{LTop}} \alpha$, *if there exists a derivation of it in* **LTop**.
(2) Let $\Gamma \cup \{\alpha\}$ *be a set of formulas. We say that* α *is* derivable in **LTop** from Γ, *and we write* $\Gamma \vdash_{\mathbf{LTop}} \alpha$, *if either* $\vdash_{\mathbf{LTop}} \alpha$ *or there exists a finite, non-empty subset* $\{\gamma_1, \ldots, \gamma_n\}$ *of* Γ *such that* $(\gamma_1 \wedge (\gamma_2 \wedge (\ldots \wedge (\gamma_{n-1} \wedge \gamma_n) \ldots))) \to \alpha$ *is derivable in* **LTop**.

By the very definition, $\emptyset \vdash_{\mathbf{LTop}} \alpha$ iff $\vdash_{\mathbf{LTop}} \alpha$.

Remark 7.4 *The consequence relation* $\vdash_{\mathbf{LTop}}$ *for* **LTop** *given in the previous definition is Tarskian and finitary, that is, it satisfies the following properties: (i) if* $\alpha \in \Gamma$ *then* $\Gamma \vdash_{\mathbf{LTop}} \alpha$*; (ii) if* $\Gamma \vdash_{\mathbf{LTop}} \alpha$ *and* $\Gamma \subseteq \Delta$

then $\Delta \vdash \alpha$*; (iii) if* $\Gamma \vdash_{\mathbf{LTop}} \Delta$ *and* $\Delta \vdash_{\mathbf{LTop}} \alpha$ *then* $\Gamma \vdash_{\mathbf{LTop}} \alpha$*, where* $\Gamma \vdash \Delta$ *means that* $\Gamma \vdash \delta$ *for every* $\delta \in \Delta$*; and (iv) if* $\Gamma \vdash_{\mathbf{LTop}} \alpha$ *then* $\Gamma_0 \vdash_{\mathbf{LTop}} \alpha$ *for some finite* Γ_0 *contained in* Γ*. This can be proven by adapting to the logic* **LTop***, together with the pair of connectives* $(\rightarrow, \wedge)$*, a general result (Theorem 2.10.2) concerning entailment systems found in [20].*

Remark 7.5 Note that (**Ax1**), (**Ax2**) and (**Ax3**) plus (**MP**) contitute an axiomatization of propositional classical logic **CPL** over the signature $\{\sim, \rightarrow\}$.

7.3 Basic Propositions Derivable in LTop

In this section some basic propositions will be derived in **LTop**.

Definition 7.6 *Define the relation* $\equiv$ *in* $\mathbb{L}$ *as follows:*

$$\alpha \equiv \beta \qquad \textit{iff} \qquad \vdash_{\mathbf{LTop}} \alpha \rightarrow \beta \ \ \textit{and} \ \ \vdash_{\mathbf{LTop}} \beta \rightarrow \alpha$$

An immediate consequence of the rules **CR** and **MP** is the following:

Proposition 7.7 *If* $\alpha \equiv \beta$ *then* $\neg\alpha \equiv \neg\beta$*.*

From Proposition 7.7 and Remark 7.5 it follows:

Corollary 7.8 (Weak Replacement) *If* $\alpha_i \equiv \beta_i$ *for* $i = 1, \ldots, n$ *then, for every formula* $\varphi(p_1, \ldots, p_n)$*, it holds:*

$$\varphi[p_1/\alpha_1 \cdots p_n/\alpha_n] \equiv \varphi[p_1/\beta_1 \cdots p_n/\beta_n].$$

In the terminology introduced by Wójcicki, the latter result shows that **LTop** is a *selfextensional* logic. On the other hand, the following meta-property of **LTop** can be easily proved:

Proposition 7.9 (Deduction Theorem) *For every set of formulas* $\Gamma \cup \{\alpha, \beta\} : \Gamma, \alpha \vdash_{\mathbf{LTop}} \beta$ *iff* $\Gamma \vdash_{\mathbf{LTop}} (\alpha \rightarrow \beta)$*.*

Proof. It is an immediate consequence of the notion of derivations in **LTop** and the properties of **CPL**. The details are left to the reader. □
The relation between the two primitive negations is as follows:

Proposition 7.10 $\vdash_{\mathbf{LTop}} \sim\alpha \rightarrow \neg\alpha$.

Proof. Consider the following (meta)derivation in **LTop**:

1. $\vdash_{\mathbf{LTop}} {\sim}\alpha \to \neg{\sim}{\sim}\alpha$ (by **Ax4**)

2. $\vdash_{\mathbf{LTop}} {\sim}\alpha \to \neg\alpha$ (Replacement)

□

The converse does not hold in general (see Proposition 7.16 below). The weak negation $\neg$ satisfies the following basic properties:

Proposition 7.11 *The following holds in* **LTop***:*

(i) $\vdash_{\mathbf{LTop}} \alpha \vee \neg\alpha$.

(ii) $\neg(\alpha \wedge \beta) \equiv (\neg\alpha \vee \neg\beta)$.

(iii) $\vdash_{\mathbf{LTop}} \neg(\alpha \vee \beta) \to (\neg\alpha \wedge \neg\beta)$.

Proof. (i) Consider the following (meta)derivation in **LTop**:

1. $\vdash_{\mathbf{LTop}} {\sim}\alpha \to \neg\alpha$ (by Proposition 7.10)

2. $\vdash_{\mathbf{LTop}} (\alpha \vee {\sim}\alpha) \to (\alpha \vee \neg\alpha)$ (1, **CPL**)

3. $\vdash_{\mathbf{LTop}} \alpha \vee {\sim}\alpha$ (**CPL**)

4. $\vdash_{\mathbf{LTop}} \alpha \vee \neg\alpha$ (2,3 **MP**)

(ii) By (**Ax6**) it is enough to show that $\vdash_{\mathbf{LTop}} (\neg\alpha \vee \neg\beta) \to \neg(\alpha \wedge \beta)$. Thus, consider the following (meta)derivation in **LTop**:

1. $\vdash_{\mathbf{LTop}} (\alpha \wedge \beta) \to \alpha$ (**CPL**)

2. $\vdash_{\mathbf{LTop}} \neg\alpha \to \neg(\alpha \wedge \beta)$ (**CR**,1)

3. $\vdash_{\mathbf{LTop}} (\alpha \wedge \beta) \to \beta$ (**CPL**)

4. $\vdash_{\mathbf{LTop}} \neg\beta \to \neg(\alpha \wedge \beta)$ (**CR**,3)

5. $\vdash_{\mathbf{LTop}} (\neg\alpha \vee \neg\beta) \to \neg(\alpha \wedge \beta)$ (2,4, **CPL**)

(iii) By **CPL**, $\vdash_{\mathbf{LTop}} \alpha \to (\alpha \vee \beta)$ whence $\vdash_{\mathbf{LTop}} \neg(\alpha \vee \beta) \to \neg\alpha$, by (**CR**). Analogously, $\vdash_{\mathbf{LTop}} \neg(\alpha \vee \beta) \to \neg\beta$. The result follows ny **CPL**.
□

The converse of item (iii) of the latter proposition does not hold in general. It will be proven in the next section by using the topological semantics for **LTop** (see Proposition 7.16).

7.4 Topological Semantics for LTop

In this section, an intuitive semantics for **LTop** over topological spaces will be given. Given a a topological space (X, τ) and $A \subseteq X$, the interior and the closure of A in the given topology, as well as its complement (relative to X), will be denoted by $Int(A)$, $\overline{A}$ and A^c, respectively. The powerset of X will be denoted by $\wp(X)$.

Definition 7.12 *A* Topological structure *for* **LTop** *is a topological space* $\mathcal{T} = \langle X, \tau \rangle$. *A* Topological model *for* **LTop** *is a pair* $\mathcal{M} = \langle \mathcal{T}, v \rangle$ *such that* $\mathcal{T}$ *is a topological structure* $\langle X, \tau \rangle$ *for* **LTop** *and* $v : \mathbb{L} \to \wp(X)$ *is a function, called* valuation, *satisfying the following conditions:*

1. $v(\alpha \to \beta) = v(\alpha)^c \cup v(\beta)$;
2. $v({\sim}\alpha) = v(\alpha)^c$;
3. $v(\neg\alpha) = \overline{v(\alpha)^c}$.

Note that, because of (conj) and (disj):

(v-conj) $v(\alpha \wedge \beta) = v({\sim}(\alpha \to {\sim}\beta)) = (v(\alpha)^c \cup v(\beta)^c)^c = v(\alpha) \cap v(\beta)$;

(v-disj) $v(\alpha \vee \beta) = v({\sim}\alpha \to \beta) = (v(\alpha)^c)^c \cup v(\beta) = v(\alpha) \cup v(\beta)$.

Definition 7.13 (Semantical consequence in LTop)
(1) A formula α *in* $\mathbb{L}$ *is* true *in a topological model* $\mathcal{M} = \langle\langle X, \tau \rangle, v\rangle$, *written as* $\mathcal{M} \models \alpha$, *if* $v(\alpha) = X$.
(2) A formula α *in* $\mathbb{L}$ *is* valid *in* **LTop**, *denoted by* $\models_{\mathbf{LTop}} \alpha$, *if* $\mathcal{M} \models \alpha$ *for every topological model* $\mathcal{M}$.
(3) Let $\Gamma \cup \{\alpha\}$ *be a set of formulas. We say that* α *is a* semantical consequence of Γ in **LTop**, *denoted by* $\Gamma \models_{\mathbf{LTop}} \alpha$, *if either* $\models_{\mathbf{LTop}} \alpha$ *or there exists a finite non-empty set* $\Gamma_0 \subseteq \Gamma$ *such that* $\bigcap_{\gamma \in \Gamma_0} v(\gamma) \subseteq v(\alpha)$, *for every topological model* $\langle\langle X, \tau \rangle, v\rangle$.

Note that, by the very definition, $\emptyset \models_{\mathbf{LTop}} \alpha$ iff $\models_{\mathbf{LTop}} \alpha$.

Proposition 7.14 *[Soundness] The logic* **LTop** *is sound with respect to the topological semantics, that is:* $\Gamma \vdash_{\mathbf{LTop}} \alpha$ *implies* $\Gamma \models_{\mathbf{LTop}} \alpha$, *for every set of formulas* $\Gamma \cup \{\alpha\}$. *In particular,* $\vdash_{\mathbf{LTop}} \alpha$ *implies that* $\models_{\mathbf{LTop}} \alpha$.

Proof. We first prove that $\vdash_{\mathbf{LTop}} \alpha$ implies $\models_{\mathbf{LTop}} \alpha$, since **LTop** is finitary and by Proposition 7.9. But the latter is easily proved by observing that every axiom is valid in every topological space, and that inference rules preserve validity. The case for $\Gamma \neq \emptyset$ follows from the very definitions. □

Corollary 7.15 *If $\alpha \equiv \beta$ then $v(\alpha) = v(\beta)$ for every topological model $\langle \mathcal{T}, v\rangle$.*

Proof. Suppose that $\alpha \equiv \beta$, and let $\langle\langle X, \tau\rangle, v\rangle$ be a topological model for **LTop**. The result follows by observing that $v(\alpha \to \beta) = X$ iff $v(\alpha) \subseteq v(\beta)$. □

The converse of the last corollary will follow from the completeness theorem (see Theorem 7.36 below). Thanks to that, and by the definition of the topological semantics, it is easy to see that Proposition 7.11(ii) corresponds to one of the Kuratowski axioms for the closure operator, namely: $\overline{A} \cup \overline{B} = \overline{A \cup B}$ for every $A, B \subseteq X$.

By using again the soundness theorem of **LTop** w.r.t. topological semantics, it can be seen that some properties of the classical negation $\sim$ fail for the negation $\neg$:

Proposition 7.16 *Let p and q be two different propositional variables. Then:*

(i) $\nvdash_{\mathbf{LTop}} (\neg p \wedge \neg q) \to \neg(p \vee q)$.

(ii) $\nvdash_{\mathbf{LTop}} p \to \neg\neg p$.

(iii) $\nvdash_{\mathbf{LTop}} \neg\neg p \to p$.

(iv) $\nvdash_{\mathbf{LTop}} \neg p \to {\sim} p$.

Proof. Consider $\mathcal{M} = \langle\langle \mathbb{R}, \tau\rangle, v\rangle$ such that τ is the usual topology on the set $\mathbb{R}$ of real numbers.

(i) Let $v(p) = (-\infty, 0] \cup [1, +\infty)$ and $v(q) = (-\infty, 1] \cup [2, +\infty)$. Then $v(\neg p) = \overline{(0,1)} = [0,1]$ and $v(\neg q) = \overline{(1,2)} = [1,2]$ and so $v(\neg p \wedge \neg q) = v(\neg p) \cap v(\neg q) = \{1\}$. On the other hand, $v(p \vee q) = \mathbb{R}$ and so $v(\neg(p \vee q)) = \overline{\emptyset} = \emptyset$. Since $v(\neg p \wedge \neg q) \nsubseteq v(\neg(p \vee q))$ it follows that $v((\neg p \wedge \neg q) \to \neg(p \vee q)) \neq \mathbb{R}$. By soundness of **LTop** w.r.t. topological models, $\nvdash_{\mathbf{LTop}} (\neg p \wedge \neg q) \to \neg(p \vee q)$.

(ii), (iii) and (iv) Let $v(p) = (0,1) \cup \{2\}$. Then $v(\neg\neg p) = [0,1]$, which is incomparable with $v(p)$. On the other hand, $v(\neg p) \nsubseteq v({\sim} p)$. □

7.5 LTop as a Modal Logic

Consider the following abbreviations:

1. $\Box\alpha \stackrel{\text{def}}{=} {\sim}\neg\alpha$

2. $\Diamond\alpha \stackrel{\text{def}}{=} \neg{\sim}\alpha$

Semantically, it means the following:

1. $v(\Box\alpha) = v({\sim}\neg\alpha) = (\overline{v(\alpha)^c})^c = Int(v(\alpha))$

2. $v(\Diamond\alpha) = v(\neg{\sim}\alpha) = \overline{(v(\alpha)^c)^c} = \overline{v(\alpha)}$

The relationship between $\Box$ and $\Diamond$ is as expected.

Proposition 7.17 *The following holds in* **LTop***:*

(i) $\Box\alpha \equiv {\sim}\Diamond{\sim}\alpha$.

(ii) $\Diamond\alpha \equiv {\sim}\Box{\sim}\alpha$.

Proof. It is immediate from the axioms and rules of **LTop**. $\Box$

The following properties of $\Diamond$ can be proven in **LTop**:

Proposition 7.18 *The operator $\Diamond$ satisfies the following properties in* **LTop***:*

(i) $\vdash_{\mathbf{LTop}} \alpha \to \Diamond\alpha$.

(ii) $\Diamond\alpha \equiv \Diamond\Diamond\alpha$.

(iii) $\Diamond(\alpha \vee \beta) \equiv \Diamond\alpha \vee \Diamond\beta$.

(iv) If $\vdash_{\mathbf{LTop}} (\alpha \to \beta)$ *then* $\vdash_{\mathbf{LTop}} (\Diamond\alpha \to \Diamond\beta)$.

Proof. (i) and (ii): It follows easily from the axioms and rules of **LTop**.
(iii) Since, for any α and β,

$$(\neg\alpha \vee \neg\beta) \equiv \neg(\alpha \wedge \beta)$$

then, in particular,

$$(\neg{\sim}\alpha \vee \neg{\sim}\beta) \equiv \neg({\sim}\alpha \wedge {\sim}\beta).$$

But $\sim\alpha \wedge \sim\beta \equiv \sim(\alpha \vee \beta)$ (by **CPL**) and then, using Replacement, we have that

$$(\neg\sim\alpha \vee \neg\sim\beta) \equiv \neg\sim(\alpha \vee \beta)$$

namely $\Diamond\alpha \vee \Diamond\beta \equiv \Diamond(\alpha \vee \beta)$.

(iv) Consider the following (meta)derivation in **LTop**:

1. $\vdash_{\mathbf{LTop}} (\alpha \to \beta)$ (by Hypothesis)
2. $\vdash_{\mathbf{LTop}} (\sim\beta \to \sim\alpha)$ (**CPL**)
3. $\vdash_{\mathbf{LTop}} (\neg\sim\alpha \to \neg\sim\beta)$ (**CR**)

□

Remark 7.19 *Observe that the logical properties of the connective* $\Diamond$ *in* **LTop** *stated in Proposition 7.18 reflect, when interpreted in topological structures, the basic properties of a closure operator. Indeed, they represent the following properties, for all subsets* A, B *of a topological space* X*: (i)* $A \subseteq \overline{A}$*; (ii)* $\overline{A} = \overline{\overline{A}}$*; (iii)* $\overline{A \cup B} = \overline{A} \cup \overline{B}$*; and (iv) if* $A \subseteq B$ *then* $\overline{A} \subseteq \overline{B}$.

Dually, the following properties of □, seeing as an interior operator, can be proved in **LTop**:

Proposition 7.20 *The following holds in* **LTop***:*

(i) $\vdash_{\mathbf{LTop}} \Box(\alpha \to \alpha)$

(ii) $\vdash_{\mathbf{LTop}} \Box\alpha \to \alpha$.

(iii) $\Box\alpha \equiv \Box\Box\alpha$.

(iv) $\Box(\alpha \wedge \beta) \equiv \Box\alpha \wedge \Box\beta$.

(v) If $\vdash_{\mathbf{LTop}} (\alpha \to \beta)$ *then* $\vdash_{\mathbf{LTop}} (\Box\alpha \to \Box\beta)$.

(vi) $\vdash_{\mathbf{LTop}} \Box(\alpha \to \beta) \to (\Box\alpha \to \Box\beta)$.

Proof. Items (i)-(iv) are left to the reader.

(v) Consider the following (meta) derivation in **LTop**:

1. $\vdash_{\mathbf{LTop}} (\alpha \to \beta)$ (Hypothesis)
2. $\vdash_{\mathbf{LTop}} (\neg\beta \to \neg\alpha)$ (**CR**)
3. $\vdash_{\mathbf{LTop}} (\neg\beta \to \neg\alpha) \to (\sim\neg\alpha \to \sim\neg\beta)$ (**CPL**)

4. $\vdash_{\mathbf{LTop}} (\sim\neg\alpha \to \sim\neg\beta)$ (2,3 **MP**)

(vi) Since $\vdash_{\mathbf{LTop}} ((\alpha \to \beta) \wedge \alpha) \to \beta$ then $\vdash_{\mathbf{LTop}} \Box((\alpha \to \beta) \wedge \alpha) \to \Box\beta$, by (v). The result is an immediate consequence of $\Box((\alpha \to \beta) \wedge \alpha) \equiv \Box((\alpha \to \beta)) \wedge \Box\alpha$ (by item (iv)), Replacement and the properties of **CPL**. □

Remark 7.21 *Observe that properties (ii), (iii) and (vi) of the last proposition correspond to the well-known modal axioms (**T**), (4) and (**K**), respectively. By its turn, property (v), together with (i), produce the modal necessitation rule. In fact, as we shall see in Section 7.9,* **LTop** *coincides with modal logic* **S4** *up to language. Additionally, the logical properties (i)-(v) of the connective □ stated in the last proposition reflect, when interpreted in topological structures, the basic properties of an interior operator. Specifically, they state the following properties, for all subsets A, B of a topological space X: (i) $Int(X) = X$; (ii) $Int(A) \subseteq A$; (iii) $Int(A) = Int(Int(A))$; (iv) $Int(A \cap B) = Int(A) \cap Int(B)$; and (v) if $A \subseteq B$ then $Int(A) \subseteq Int(B)$.*

7.6 LTop as a Logic of Formal Inconsistency

The *Logics of Formal Inconsistency* (**LFIs**) where introduced by W. Carnielli and J. Marcos in [8], and additionally studied in [7, 6] (among others). They are paraconsistent logics (that is, logics with a non-explosive negation ¬) and with a *consistency* operator which allows to recover the explosion law w.r.t. ¬ in a non-trivial way. In formal terms:[1]

Definition 7.22 *Let $\mathbf{L} = \langle \Theta, \vdash \rangle$ be a Tarskian, finitary and structural logic defined over a propositional signature Θ, which contains a negation ¬, and let ∘ be a (primitive or defined) unary connective. Then,* **L** *is said to be a* Logic of Formal Inconsistency *(an* **LFI**, *for short) with respect to ¬ and ∘ if the following holds:*

(i) $\varphi, \neg\varphi \nvdash \psi$ *for some* φ *and* ψ*;*

(ii) there are two formulas α *and* β *such that*

(ii.a) $\circ(\alpha), \alpha \nvdash \beta$*;*

(ii.b) $\circ(\alpha), \neg\alpha \nvdash \beta$*;*

[1]This is a simplified version of the definition of **LFIs**. In the general definition, a non-empty set $\bigcirc(p)$ of formulas depending exactly on the propositional variable p is considered, instead of a connective $\circ(p)$.

(iii) $\circ(\varphi), \varphi, \neg\varphi \vdash \psi$ *for every* φ *and* ψ.

Condition (ii) of the definition of **LFIs** is required in order to satisfy condition (iii) (called *gentle explosion law*) in a non-trivial way. Examples of consistency operators defined in **LTop** violating condition (ii) will be given in Remark 7.30.

Consider the following proposal for a consistency operator in **LTop** w.r.t. the negation $\neg$:

$$\text{(cons)} \quad \circ\alpha \overset{\text{def}}{=} \Diamond\alpha \to \Box\alpha \equiv \Box{\sim}\alpha \vee \Box\alpha = {\sim}\neg{\sim}\alpha \vee {\sim}\neg\alpha.$$

Let **L** be an **LFI**. An *inconsistency* operator $\bullet$ can also be defined, which satisfies the following: $\vdash (\alpha \wedge \neg\alpha) \to \bullet\alpha$. In general, $\bullet$ can be defined in **L** in two ways: $\bullet\alpha \overset{\text{def}}{=} \neg\circ\alpha$, or $\bullet\alpha \overset{\text{def}}{=} {\sim}\circ\alpha$, where $\sim$ is a classical negation definable in **L**. In the case of **LTop**, these alternatives produce the following:

$$\text{(incons)} \quad \bullet\alpha \overset{\text{def}}{=} {\sim}\circ\alpha = {\sim}(\Diamond\alpha \to \Box\alpha) = {\sim}(\Diamond\alpha \to {\sim}\neg\alpha) = \Diamond\alpha \wedge \neg\alpha;$$

$$\text{(incons')} \quad \bullet'\alpha \overset{\text{def}}{=} \neg\circ\alpha \equiv \neg({\sim}\neg{\sim}\alpha \vee {\sim}\neg\alpha) \equiv \neg{\sim}(\neg{\sim}\alpha \wedge \neg\alpha) = \Diamond(\Diamond\alpha \wedge \neg\alpha).$$

Let $\langle\langle X, \tau\rangle, v\rangle$ be a topological model for **LTop**. Recall that $\partial A = \overline{A} \cap \overline{A^c}$ and $Ext(A) = Int(A^c)$ denote the *boundary* and the *exterior* of a set $A \subseteq X$. Thus, the operators $\circ$, $\bullet$ and $\bullet'$ for consistency and inconsistency in **LTop** are semantically characterized as follows (by using Corollary 7.15):

$$\text{(v-cons)} \quad v(\circ\alpha) = v(\Box{\sim}\alpha \vee \Box\alpha) = Ext(v(\alpha)) \cup Int(v(\alpha));$$

$$\text{(v-incons)} \quad v(\bullet\alpha) = v(\Diamond\alpha \wedge \neg\alpha) = \overline{v(\alpha)} \cap \overline{v(\alpha)^c} = \partial v(\alpha);$$

$$\text{(v-incons')} \quad v(\bullet'\alpha) = v(\Diamond(\Diamond\alpha \wedge \neg\alpha)) = \overline{\overline{v(\alpha)} \cap \overline{v(\alpha)^c}}.$$

Given that the intersection of closed sets is a closed set, it follows that $\overline{v(\alpha)} \cap \overline{v(\alpha)^c} = \overline{\overline{v(\alpha)} \cap \overline{v(\alpha)^c}}$ and so $v(\bullet\alpha) = v(\bullet'\alpha)$. This means that, given the consistency operator proposed for **LTop**, there is just one inconsistency operator generated from it by means of the usual definitions. It is worth noting that $\circ$ and $\bullet$, as well as the negation $\neg$, have a nice interpretation in topological terms.

Remark 7.23 *Note that, for every topological model* $\mathcal{M}$,

$$v(\alpha \wedge \neg\alpha) = v(\alpha) \cap \overline{v(\alpha)^c} \subseteq \overline{v(\alpha)} \cap \overline{v(\alpha)^c} = v(\bullet\alpha)$$

and so $\bullet$ *satisfies the basic requirement for an inconsistency operator. The inclusion above is strict in general: let* $\mathcal{M} = \langle\langle \mathbb{R}, \tau\rangle, v\rangle$ *such that*

τ is the usual topology on the set $\mathbb{R}$ of real numbers, and $v(p) = (0,1)$, where p is a propositional variable. Then $v(\neg p) = (-\infty, 0] \cup [1, +\infty)$ and so $v(p \wedge \neg p) = v(p) \cap v(\neg p) = \emptyset$. On the other hand $\overline{v(p)} = [0,1]$ and so $v(\bullet p) = \{0,1\} \not\subseteq \emptyset = v(p \wedge \neg p)$. This means that **LTop** *separates the notions of $\neg$-contradiction and inconsistency: every $\neg$-contradiction is an inconsistency but the converse is not always true. In logical terms:*[2]

$$\vdash_{\mathbf{LTop}} (\alpha \wedge \neg\alpha) \rightarrow \bullet\alpha \quad \textit{but} \quad \nvdash_{\mathbf{LTop}} \bullet\alpha \rightarrow (\alpha \wedge \neg\alpha).$$

In order to prove that **LTop** is an **LFI** w.r.t. $\neg$ and $\circ$, we begin by proving the following:

Lemma 7.24 *Let p and q be two different propositional variables. Then:*

(i) $p, \neg p \nvdash_{\mathbf{LTop}} q$;

(ii) $\circ p, p \nvdash_{\mathbf{LTop}} q$;

(iii) $\circ p, \neg p \nvdash_{\mathbf{LTop}} q$.

Proof. Let $\mathcal{M} = \langle\langle \mathbb{R}, \tau\rangle, v\rangle$ such that τ is the usual topology on the set $\mathbb{R}$ of real numbers, $v(p) = [0,1)$ and $v(q) = (2,3)$.
(i) Since $v(p)^c = (-\infty, 0) \cup [1, +\infty)$ it follows that $v(\neg p) = (-\infty, 0] \cup [1, +\infty)$. Hence $v(p \wedge \neg p) = v(p) \cap v(\neg p) = \{0\} \not\subseteq v(q)$.
(ii) Note that $Ext(v(p)) = (-\infty, 0) \cup (1, +\infty)$. Given that $Int(v(p)) = (0,1)$ it follows that $v(\circ p) = Ext(v(p)) \cup Int(v(p)) = \mathbb{R} \setminus \{0,1\}$. Hence, $v(\circ p) \cap v(p) = (0,1) \not\subseteq v(q)$.
(iii) From (i) and (ii) it follows that $v(\circ p) \cap v(\neg p) = (\mathbb{R} \setminus \{0,1\}) \cap ((-\infty, 0] \cup [1, +\infty)) = (-\infty, 0) \cup (1, +\infty) \not\subseteq v(q)$. □

The previous lemma shows that $\neg$ and $\circ$ satisfy in **LTop** properties (i) and (ii) of Definition 7.22. In order to prove that condition (iii) of such definition is also satisfied, it is necessary to prove some previous results in **LTop**.

Lemma 7.25 *If* $\vdash_{\mathbf{LTop}} \alpha \rightarrow \gamma$ *and* $\vdash_{\mathbf{LTop}} \beta \rightarrow \gamma$ *then* $\vdash_{\mathbf{LTop}} (\alpha \vee \beta) \rightarrow \gamma$.

Proof. Assume that $\vdash_{\mathbf{LTop}} \alpha \rightarrow \gamma$ and $\vdash_{\mathbf{LTop}} \beta \rightarrow \gamma$. By using **CPL**, we also have that $\vdash_{\mathbf{LTop}} (\alpha \rightarrow \gamma) \rightarrow ((\beta \rightarrow \gamma) \rightarrow ((\alpha \vee \beta) \rightarrow \gamma)$. The result follows by (**MP**). □

[2] Of course this holds after proving the completeness of **LTop** w.r.t. topological models, see Theorem 7.36 below.

Lemma 7.26 $\vdash_{\mathbf{LTop}} \Box\alpha \to (\alpha \to (\neg\alpha \to \beta))$.

Proof. By **CPL**, it is enough to shown that $\vdash_{\mathbf{LTop}} (\Box\alpha \wedge \alpha \wedge \neg\alpha) \to \beta$. Thus, it will shown that $\vdash_{\mathbf{LTop}} ({\sim}\neg\alpha \wedge \alpha \wedge \neg\alpha) \to \beta$. Consider the following (meta)derivation in **LTop**:

1. $\vdash_{\mathbf{LTop}} ({\sim}\neg\alpha \wedge \alpha \wedge \neg\alpha) \to ({\sim}\neg\alpha \wedge \neg\alpha)$ (**CPL**)
2. $\vdash_{\mathbf{LTop}} ({\sim}\neg\alpha \wedge \neg\alpha) \to \beta$ (**CPL**)
3. $\vdash_{\mathbf{LTop}} ({\sim}\neg\alpha \wedge \alpha \wedge \neg\alpha) \to \beta$ (1, 2, **CPL**)

□

Similarly, it can be proven the following:

Lemma 7.27 $\vdash_{\mathbf{LTop}} \Box{\sim}\alpha \to (\alpha \to (\neg\alpha \to \beta))$.

Finally:

Lemma 7.28 $\vdash_{\mathbf{LTop}} \circ\alpha \to (\alpha \to (\neg\alpha \to \beta))$.

Proof. Recall that $\circ\alpha \equiv \Box{\sim}\alpha \vee \Box\alpha$. The result follows from lemmas 7.26, 7.27 and 7.25, and by Replacement. □

Proposition 7.29 **LTop** *is an* **LFI** *w.r.t.* $\neg$ *and* $\circ$.

Proof. It follows from lemmas 7.24 and 7.28. □

Remark 7.30 *Let* $\circ'\alpha \stackrel{\text{def}}{=} \Box\alpha$ *and* $\circ''\alpha \stackrel{\text{def}}{=} \Box{\sim}\alpha$. *Then, lemmas 7.26 and 7.27 state that these unary operators satisfy item (iii) of Definition 7.22 of* **LFI***s. Since item (i) of that definition is also satisfied (because* **LTop** *is* $\neg$*-paraconsistent) one wonders if* $\circ'$ *and* $\circ'$ *could be considered as alternative consistency operators in* **LTop** *w.r.t.* $\neg$. *However, it is easy to see that they are trivial, in the sense that property (ii) of Definition 7.22 fails for both of them. Indeed,* $v(\circ'\alpha \wedge \neg\alpha) = \emptyset$ *in every topological model and so condition (ii.b) fails for* $\circ'$. *On the other hand,* $v(\circ''\alpha \wedge \alpha) = \emptyset$ *in every topological model, hence condition (ii.a) fails for* $\circ''$. *Being so, the gentle exlosion law is satisfied by both operators in a trivial way.*

It is worth noting that Marcos suggested in [13] the following definition of a consistency operator inside a modal logic: $\oslash\alpha \stackrel{\text{def}}{=} \alpha \to \Box\alpha$. *Considered in* **LTop**, *this formula defines a consistency operator in the sense of Definition 7.22. The inconsistency operators naturally associated to it are* $\bullet\alpha = {\sim}\oslash\alpha = \alpha \wedge \neg\alpha$ *(where inconsistency is identified with contradiction) and* $\bullet'\alpha = \neg\oslash\alpha = \overline{\alpha \wedge \neg\alpha}$ *(where inconsistency and contradiction are diferent notions).*

7.7 *Intermezzo*: a Problem on Kuratowski Operators

In this section a technical result (Proposition 7.31) concerning the definition of a closure operator from a given colletion of sets enjoying some properties will be given. This result will be used in the proof of completeness of **LTop** w.r.t. topological models, in Section 7.8.

Let $X \neq \emptyset$ be a set. Let $\mathcal{B} \subseteq \wp(X)$ be a collection of subsets of X such that: (i) $\emptyset \in \mathcal{B}$ and $X \in \mathcal{B}$; and (ii) if $F, G \in \mathcal{B}$ then $F \cup G \in \mathcal{B}$. Now, let $\widehat{(\cdot)} : \mathcal{B} \to \mathcal{B}$ be a mapping such that:

1. $\widehat{\emptyset} = \emptyset$;
2. $F \subseteq \widehat{F}$ for every $F \in \mathcal{B}$;
3. $\widehat{F \cup G} = \widehat{F} \cup \widehat{G}$ for every $F, G \in \mathcal{B}$;
4. $\widehat{\widehat{F}} = \widehat{F}$ for every $F \in \mathcal{B}$.

Note that $\widehat{X} = X$, from 2. From 3 it follows that $\widehat{(\cdot)}$ is monotonic: $F \subseteq G$ implies $\widehat{F} \subseteq \widehat{G}$, for every $F, G \in \mathcal{B}$. We ask wether it is possible to extend $\widehat{(\cdot)}$ to a Kuratowski closure operator $\overline{(\cdot)} : \wp(X) \to \wp(X)$ over X (that is, satisfying properties 1-4 above for every $F, G \in \wp(X)$). Next result give us a positive answer.

Proposition 7.31 *Let $\mathcal{B}$ and $\widehat{(\cdot)}$ as above. The operator $\overline{(\cdot)} : \wp(X) \to \wp(X)$ given by $\overline{A} = \bigcap\{\widehat{F} \ : \ F \in \mathcal{B}$ and $A \subseteq \widehat{F}\}$, is a Kuratowski closure operator over X such that $\overline{F} = \widehat{F}$ if $F \in \mathcal{B}$.*

Proof. First, it is easy to see that $\overline{F} = \widehat{F}$ if $F \in \mathcal{B}$. In fact, given $F \in \mathcal{B}$ then $F \subseteq \widehat{F}$ and so $\overline{F} = \bigcap\{\widehat{G} \ : \ G \in \mathcal{B}$ and $F \subseteq \widehat{G}\} \subseteq \widehat{F}$. On the other hand, if $G \in \mathcal{B}$ and $F \subseteq \widehat{G}$ then $\widehat{F} \subseteq \widehat{\widehat{G}} = \widehat{G}$, and so $\widehat{F} \subseteq \overline{F}$.

This means that $\overline{(\cdot)}$ extends the operator $\widehat{(\cdot)}$ to $\wp(X)$. We will prove now that $\overline{(\cdot)}$ is a Kuratowski closure operator.
(1) From the observation above, $\overline{\emptyset} = \widehat{\emptyset} = \emptyset$.
(2) Clearly $A \subseteq \overline{A}$.
(3) Let $A, B \subseteq X$. Then

$$\overline{A} \cup \overline{B} = \left(\bigcap_{A \subseteq \widehat{F}} \widehat{F}\right) \cup \left(\bigcap_{B \subseteq \widehat{G}} \widehat{G}\right) = \bigcap \mathcal{F}$$

such that $\mathcal{F} = \{\widehat{F} \cup \widehat{G} \ : \ F, G \in \mathcal{B}; \ A \subseteq \widehat{F} \text{ and } B \subseteq \widehat{G}\}$. On the other hand, $\overline{A \cup B} = \bigcap \mathcal{G}$ such that $\mathcal{G} = \{\widehat{H} \ : \ H \in \mathcal{B} \text{ and } A \cup B \subseteq \widehat{H}\}$. Since $\widehat{F \cup G} = \widehat{F} \cup \widehat{G}$ then $\mathcal{F} \subseteq \mathcal{G}$. On the other hand, let $\widehat{H} \in \mathcal{G}$. Then $\widehat{H} \in \mathcal{B}$ is such that $A \subseteq \widehat{H}$ and $B \subseteq \widehat{H}$ and so $\widehat{H} = \widehat{H} \cup \widehat{H} \in \mathcal{F}$. From this, $\mathcal{G} \subseteq \mathcal{F}$. Therefore, $\overline{A \cup B} = \overline{A} \cup \overline{B}$.
(4) By definition, $\overline{\overline{A}} = \bigcap \mathcal{F}$ such that $\mathcal{F} = \{\widehat{F} \ : \ F \in \mathcal{B} \text{ and } \overline{A} \subseteq \widehat{F}\}$. By its turn, $\overline{A} = \bigcap \mathcal{G}$ such that $\mathcal{G} = \{\widehat{G} \ : \ G \in \mathcal{B} \text{ and } A \subseteq \widehat{G}\}$. Let $\widehat{G} \in \mathcal{G}$. Then $G \in \mathcal{B}$ such that $\overline{A} \subseteq \widehat{G}$ and so $\widehat{G} \in \mathcal{F}$. This means that $\mathcal{G} \subseteq \mathcal{F}$ and so $\overline{\overline{A}} = \bigcap \mathcal{F} \subseteq \bigcap \mathcal{G} = \overline{A}$. □

7.8 Completeness Theorem for the Logic LTop

Recall that, given a logic **L** and a formula α in the language $\mathbb{L}$ of **L**, a set $\Delta \subseteq \mathbb{L}$ is α-saturated in **L** if: (1) $\Delta \nvdash_{\mathbf{L}} \alpha$; and (2) if $\beta \notin \Delta$ then $\Delta, \beta \vdash_{\mathbf{L}} \alpha$. It follows that α-saturated sets are closed non-trivial theories of **L**, that is: $\Delta \neq \mathbb{L}$, and $\beta \in \Delta$ iff $\Delta \vdash_{\mathbf{L}} \beta$. The following classical result will be useful:

Theorem 7.32 (Lindenbaum-Łos) *Let* **L** *be a Tarskian and finitary logic over a language* $\mathbb{L}$. *Let* $\Gamma \cup \{\alpha\}$ *be a set of formulas of* $\mathbb{L}$ *such that* $\Gamma \nvdash_{\mathbf{L}} \alpha$. *Then, it is possible to extend* Γ *to an* α*-saturated set* Δ *in* **L**.

Proof. See [19, Theorem 22.2]. □

Since the logic **LTop** is Tarskian and finitary (see Remark 7.4) then the last theorem holds for it. Additionally, α-saturated sets in **LTop** satisfy the following properties:

Proposition 7.33 *Let* Δ *be an* α*-saturated set in* **LTop**. *Then:*
(i) $(\beta \to \beta) \in \Delta$ *and* ${\sim}\neg(\beta \to \beta) \in \Delta$ *for every* β;
(ii) $(\beta \to \gamma) \in \Delta$ *iff* $\beta \notin \Delta$ *or* $\gamma \in \Delta$;
(iii) ${\sim}\beta \in \Delta$ *iff* $\beta \notin \Delta$;
(iv) $(\beta \wedge \gamma) \in \Delta$ *iff* $\beta \in \Delta$ *and* $\gamma \in \Delta$;
(v) $(\beta \vee \gamma) \in \Delta$ *iff* $\beta \in \Delta$ *or* $\gamma \in \Delta$;
(vi) $\beta \notin \Delta$ *implies* $\neg\beta \in \Delta$.

Now, let $X_c = \{\Delta \subseteq \mathbb{L} \ : \ \Delta$ is an α-saturated set in **LTop** for some $\alpha \in \mathbb{L}\}$. For every $\varphi \in \mathbb{L}$ let $F_\varphi = \{\Delta \in X_c \ : \ \varphi \notin \Delta\}$. Observe that $F_\varphi = F_\psi$ whenever $\varphi \equiv \psi$. Let $\mathcal{B} = \{F_\varphi \ : \ \varphi \in \mathbb{L}\}$. Clearly:

(i) $\emptyset \in \mathcal{B}$ since $\emptyset = F_{(\varphi\to\varphi)}$, by Proposition 7.33(i); additionally, $X_c \in \mathcal{B}$ since $X_c = F_{\sim(\varphi\to\varphi)}$ (recalling that α-saturated sets are non-trivial theories);

(ii) $F_\varphi \cup F_\psi = F_{(\varphi\wedge\psi)} \in \mathcal{B}$, by Proposition 7.33(iv).

Notice that $X_c \setminus F_\varphi = F_{\sim\varphi}$, by Proposition 7.33(iii). Consider now a mapping $\widehat{(\cdot)} : \mathcal{B} \to \mathcal{B}$ defined as follows: $\widehat{F_\varphi} = F_{\sim\neg\varphi} = \{\Delta \in X_c : \neg\varphi \in \Delta\}$. Observe the following:

(i) $\widehat{\emptyset} = \emptyset$. In fact: $\widehat{\emptyset} = \widehat{F_{(\varphi\to\varphi)}} = F_{\sim\neg(\varphi\to\varphi)} = \emptyset$, by Proposition 7.33(i).

(ii) $F_\varphi \subseteq \widehat{F_\varphi}$. In fact: if $\Delta \in F_\varphi$ then $\varphi \notin \Delta$ and so $\neg\varphi \in \Delta$, by Proposition 7.33(vi). Thus $\Delta \in \widehat{F_\varphi}$.

(iii) $\widehat{F_\varphi \cup F_\psi} = \widehat{F_\varphi} \cup \widehat{F_\psi}$. In fact: $\widehat{F_\varphi} \cup \widehat{F_\psi} = F_{\sim\neg\varphi} \cup F_{\sim\neg\psi} = F_{(\sim\neg\varphi\wedge\sim\neg\psi)} = F_{\sim\neg(\varphi\wedge\psi)} = \widehat{F_{(\varphi\wedge\psi)}} = \widehat{F_\varphi \cup F_\psi}$, by Proposition 7.20(iv).

(iv) $\widehat{\widehat{F_\varphi}} = \widehat{F_\varphi}$. In fact: $\widehat{\widehat{F_\varphi}} = \widehat{F_{\sim\neg\varphi}} = F_{\sim\neg\sim\neg\varphi} = F_{\sim\neg\varphi} = \widehat{F_\varphi}$, by Proposition 7.20(iii).

Then, by Proposition 7.31, the operator $\overline{(\cdot)} : \wp(X_c) \to \wp(X_c)$ given by

$$\overline{A} = \bigcap\{F_{\sim\neg\varphi} \ : \ A \subseteq F_{\sim\neg\varphi}\}$$

is a Kuratowski closure operator over X_c such that $\overline{F_\varphi} = \widehat{F_\varphi} = F_{\sim\neg\varphi}$.

Definition 7.34 *The* canonical structure *for* **LTop** *is the topological space $\mathcal{M}_c = \langle X_c, \tau_c\rangle$ such that τ_c is the topology over X_c generated by the Kuratowski closure $\overline{(\cdot)}$ defined above. The* canonical model *for* **LTop** *is the model $\langle \mathcal{M}_c, v_c\rangle$ such that $v_c(p) = F_{\sim p} = \{\Delta \in X_c \ : \ p \in \Delta\}$, for every $p \in \mathcal{V}$.*

Notice that, since $X_c \setminus F_\varphi = F_{\sim\varphi}$, then each F_φ is clopen in $\mathcal{M}_c$.

Lemma 7.35 (Truth Lemma) *For every $\varphi \in \mathbb{L}$ it holds:*

$$v_c(\varphi) = F_{\sim\varphi} = \{\Delta \in X_c \ : \ \varphi \in \Delta\}.$$

Proof. Since it was observed, $F_{\sim\varphi} = \{\Delta \in X_c \ : \ \varphi \in \Delta\}$. By induction on the complexity of φ it will be proven that $v_c(\varphi) = F_{\sim\varphi}$.
If $\varphi = p \in \mathcal{V}$ then the result holds by definition of v_c.
If $\varphi = (\psi \to \gamma)$ then $v_c(\varphi) = v_c(\psi \to \gamma) = (X_c \setminus v_c(\psi)) \cup v_c(\gamma)$, by definition of v_c. By induction hypothesis and Proposition 7.33, the latter is equal to $(X_c \setminus F_{\sim\psi}) \cup F_{\sim\gamma} = F_{\sim\sim\psi} \cup F_{\sim\gamma} = F_\psi \cup F_{\sim\gamma} = F_{(\psi\wedge\sim\gamma)} = F_{\sim(\psi\to\gamma)} = F_{\sim\varphi}$.

If $\varphi = {\sim}\psi$ then $v_c(\varphi) = v_c({\sim}\psi) = (X_c \setminus v_c(\psi))$, by definition of v_c. By induction hypothesis it is equal to $(X_c \setminus F_{\sim\psi}) = F_{\sim\sim\psi} = F_{\sim\varphi}$.
If $\varphi = \neg\psi$ then $v_c(\varphi) = v_c(\neg\psi) = \overline{(X_c \setminus v_c(\psi))}$, by definition of v_c. By induction hypothesis and the definition of closure it is equal to $\overline{(X_c \setminus F_{\sim\psi})} = \overline{F_{\sim\sim\psi}} = \overline{F_\psi} = \widehat{F_\psi} = F_{\sim\neg\psi} = F_{\sim\varphi}$. □

Theorem 7.36 (Completeness of LTop) *For every set of formulas* $\Gamma \cup \{\varphi\}$ *it holds:* $\Gamma \models_{\mathbf{LTop}} \varphi$ *implies* $\Gamma \vdash_{\mathbf{LTop}} \varphi$.

Proof. Suppose that $\Gamma \nvdash_{\mathbf{LTop}} \varphi$, and assume that $\Gamma \neq \emptyset$. By Theorem 7.32, there exists a φ-saturated set Δ in **LTop** such that $\Gamma \subseteq \Delta$. Consider the canonical model $\langle \mathcal{M}_c, v_c \rangle$. Since $\Gamma \subseteq \Delta$ then $\Delta \in v_c(\gamma)$ for every $\gamma \in \Gamma$, by Lemma 7.35, and so $\Delta \in \bigcap_{\gamma\in\Gamma} v_c(\gamma)$. On the other hand $\varphi \notin \Delta$ and so $\Delta \notin v_c(\varphi)$, by Lemma 7.35. Thus $\bigcap_{\gamma\in\Gamma} v_c(\gamma) \not\subseteq v_c(\varphi)$. This shows that $\Gamma \not\models_{\mathbf{LTop}} \varphi$. The case $\Gamma = \emptyset$ is proved analogously. □

The latter result can be applied to the question of defining unary connectives in **LTop**, which can be solved semantically in terms of the Kuratowski's closure-complement problem. Indeed, a theorem due to Kuratowski states that there are at most 14 distinct sets obtainable by iterations of closure (k) and complement (c) to a given subset of X. As a consequence of this, and by soundness and completeness of **LTop** w.r.t. topological models, there are only 14 unary conectives definable in **LTop** by iterations of $\neg$ and $\sim$. They are listed in the table below (as usual, the composition of operators must be read from right to left, and so, for instance, kc stands for c followed by k).

Operator	Connective
id	$\sim\sim$
c	$\sim$
k	$\neg\sim$
ck	$\sim\neg\sim$
kc	$\neg$
ckc	$\sim\neg$
kck	$\neg\neg\sim$
ckck	$\sim\neg\neg\sim$
kckc	$\neg\neg$
ckckc	$\sim\neg\neg$
kckck	$\neg\neg\neg\sim$
ckckck	$\sim\neg\neg\neg\sim$
kckckc	$\neg\neg\neg$
ckckckc	$\sim\neg\neg\neg$

7.9 LTop is S4 up to Language

In this section it will be shown that **LTop** is nothing more than the well-known modal logic **S4**, presented in a different (non-modal) language. In order to do this, two translation mappings $* : \mathbb{L}_4 \to \mathbb{L}$ and $\otimes : \mathbb{L} \to \mathbb{L}_4$ will be defined, where $\mathbb{L}_4$ is the set of formulas of **S4**, satisfying the following: (1) $\Theta \vdash_{\mathbf{S4}} \beta$ iff $\Theta^* \vdash_{\mathbf{LTop}} \beta^*$ for every $\Theta \cup \{\beta\} \subseteq \mathbb{L}_4$; and (2) $\Gamma \vdash_{\mathbf{LTop}} \alpha$ iff $\Gamma^\otimes \vdash_{\mathbf{S4}} \alpha^\otimes$ for every $\Gamma \cup \{\alpha\} \subseteq \mathbb{L}$, where $\vdash_{\mathbf{S4}}$ denotes the consequence relation of **S4** (observe that derivations in **S4** from non-empty sets of premisses are defined in a similar way as in **LTop**). Here, $\Theta^* = \{\gamma^* \ : \ \gamma \in \Theta\}$ and γ^* denotes $*(\gamma)$, for every $\gamma \in \mathbb{L}_4$. A similar notation is adopted for the translation mapping $\otimes$. Moreover, $\beta \equiv_4 (\beta^*)^\otimes$ and $\alpha \equiv (\alpha^\otimes)^*$ for every $\beta \in \mathbb{L}_4$ and $\alpha \in \mathbb{L}$, where $\beta \equiv_4 \gamma$ means that $\vdash_{\mathbf{S4}} \beta \to \gamma$ and $\vdash_{\mathbf{S4}} \gamma \to \beta$. From this result, it can be said that **LTop** coincides with **S4** 'up to translations' or 'up to language'.

Previous to prove this result, it will be shown that the modal *Necessitation rule* ('if α is a theorem then $\Box\alpha$ is a theorem') is admissible in **LTop** (where $\Box\varphi$ denotes, as stated above, the formula ${\sim}\neg\varphi$ of $\mathbb{L}$). This means that adding this rule to **LTop** does not add any new theorem to the resulting logic.

Theorem 7.37 (Admissibility in LTop of the Necessitation rule) *Consider, in the language* $\mathbb{L}$ *of* **LTop**, *the Necessitation rule:*

$$\textbf{(NEC)}\ \frac{\alpha}{\Box\alpha}$$

(where, as stated above, $\Box\alpha$ *denotes* ${\sim}\neg\alpha$*). Then,* **(NEC)** *is admissible in* **LTop**, *that is: if* $\vdash_{\mathbf{LTop}} \varphi$ *then* $\vdash_{\mathbf{LTop}} \Box\varphi$.

Proof. Suppose that $\vdash_{\mathbf{LTop}} \varphi$. By soundness of **LTop** with respect to topological structures, it follows that $\models_{\mathbf{LTop}} \varphi$. This means that $v(\varphi) = X$ for every model $\langle\langle X, \tau\rangle, v\rangle$. But then $v(\neg\varphi) = \overline{v(\varphi)^c} = \overline{\emptyset} = \emptyset$ and so $v(\Box\varphi) = v({\sim}\neg\varphi) = v(\neg\varphi)^c = X$, for every model $\mathcal{M}$. Thus $\models_{\mathbf{LTop}} \Box\varphi$ and so $\vdash_{\mathbf{LTop}} \Box\varphi$, by the completeness theorem. $\Box$

Now, the logic **S4** will be formally analyzed. Recall from Remark 7.5 that the Hilbert calculus formed by axioms **(Ax1)**, **(Ax2)** and **(Ax3)** of **LTop** plus the rule **(MP)** constitutes a Hilbert calculus $\mathbf{H_{CPL}}$ for **CPL** over the signature $\{\sim, \to\}$.

Definition 7.38 (Modal logic S4) *Consider the signature*

$$\Sigma_\Box = \{\Box, \sim, \to\}$$

and let $\mathbb{L}_4$ be the language generated by the set $\mathcal{V}$ of propositional variables (recall Definition 7.1) over the signature $\Sigma_\Box$. The modal logic **S4** *is defined over the language $\mathbb{L}_4$ by adding to* $\mathbf{H_{CPL}}$ *the following axioms and rules:*

$$\Box(\alpha \to \beta) \to (\Box\alpha \to \Box\beta) \tag{K}$$

$$\Box\alpha \to \alpha \tag{T}$$

$$\Box\alpha \to \Box\Box\alpha \tag{4}$$

$$\frac{\alpha}{\Box\alpha} \quad \textit{(NEC)}$$

Derivations (without and with premisses) are defined in **S4** in a similar way as in **LTop** (recall Definition 7.3), as usual in modal systems. It is well-known (consult, for instance, [5]) that the Hilbert calculus presented in Definition 7.38, together with its notion of derivations, is adequate for the modal logic **S4**.

If $\beta, \gamma \in \mathbb{L}_4$ then $\beta \equiv_4 \gamma$ will mean that $\vdash_{\mathbf{S4}} \beta \to \gamma$ and $\vdash_{\mathbf{S4}} \gamma \to \beta$.

Now, the two translation mappings will be defined.

Definition 7.39 *Let $* : \mathbb{L}_4 \to \mathbb{L}$ be a mapping defined recursively as follows (here, $*(\gamma)$ will be denoted by γ^*, for every $\gamma \in \mathbb{L}_4$):*

(i) $p^* = p$ *if p is a propositional variable;*

(ii) $(\Box\gamma)^* = {\sim}\neg(\gamma^*)$;

(iii) $({\sim}\gamma)^* = {\sim}(\gamma^*)$;

(iv) $(\gamma \to \delta)^* = (\gamma^*) \to (\delta^*)$.

Definition 7.40 *Let $\otimes : \mathbb{L} \to \mathbb{L}_4$ be a mapping defined recursively as follows (here, $\otimes(\gamma)$ will be denoted by $\gamma^{\otimes}$, for every $\gamma \in \mathbb{L}$):*

(i) $p^{\otimes} = p$ *if p is a propositional variable;*

(ii) $(\neg\gamma)^{\otimes} = {\sim}\Box(\gamma^{\otimes})$;

(iii) $({\sim}\gamma)^{\otimes} = {\sim}(\gamma^{\otimes})$;

(iv) $(\gamma \to \delta)^{\otimes} = (\gamma^{\otimes}) \to (\delta^{\otimes})$.

Proposition 7.41 *For every $\beta \in \mathbb{L}_4$, $\beta \equiv_4 (\beta^*)^{\otimes}$.*

Proof. The proof is done by induction on the complexity of $\beta \in \mathbb{L}_4$ (which is defined as usual). By definition of $*$, it is enough to analyze the induction step when $\beta = \Box\gamma$. Then, we have that $\beta^* = {\sim}\neg(\gamma^*)$. From this, by definition of $\otimes$, by Replacement of **S4** and by induction hypothesis (from which $\gamma \equiv_4 (\gamma^*)^{\otimes}$),

$$(\beta^*)^{\otimes} = {\sim}{\sim}\Box((\gamma^*)^{\otimes}) \equiv_4 \Box((\gamma^*)^{\otimes}) \equiv_4 \Box\gamma = \beta.$$

That is, $\beta \equiv_4 (\beta^*)^{\otimes}$. □

Proposition 7.42 *For every* $\alpha \in \mathbb{L}$, $\alpha \equiv (\alpha^{\otimes})^*$.

Proof. The proof is analogous to that of Proposition 7.41, but now the only case to be analyzed is when $\alpha = \neg\delta$. The details are left to the reader □

Proposition 7.43 *For every* $\beta \in \mathbb{L}_4$, $\vdash_{\mathbf{S4}} \beta$ *implies that* $\vdash_{\mathbf{LTop}} \beta^*$.

Proof. Consider the Hilbert calculus for **S4** presented above. It is immediate to see, by the results about **LTop** proved in the previous sections, that any instance β of an axiom of **S4** is such that β^* is a theorem of **LTop**. On the other hand, by definition of $*$, it is immediate that (**MP**) is translated as itself by $*$. Finally, the translation of (**NEC**) by $*$ is an admissible rule in **LTop**, by Theorem 7.37. From this, it is easy to prove, by induction on the length of a derivation of β in **S4**, that $\vdash_{\mathbf{S4}} \beta$ implies that $\vdash_{\mathbf{LTop}} \beta^*$. The details are left to the reader. □

Proposition 7.44 *For every* $\alpha \in \mathbb{L}$, $\vdash_{\mathbf{LTop}} \alpha$ *implies that* $\vdash_{\mathbf{S4}} \alpha^{\otimes}$.

Proof. The proof uses an argument analogous to that of Proposition 7.43. Let us begin by observing that the translation by $\otimes$ of axioms (**Ax1**), (**Ax2**) and (**Ax3**) of **LTop** are instances of the same axioms in **S4**. The same happens with the rule (**MP**). Now, observe that the translation by $\otimes$ of axioms (**Ax4**) and (**Ax5**) of **LTop** corresponds to axioms (**T**) and (**4**) of **S4**, but written in terms of the possibility operator. Then, they are derivable in **S4**. The translation by $\otimes$ of any instance axiom (**Ax6**) of **LTop** has the form

$${\sim}\Box(\gamma \wedge \delta) \rightarrow ({\sim}\Box\gamma \vee {\sim}\Box\delta).$$

The latter is equivalent, by **CPL**, to $(\Box\gamma \wedge \Box\delta) \rightarrow \Box(\gamma \wedge \delta)$, which is a theorem of **S4**. The translation by $\otimes$ of any instance of axiom (**Ax7**) of **LTop** has the form $\Box(\gamma \rightarrow \gamma)$, which is derivable in **S4**. The rule (**DR**) is translated by $\otimes$ as itself, being clearly a derived rule in the Hilbert calculus of **S4**. Finally, the translation by $\otimes$ of the rule (**CR**) has the form $\gamma \to \delta / {\sim}\Box\delta \to {\sim}\Box\gamma$. Suppose that $\vdash_{\mathbf{S4}} \gamma \to \delta$. Then $\vdash_{\mathbf{S4}} \Box\gamma \to \Box\delta$ (this is a well known fact of **S4**) and so, by **CPL**, $\vdash_{\mathbf{S4}} {\sim}\Box\delta \to {\sim}\Box\gamma$. This proves that the translation by $\otimes$ of the rule (**CR**) is an admissible rule in **S4**, concluding the proof. □

Corollary 7.45 *Let* $\alpha \in \mathbb{L}$ *and* $\beta \in \mathbb{L}_4$. *Then:*

(i) $\vdash_{\mathbf{LTop}} \alpha$ *iff* $\vdash_{\mathbf{S4}} \alpha^{\otimes}$.

(ii)) $\vdash_{\mathbf{S4}} \beta$ *iff* $\vdash_{\mathbf{LTop}} \beta^{*}$.

Proof.

(i) If $\vdash_{\mathbf{LTop}} \alpha$ then $\vdash_{\mathbf{S4}} \alpha^{\otimes}$, by Proposition 7.44. Conversely, if $\vdash_{\mathbf{S4}} \alpha^{\otimes}$ then $\vdash_{\mathbf{LTop}} (\alpha^{\otimes})^{*}$, by Proposition 7.43. The result follows by Proposition 7.42.

(ii) It is proved analogously, by using Proposition 7.41 in the last step.

□

Corollary 7.46 *Let* $\Gamma \cup \{\alpha\} \subseteq \mathbb{L}$ *and* $\Theta \cup \{\beta\} \subseteq \mathbb{L}_4$. *Then:*

(i) $\Gamma \vdash_{\mathbf{LTop}} \alpha$ *iff* $\Gamma^{\otimes} \vdash_{\mathbf{S4}} \alpha^{\otimes}$.

(ii) $\Theta \vdash_{\mathbf{S4}} \beta$ *iff* $\Theta^{*} \vdash_{\mathbf{LTop}} \beta^{*}$.

Proof. It is an immediate consequence from Corollary 7.45, the definition of the respective consequence relation in both logics, and the fact that both translations preserve the connectives of **CPL**. The details are left to the reader. □

The last result, together with propositions 7.41 and 7.42, justify the claim that **LTop** and **S4** define the same logic, up to language.

7.10 Encoding Intuitionistic Logic inside LTop

Finally, it will be shown that **LTop** has the expressive power to encode Intuitionistic Propositional Logic **IPL** by means of a conservative translation. This is not surprising, since **LTop** is **S4** up to translations, as it was proven in the previous section. On the other and, it was proved by McKinsey-Tarski in [15] that Gödel's translation T of **IPL** into **S4** is conservative for valid formulas, that is: α is intuitionistically valid iff $T(\alpha)$ is derivable in **S4**.

In order to define a conservative translation from **IPL** to **LTop**, consider the signature $\Sigma_{int} = \{\sim, \rightarrow, \wedge, \vee\}$ for **IPL** and let $\mathbb{L}_{\mathbf{IPL}}$ be the language generated by the set $\mathcal{V}$ of propositional variables (recall Definition 7.1) over the signature Σ_{int}. The consequence relation of **IPL**, denoted by $\vdash_{\mathbf{IPL}}$, can be axiomatized by a Hilbert calculi over $\mathbb{L}_{\mathbf{IPL}}$ (see, for instance, [15]). It is well known that $\vdash_{\mathbf{IPL}}$ is semantically characterized by a topological semantics given by topological models $\mathcal{M} = \langle \mathcal{T}, v_0 \rangle$ such that $\mathcal{T} = \langle X, \tau \rangle$ is a topological space and $v_0 \colon \mathbb{L}_{\mathbf{IPL}} \rightarrow \tau$ is a valuation satisfying the following conditions:

1. $v_0(\sim\alpha) = Int(v_0(\alpha)^c)$;
2. $v_0(\alpha \rightarrow \beta) = Int(v_0(\alpha)^c \cup v_0(\beta))$;
3. $v_0(\alpha \wedge \beta) = v_0(\alpha) \cap v_0(\beta)$;
4. $v_0(\alpha \vee \beta) = v_0(\alpha) \cup v_0(\beta)$.

It generates a semantical consequence as follows: $\models_{\mathbf{IPL}} \alpha$ iff $v_0(\alpha) = X$ for every topological model $\langle \langle X, \tau \rangle, v_0 \rangle$. Moreover, $\Gamma \models_{\mathbf{IPL}} \alpha$ iff either $\models_{\mathbf{IPL}} \alpha$ or there exists a finite non-empty set $\Gamma_0 \subseteq \Gamma$ such that $\bigcap_{\gamma \in \Gamma_0} v_0(\gamma) \subseteq v_0(\alpha)$, for every topological model $\langle \langle X, \tau \rangle, v_0 \rangle$. By definition, $\emptyset \models_{\mathbf{IPL}} \alpha$ iff $\models_{\mathbf{IPL}} \alpha$. Consider now the following translation mapping:

Definition 7.47 *Let* $\odot : \mathbb{L}_{\mathbf{IPL}} \rightarrow \mathbb{L}$ *be a mapping defined recursively as follows:*

(i) $p^{\odot} = \Box p = \sim\neg p$ *if* p *is a propositional variable;*

(ii) $(\sim\gamma)^{\odot} = \Box\sim(\gamma^{\odot}) = \sim\neg\sim(\gamma^{\odot})$;

(iii) $(\gamma \rightarrow \delta)^{\odot} = \Box((\gamma^{\odot}) \rightarrow (\delta^{\odot})) = \sim\neg((\gamma^{\odot}) \rightarrow (\delta^{\odot}))$;

(iv) $(\gamma \wedge \delta)^{\odot} = (\gamma^{\odot}) \wedge (\delta^{\odot})$;

(v) $(\gamma \vee \delta)^{\odot} = (\gamma^{\odot}) \vee (\delta^{\odot})$.

Now, the desired result is the following:

Theorem 7.48 *Let* $\Gamma \cup \{\varphi\} \subseteq \mathbb{L}_{\mathbf{IPL}}$. *Then:*

$$\Gamma \vdash_{\mathbf{IPL}} \varphi \quad \textit{iff} \quad \Gamma^{\odot} \vdash_{\mathbf{LTop}} \varphi^{\odot}.$$

Proof. ('Only if' part) Suppose that $\Gamma \vdash_{\mathbf{IPL}} \varphi$ such that $\Gamma \neq \emptyset$, and let $\langle\langle X, \tau\rangle, v\rangle$ be a topological model for **LTop**. It is easy to prove that, for every $\alpha \in \mathbb{L}_{\mathbf{IPL}}$, $v(\alpha^{\odot}) \in \tau$. Thus, there is a mapping $v_0 : \mathbb{L}_{\mathbf{IPL}} \to \tau$ given by $v_0(\alpha) \stackrel{\text{def}}{=} v(\alpha^{\odot})$ for every α. It can be seen that v_0 is a valuation for **IPL** whence $\langle\langle X, \tau\rangle, v_0\rangle$ is a topological model for **IPL**. By soundness of **IPL** w.r.t. topological models it follows that $\bigcap_{\gamma\in\Gamma} v_0(\gamma) \subseteq v_0(\varphi)$. That is, $\bigcap_{\gamma\in\Gamma} v(\gamma^{\odot}) \subseteq v(\varphi^{\odot})$. This means that $\Gamma^{\odot} \models_{\mathbf{LTop}} \varphi^{\odot}$ and so $\Gamma^{\odot} \vdash_{\mathbf{LTop}} \varphi^{\odot}$, by completeness of **LTop** w.r.t. topological models. The case when $\vdash_{\mathbf{IPL}} \varphi$ is proved analogously.

('If' part) Suppose that $\Gamma^{\odot} \vdash_{\mathbf{LTop}} \varphi^{\odot}$ such that $\Gamma \neq \emptyset$, and let $\langle\langle X, \tau\rangle, v_0\rangle$ be a topological model for **IPL**. Define a mapping $v_1 : \mathcal{V} \to \wp(X)$ such that $v_1(p) = v_0(p)$ for every propositional variable p. Extend v_1 to a valuation $v : \mathbb{L} \to \wp(X)$ for **LTop** by using the clauses of Definition 7.12. It is easy to prove that $v_0(\alpha) = v(\alpha^{\odot})$ for every $\alpha \in \mathbb{L}_{\mathbf{IPL}}$. Let $\langle\langle X, \tau\rangle, v\rangle$ be the resulting topological model for **LTop**. By hypothesis and soundness of **LTop** w.r.t. topological models, $\bigcap_{\gamma\in\Gamma} v(\gamma^{\odot}) \subseteq v(\varphi^{\odot})$ and so $\bigcap_{\gamma\in\Gamma} v_0(\gamma) \subseteq v_0(\varphi)$. This shows that $\Gamma \models_{\mathbf{IPL}} \varphi$, and then $\Gamma \vdash_{\mathbf{IPL}} \varphi$ by completeness of **IPL** w.r.t. topological models. The case when $\vdash_{\mathbf{LTop}} \varphi^{\odot}$ is proved analogously. □

The latter result shows that $\odot$ is a conservative translation from **IPL** to **LTop**.

Acknowledgments

The first author has been supported by FAPESP (Thematic Project LogCons 2010/51038-0), and by a research grant from CNPq (PQ 308524/2014-4).

Bibliography

[1] Can Başkent. Some topological properties of paraconsistent models. *Synthese*, **190**(18):4023–4040, 2013.

[2] Jean-Yves Beziau. **S5** is a paraconsistent logic and so is first-order classical logic. *Logical Studies*, **9**:301–309, 2002.

[3] Jean-Yves Beziau. Paraconsistent Logic from a Modal Viewpoint. *Journal of Applied Logic*, **3**(1):7–14, 2005.

[4] Jean-Yves Beziau. The paraconsistent logic **Z**: a possible solution to Jaśkowski's problem. *Logic and Logical Philosophy*, **15**:99–111, 2006.

[5] Patrick Blackburn, Maarten de Rijke and Yde Venema. *Modal Logic.* Volume 53 of *Cambridge Tracts in Theoretical Computer Science* series. Cambridge University Press, 2002.

[6] Walter A. Carnielli and Marcelo E. Coniglio. *Paraconsistent Logic: Consistency, Contradiction and Negation.* Volume 40 of *Logic, Epistemology, and the Unity of Science* series. Springer, 2016.

[7] Walter A. Carnielli, Marcelo E. Coniglio, and João Marcos. Logics of Formal Inconsistency. In: D. M. Gabbay and F. Guenthner, editors, *Handbook of Philosophical Logic (2nd. edition)*, volume 14, pages 1–93. Springer, 2007.

[8] Walter A. Carnielli and João Marcos. A taxonomy of C-systems. In: W. A. Carnielli, M. E. Coniglio, and I. M. L. D'Ottaviano, editors, *Paraconsistency: The Logical Way to the Inconsistent*, volume 228 of *Lecture Notes in Pure and Applied Mathematics*, pages 1–94. Marcel Dekker, 2002.

[9] Ana L. de Araújo, Elias H. Alves, and José A. D. Guerzoni. Some relations between modal and paraconsistent logic. *The Journal of Non-Classical Logic*, **4**(2):33–44, 1987. Available at `http://www.cle.unicamp.br/jancl/`.

[10] Nicolas D. Goodman. The Logic of Contradiction. *Zeitschrift für Mathematische Logik und Grundlagen der Mathematik*, **27**(8-10):119–126, 1981.

[11] Stanisław Jaśkowski. Rachunek zdań dla systemów dedukcyjnych sprzecznych (in Polish). *Studia Societatis Scientiarun Torunesis – Sectio A*, **I**(5):55–77, 1948. Translated to English as "A propositional calculus for inconsistent deductive systems". *Logic and Logic Philosophy*, **7**:35–56, 1999. Proceedings of the Stanisław Jaśkowski's Memorial Symposium, held in Toruń, Poland, July 1998.

[12] Tamar Lando. Completeness of **S4** for the Lebesgue Measure Algebra. *Journal of Philosophical Logic*, **41**(2):287–316, 2012.

[13] João Marcos. Nearly every normal modal logic is paranormal. *Logique et Analyse*, **48**(189-192):279–300, 2005.

[14] John Charles C. McKinsey and Alfred Tarski. The Algebra of Topology. *Annals of Mathematics*, **45**:141–91, 1944.

[15] John Charles C. McKinsey and Alfred Tarski. Some theorems about the sentential calculi of Lewis and Heyting. *The Journal of Symbolic Logic*, **13**(1):1–15, 1948.

[16] Chris Mortensen. Topological Separation Principles and Logical Theories. *Synthese*, **125**(1-2):169–178, 2000.

[17] Graham Priest. Dualising Intuitionistic Negation. *Principia*, **13**(3):165–84, 2009.

[18] Johan van Benthem and Guram Bezhanishvili. Modal Logics of Space. In Marco Aiello, Ian Pratt-Hartmann and Johan Van Benthem, editors, *Handbook of Spatial Logics*, pages 217–298. Springer-Verlag, Dordrecht, 2007.

[19] Ryszard Wójcicki. *Lectures on propositional calculi.* Ossolineum, Wroclaw, Poland, 1984.

[20] Ryszard Wójcicki. *Theory of logical calculi: Basic theory of consequence operations*, volume 199 of *Synthese Library.* Kluwer Academic Publishers, Dordrecht, 1988.

8

PARACONSISTENT PROBABILITIES, THEIR SIGNIFICANCE AND THEIR USES

Walter Carnielli
Centre for Logic, Epistemology and the History of Science and
Department of Philosophy
State University of Campinas UNICAMP, Campinas, SP, Brazil
walter.carnielli@cle.unicamp.br

Juliana Bueno-Soler
Faculty of Technology
State University of Campinas UNICAMP, Limeira, SP, Brazil
juliana@ft.unicamp.br

Ao querido Amílcar, com saudades.

Abstract

This paper surveys and advances the notion of paraconsistent probability theory based on the Logics of Formal Inconsistency (LFIs), what makes possible to treat realistic probabilistic reasoning under contradictions. Some appropriate notions of conditional probability and paraconsistent updating are explored through versions of Bayes' theorem for conditionalization. Eight challenging problems with different degrees of difficulty and different motivations are proposed here as an invitation to further investigation into non-standard probability theories.

8.1 Where logic meets probability, and how

The question of the relationship between logic and probability has been debated for more than three centuries. Leibniz believed that probabilities could help the legal standpoint, in what he called a "new kind of logic", but he did not have the relevant mathematical tools to develop it at his time. To this tradition, one can add Jacobus Bernoulli, Johann Heinrich Lambert, Bernard Bolzano, and Augustus De Morgan among others. George Boole presented a calculus of probabilities based on his logical calculus in his book *The Laws of Thought*. The famous economist John M. Keynes pursued this tradition in [Key21] with his logical theory of probability and even claimed that some probabilities are not measurable (that is, fail to have a numeric value), thus opening the possibility for a non-classical logic about probability (a way, however, Keynes did not take). Frank Ramsey considered the theory of probability as a branch of logic, but enough different from Keynes' viewpoint; in [Ram26] echoes of Leibniz's ideas are found. Rudolf Carnap ([Car50]) was also interested in assigning probabilities to sentences in a formal language, having conditional probabilities in sight. This tradition justifies the view of regarding probabilities as a generalization of standard logic, as R. Jeffrey argues in [Jef92]. From the purely logical point of view, a probability assignment can be regarded as the measure of the degree to which a set of sentences support a conclusion, thus committing the concept of probability to the notions of deduction and argumentation.

The connections between logic and probability were also emphasized by Bruno de Finetti in [dF37], who professed a notion of probabilistic coherence very close to (classical) logic consistency. Indeed, as shown in [DES09] the logical consistency of a collection of assertions about events can be viewed as a particular case of coherent probability assessments in the sense of de Finetti.

A new approach to the probabilization of propositional logic is given in [RSS16], where a novel notion of stochastic valuation and its main properties are studied. The main result is the proof of the equivalence of assigning probabilities to formulas and assigning probabilities to valuations. Such novel approach can, in principle, be extended to the probabilization of non-standard propositional logics.

The view of probability as linked to deduction and argumentation has more philosophical appeal, but it clashes with the approach of probability measures, or probability on sets, a more recent tradition followed by physicists and engineers. In 1933 Andrei Kolmogorov introduced the idea of probabilities as defined by mathematical structures, akin for instance to topological spaces, in his classic book in German translated as [Kol56], an approach he called "elementary theory of probability". This approach is widely used by mathematicians, statisticians and engineers.

One of the difficulties of the Kolmogorovian approach is the axiom of countable additivity for probability, introduced by Émile Borel, which has always been controversial: If $\{A_i\}$ is a countable (finite or infinite) collection of (pairwise) disjoint events, then the probability of $\cup\{A_i\}$ is the sum of the individual probabilities of $\{A_i\}$. Even if used in measure-theoretic probability by Kolmogorov, he admitted it to be merely a mathematical convenience, as explained in his *Grundbegriffe*: he viewed countable additivity not as meaningful applications, but just as a mathematical aid.

For de Finneti (cf. [dF74]) as well, infinite additivity was considered a mathematical expedient, not strictly justified by the concept of probability. As it is well known, de Finetti regarded the notion of countable additivity to be circular (see [dF72], pp. 91-92). On the other hand, for probability as linked to deductions, this is not any significant restriction given the compactness property of classical logic.

The divide between probability on sentences and probability on sets is also furthered by certain hidden philosophical principles that engineers and other users sometimes apply unconsciously. For instance, the *Principle of Insufficient Reason* proposed by Jacob Bernoulli and Pierre Simon Laplace in the 18th century says that if there are a finite number of propositions and a state of knowledge according to which none of the propositions is more plausible than any other, then all of them have the same probability summing 1 (conditional on that state of knowledge). This principle is referred to by Keynes in chapter 4 of [Key21] as the *Principle of Indifference.* So, for instance, if you do not know the color of a certain book, you are entitled, according to the Principle of Indifference, to assign probability 1/2 to both propositions "The book is blue" and "The book is not blue", and the same for red or yellow. No contradictions arise between such assumptions because different colors mean different states of knowledge.

As a rule for assigning epistemic probabilities, the Principle of Indifference becomes more debatable if there is an infinite number of events, and if it is conjugated with the axiom of countable additivity for probability.

Another metaphysical principle that is taken for granted in the probabilistic praxis, this one emphasized in the philosophy of the 17th- and 18th-century by Leibniz, is the *Principle of Sufficient Reason.* It claims that necessarily every contingent truth has an explanation or simply 'nothing is without reason'. Reasons are needed for calculating chances: if there are no reasons, then there are no explanations, and *a fortiori* no way of calculating probabilities in any non-arbitrary way. As Leibniz is known to have said, "My metaphysics is all mathematics". This view, combined with his views on probability as a 'new kind of logic', makes entirely feasible to think on probability pluralism, in much a parallel way as we think of logical pluralism.

But again, if probability pluralism makes perfect sense when considering probabilities on sentences, as done here, it makes less sense, if any, when considering probabilities on sets or random fields: the latter depends much less on logic directly, and much more on mathematical structures. Of course, it is not impossible to imagine new mathematical theories based on non-standard logics (we already have available paraconsistent set theories, see e.g [CC16], chapter 8), but this is far from the mathematical reality.

For instance, a stochastic model that describes phenomena evolving over time or space can be directly generalized by parameterizing time and space variables by elements of a topological structure. Mathematicians do not need (yet) any non-classical topological spaces–what it all means is that, if we expand the notions of probabilities on sentences, what can be done quite naturally as will be sketched in the next section, the companion theory of probabilities on sets (a.k.a. Kolmogorov probability spaces) will not be naturally generalizable. If, in the classical case, they are equivalent under certain conditions, in the non-standard cases the divide will be deepened.

8.2 Evidence: an epistemological key to understanding paraconsistency

Paraconsistency is the investigation of logic systems endowed with a negation $\neg$ such that contradictions of the form α and $\neg\alpha$ do not entail everything ("explode", in pictorial terms). In other words, a paraconsistent logic does not suffer from trivialism, the undesired effects of contradictions in trivializing the deductive machinery of the system. Trivialism can be regarded as an allergic reaction to contradictions, an exaggerated 'genetically determined' response to an 'offending' agent represented by a pair of contradictory statements. The medical analogy makes sense since even an irrelevant contradiction in traditional logic obliges a classical reasoner to derive anything from a contradictory pair $\{\alpha, \neg\alpha\}$, by abiding the so-called Principle of Explosion:

(PEx) $\alpha, \neg\alpha, \vdash \beta$, for arbitrary β, while a paraconsistent logician, by using a more cautious way of reasoning, is free of the burden of (PEx) and could pause to investigate the causes for the contradiction.

Paraconsistent logic, following the biological analogy, is like a beneficial genetic modification of logic [1] or like the immune response of a vaccinated against specific diseases: the virus of contradiction do not cause the fever of trivialism. This does not mean, however, that the virus of contradiction is welcome, or desirable, or logically derivable: paraconsistent logics do not derive any contradictions, they just can stand contradictory hypothesis and pause to investigate their causes instead of foolishly deriving unwanted consequences from them.

The Principle of Explosion may be rejected for good epistemological reasons, rejecting the thesis that there are true contradictions. Indeed, the plethora of contradictions that occur in several contexts of reasoning do not mean necessarily that a given proposition α and its negation $\neg\alpha$ are true, nor that α is both true and false. Contradictions in empirical sciences are routine and considered to be almost unavoidable in empirical sciences. But no scientist falls down into trivialism.

An epistemic view on contradictions, following [CR16] and [CCR], is supported by the notion of *evidence*, where 'there is evidence that α is true' means that 'there are some reasons for believing that α is true'. Among them, the notions of *conflicting evidence* and *conclusive evidence* deserves particular attention as illuminating cases for understanding paraconsistent logics, and especially paraconsistent probabilities.

If we understand 'evidence that α is true' as reasons for believing in the truth of α, and 'evidence that α is false' as reasons for believing in the falsity of α, then:

1. 'α holds' means 'there is evidence that α is true';

2. 'α does not hold' means 'there is no evidence that α is true';

3. '$\neg\alpha$ holds' means 'there is evidence that α is false';

4. '$\neg\alpha$ does not hold' means 'there is no evidence that α is false'.

A *Basic Logic of Evidence*, *BLE*, proposed in [CR16] (see also [CCR]), suggests the following scenarios describing a proposition α on what concerns its evidence content:

1. There is only evidence that α is true: α holds, $\neg\alpha$ does not hold.

[1]Without the dubious effect of GMO in human nutrition, that are suspicious of causing new allergic reactions.

2. There is only evidence that α is false: $\neg\alpha$ holds, α does not hold.

3. There is conflicting evidence: both α and $\neg\alpha$ hold.

The case of total lack of evidence (neither α nor $\neg\alpha$ hold), though relevant in some cases, does not apply in our analysis: the logic **Ci** a member of the hierarchy of the Logics of Formal Inconsistency (LFIs) which will be analyzed below, is precisely the logic of rational situations where there is always evidence for or against something. As we will see, this interpretation fits well with the notions of 'assertion' and 'denial' to be defined within **Ci**.

Another feature of **Ci** is to represent the fact that some contradictions cannot be tolerated because they refer to something so basic that it would destroy the very act of reasoning (that is, lead to trivialization). This amounts to recognizing that not all contradictions are equivalent. **Ci** as an LFI, is designed to express the notion of consistency (and inconsistency as well) within the object language by employing a connective $\circ$ (reading $\circ\alpha$ as "α is consistent"), embodies the above intuitions.

LFIs extend classical logic in the sense that classicality may be recovered in the presence of consistency: contradictions involving consistent sentences will lead to explosive triviality. Consistency in the LFIs is not regarded as synonymous with freedom from contradiction (the traditional notion of consistency of a theory T, for instance, is taken to mean that there is no sentence α such that $T \vdash \alpha$ and $T \vdash \neg\alpha$, where $\vdash$ is a specific consequence relation in the language of T). Such a notion of consistency defined over negation may be suitable for mathematical purposes, but not for the whole enterprise of reasoning (see [Car11], and [Wil81]). In the LFIs the notions of consistency and non-contradiction are not coincident, nor are the notions of inconsistency and contradiction the same.

What grants to the LFIs their characteristics of generality is that the Principle of Explosion is not valid in general, although this principle is not abolished, but restricted to consistent sentences. Therefore, a contradictory theory may be non-trivial, unless the contradiction refers to something consistent.

Such features of the LFIs are condensed in the following law, which is referred to as the "Principle of Gentle Explosion":

(PGE) $\quad \circ\alpha, \alpha, \neg\alpha \vdash \beta$, for every β , although $\alpha, \neg\alpha \nvdash \beta$, for some β.

8.3 On Ci and its expressability

This section delineates some relevant properties of a particular LFI, the system **Ci**, that will be employed to establish a useful paraconsistent probability theory.

The logic **Ci** is defined by the following propositional axioms and the rule of Modus Ponens (**MP**):

Definition 8.1 *Let Σ be a propositional signature. The logic* **Ci** *(over Σ) is defined by the following Hilbert calculus:*

Axiom Schemas

Ax1. $\alpha \to (\beta \to \alpha)$

Ax2. $(\alpha \to \beta) \to ((\alpha \to (\beta \to \gamma)) \to (\alpha \to \gamma))$

Ax3. $\alpha \to (\beta \to (\alpha \wedge \beta))$

Ax4. $(\alpha \wedge \beta) \to \alpha$

Ax5. $(\alpha \wedge \beta) \to \beta$

Ax6. $\alpha \to (\alpha \vee \beta)$

Ax7. $\beta \to (\alpha \vee \beta)$

Ax8. $(\alpha \to \gamma) \to ((\beta \to \gamma) \to ((\alpha \vee \beta) \to \gamma))$

Ax9. $\alpha \vee (\alpha \to \beta)$

Ax10. $\alpha \vee \neg\alpha$

Ax11. $\circ\alpha \to (\alpha \to (\neg\alpha \to \beta))$

Ax12. $\neg\neg\alpha \to \alpha$

Ax13. $\neg \circ \alpha \to (\alpha \wedge \neg\alpha)$

Inference Rule

MP $\alpha, \;\; \alpha \to \beta \; / \; \beta$

In [BSC16] the logic **Cie** instead of **Ci** is employed as a basis for probability theory (but it is mistakenly called **Ci**). The difference between **Ci** and **Cie** is just that $\neg\neg\alpha \to \alpha$ holds in the former, while $\neg\neg\alpha \equiv \alpha$ holds in the later. Several choices are possible, and choosing the most adequate LFI to ground a paraconsistent probability theory is a matter of preference.

Axioms **Ax1**- **Ax9** plus **MP** define a Hilbert calculus for positive propositional classical logic (see [CCM07]), and therefore all the laws concerning positive logic (as distribution of $\wedge$ over $\vee$, etc.) are valid. Properties such as, for instance, distribution of conjunction over disjunction are easily established (where $\alpha \equiv \beta$ means $\alpha \vdash_{\mathbf{Ci}} \beta$ and $\beta \vdash_{\mathbf{Ci}} \alpha$):

Theorem 8.2

1. $\alpha \wedge (\beta \vee \gamma) \equiv (\alpha \wedge \beta) \vee (\alpha \wedge \gamma)$
2. $\alpha \vee (\beta \wedge \gamma) \equiv (\alpha \vee \beta) \wedge (\alpha \vee \gamma)$

Proof. Easy consequence of axioms **Ax4**, **Ax5**, and **Ax8**. □

As proved in [CCM07], **Ci** is not truth-functional, and cannot be semantically characterized by finite matrices, but it can be characterized in terms of valuations over $\{0, 1\}$, or *bivaluations*:

Definition 8.3 *Let $\mathcal{L}$ be the collection of sentences of* **Ci**. *A function $v : \mathcal{L} \to \{0, 1\}$ is a* valuation for **Ci** *if it satisfies the following clauses:*

(Biv1) $v(\alpha \wedge \beta) = 1 \iff v(\alpha) = 1$ *and* $v(\beta) = 1$

(Biv2) $v(\alpha \vee \beta) = 1 \iff v(\alpha) = 1$ *or* $v(\beta) = 1$

(Biv3) $v(\alpha \to \beta) = 1 \iff v(\alpha) = 0$ *or* $v(\beta) = 1$

(Biv4) $v(\alpha) = 0 \implies v(\neg\alpha) = 1$

(Biv5) $v(\circ\alpha) = 1 \implies v(\alpha) = 0$ *or* $v(\neg\alpha) = 0$.

(Biv6) $v(\neg\neg\alpha) = 1 \implies v(\alpha) = 1$.

(Biv7) $v(\neg\circ\alpha) = 1 \implies v(\alpha) = 1$ *and* $v(\neg\alpha) = 1$.

The semantical consequence relation w.r.t. valuations for **Ci** is defined as expected: $\Gamma \vDash_{\mathbf{Ci}} \varphi$ iff, for every valuation v for **Ci**, if $v(\gamma) = 1$ for every $\gamma \in \Gamma$ then $v(\varphi) = 1$.

Theorem 8.4 *(Completeness for* **Ci***) For every set of sentences $\Gamma \cup \{\varphi\} \subseteq \mathcal{L}$ in* **Ci***:*

$$\Gamma \vdash_{\mathbf{Ci}} \varphi \text{ if and only if } \Gamma \vDash_{\mathbf{Ci}} \varphi .$$

Proof. See [CCM07]. □

Remark 8.5

The notion of 'α is inconsistent' is defined as 'α is not consistent', employing the weak (paraconsistent) negation, as $\bullet\alpha := \neg \circ \alpha$. Since $\alpha \wedge \neg\alpha \to \neg\circ\alpha$ is easily seen to be provable in **Ci**, *this fact together with axiom* **Ax13** *grants that $\alpha \wedge \neg\alpha \equiv \neg \circ \alpha$. This implies that '$\alpha$ is not inconsistent if and only if it is consistent': indeed, $\neg\bullet\alpha \equiv \neg\neg\circ\alpha$, but by Theorem 3.40 of [CM02] and* **Ax12**, *$\neg\neg\circ\alpha \equiv \circ\alpha$, and thus $\neg\bullet\alpha \equiv \circ\alpha$. Clearly, by definition, it also holds 'α is not consistent if and only if it is inconsistent'. This fact is recorded in Theorem 8.7, item (1), to wit: in* **Ci** *the formula $\alpha\wedge\neg\alpha$ in the classical language of $\wedge$ and $\neg$ is equivalent to a formula that expresses inconsistency, albeit there is no formula in the classical language that can express consistency in* **Ci**, *as Theorem 8.7, items (2) and (3) clarify. Further details are discussed in [CCM07].*

A strong (classical) negation[2] can be defined as $\sim\alpha := \neg\alpha \wedge \circ\alpha$, and the following table can be easily obtained using Definition 8.3 and Remark 8.5:

α	$\neg\alpha$	$\circ\alpha$	$\neg\alpha\wedge\circ\alpha$
0	1	1	1
1	0	1	0
1	1	0	0

Several meta-theorems, as the Deduction Metatheorem, can be proved in **Ci**. Some relevant results in **Ci** hold as follows (where $\alpha \equiv \beta$ denotes that $\alpha \vdash \beta$ and $\beta \vdash \alpha$):

Theorem 8.6 *(Properties of strong negation) The strong negation $\sim$ satisfies the following properties in* **Ci** *(and, therefore, also in stronger logics):*

1. $\vdash \sim\alpha \to (\alpha \to \psi)$ *for every α and ψ;*
2. $\vdash \alpha \vee \sim\alpha$
3. $\vdash \alpha \to \sim\sim\alpha$ *and* $\vdash \sim\sim\alpha \to \alpha$
4. *If* $(\Gamma \vdash \alpha \to \gamma)$ *and* $(\Delta \vdash \sim\alpha \to \gamma)$ *then* $(\Gamma, \Delta \vdash \gamma)$
5. $\vdash (\alpha \to \beta) \to (\sim\beta \to \sim\alpha)$ *and so* $\alpha \to \beta \vdash \sim\beta \to \sim\alpha$

[2] As noted in [CCM07], this negation and infinitely many strong negations of the form $\sim_\beta\alpha := \alpha \to \bot_\beta$ can be defined in **Ci** and stronger systems, where $\bot_\beta = (\beta \wedge (\neg\beta \wedge \circ\beta))$ is a bottom formula, for any sentence β. All of them are equivalent.

6. $\vdash (\sim\alpha \to \sim\beta) \to (\beta \to \alpha)$ *and so* $\sim\alpha \to \sim\beta \vdash \beta \to \alpha$
7. $\vdash (\alpha \to \sim\beta) \to (\beta \to \sim\alpha)$
8. $\vdash (\varphi \to (\alpha \to \beta)) \to (\varphi \to (\sim\beta \to \sim\alpha))$
9. $\vdash (\varphi \to (\sim\alpha \to \sim\beta)) \to (\varphi \to (\beta \to \alpha))$
10. $\vdash (\varphi \to (\alpha \to \sim\beta)) \to (\varphi \to (\beta \to \sim\alpha))$
11. $\sim\alpha \to \beta \vdash \sim\beta \to \alpha$
12. $\vdash \sim(\alpha \to \beta) \to (\alpha \wedge \sim\beta)$
13. $\vdash \bot \to \alpha$

Proof. See [CCM07]. □

Theorem 8.7

1. $\vdash \bullet\alpha \equiv \alpha \wedge \neg\alpha$
2. $\circ\alpha \vdash \neg(\alpha \wedge \neg\alpha)$, *but the converse does not hold*
3. $\circ\alpha \vdash \neg(\neg\alpha \wedge \alpha)$, *but the converse does not hold*
4. $\vdash \circ\circ\alpha$
5. $\vdash \circ\alpha \equiv \circ(\neg\alpha)$
6. $\vdash \circ\bullet\alpha$
7. $\vdash \neg\bullet\circ\alpha$
8. $\vdash \neg\bullet\bullet\alpha$
9. $\vdash \circ\alpha \vee \alpha$
10. $\vdash \circ\alpha \vee \neg\alpha$
11. $\vdash \circ\alpha \vee \alpha \wedge \neg\alpha$
12. $\vdash \alpha \equiv \alpha \wedge (\beta \vee \neg\beta)$
13. $\vdash \circ\alpha \equiv (\alpha \wedge \circ\alpha) \vee (\neg\alpha \wedge \circ\alpha)$
14. $\vdash (\alpha \wedge \circ\alpha) \vee (\neg\alpha \wedge \circ\alpha) \vee \bullet\alpha$
15. *Only consistent or inconsistent sentences can themselves be provably consistent in* **Ci** *(that is,* $\circ\alpha$ *is a theorem of* **Ci** *if, and only if,* α *is of the form* $\circ\beta$ *or* $\bullet\beta$*).*

Proof. See [CCM07] and [CM02] □

Item 15 of the previous theorem is particularly interesting as it makes clear that consistency cannot be proved based on logical principles only: judgment and philosophy are always necessary.

As remarked, it can be proved that a sentence α is consistent in **Ci** (that is, $\circ\alpha$ holds) if, and only if, α and $\neg\alpha$ are not simultaneously true. Thus $\alpha \wedge \circ\alpha$ is true, precisely, when α is true *without* its negation being false, just check the above table (and *mutatis mutandis* for $\neg\alpha$). So we define:

(i) The *assertion* of α as $\oplus\alpha := \alpha \wedge \circ\alpha$.

(ii) The *denial* of α (equivalent to its strong negation) as ${\sim}\alpha := \neg\alpha \wedge \circ\alpha$.

The following results on assertions and denial in **Ci** can be easily proved and are relevant for using in probability theory:

Theorem 8.8

1. $\vdash \oplus\alpha \vee \sim \alpha \vee \bullet\alpha$
2. $\vdash \oplus(\circ\alpha) \equiv \circ\alpha$
3. $\vdash \oplus(\bullet\alpha) \equiv \bullet\alpha$
4. $\vdash \oplus(\neg\alpha) \equiv \sim \alpha$
5. $\vdash \oplus\alpha \equiv \oplus\alpha \wedge \alpha \equiv \oplus(\circ\alpha) \wedge \alpha$
6. $\vdash \sim \alpha \wedge \neg\alpha \equiv \sim \alpha$
7. $\oplus\alpha \wedge \neg\alpha$ *is a bottom particle*
8. ${\sim}\alpha \wedge \alpha$ *is a bottom particle*
9. $\oplus\alpha \wedge \neg \oplus \alpha$ *is not a bottom particle*
10. $\oplus\alpha \wedge \oplus(\neg\alpha)$ *is a bottom particle*
11. $\circ(\oplus\alpha)$ *is not a theorem*
12. $\circ({\sim}\alpha)$ *is not a theorem*

Proof. Consequences of Theorem 8.4 and Theorem 8.7. □

After such results, one can reaccess the mentioned evidence interpretation to **Ci** as follows (notice that the evidence contents for α and for $\neg\alpha$ are independent):

1. 'α is true' if there is partial evidence for α (or against $\neg\alpha$).
2. '$\neg\alpha$ is true' if there is partial evidence for $\neg\alpha$ (or against α).
3. '$\oplus\alpha$ is true' if there is conclusive evidence for α. In this case, $\neg\alpha$ cannot be true.
4. '$\sim\alpha$ is true' if there is conclusive evidence for $\neg\alpha$. In this case, α cannot be true.

Some probabilistic models demand probability distributions to first-order, rather than to propositional sentences. This is not any impediment to our approach, since **Ci** can be extended to the first-order logic **QCi** (over an appropriate extension of Σ) by adding appropriate axioms and rules (see [CCPR15]). For more details, conceptual motivations and the main results about LFIs the reader is referred to [CM02], [CCM07], and [CC16].

8.4 Consistency in logic and in probability

The LFIs embody an expanded notion of consistency that does not necessarily depend on negation. This notion of consistency is conceived as a primitive concept, whose meaning, on the light of the epistemological view on contradictions sketched in Section 8.2 can be thought of as "conclusively established as true (or false)", by extra-logical means, depending on the subject matter. Consistency, in this sense, is a notion independent of model-theoretical and proof-theoretical means. In this sense, it is closer to the idea of regularity or something contrary to change (see [CR15] and [BSC15]).

In the traditional philosophical views on probability, consistency is connected to complying with the laws of traditional probability. For instance, F. Ramsey in [Ram26] argues that degrees of belief should satisfy the probability axioms, and defends that this is connected to a notion of consistency or coherence. In this way, the notion of consistency (at least for degrees of belief) can be regarded as the satisfaction of the probability axioms.

This paper surveys and expands a new way to define a paraconsistent theory of probability introduced in [BSC16]. Previous approaches and mentions concerning non-classical logics and probabilities appear in [Mun06] and [Wil14], and for the paraconsistent case in [Mar97], [Pri87], [Rob00], and our development elaborates on all of them, sharpening the connections between probabilities and LFIs, by arguing that LFIs (and **Ci** in particular) offer a natural and intuitive extension of standard probabilities which is useful and philosophically meaningful.

Probability functions are usually defined for a σ-algebra of subsets of a given sample space Ω, but it is also natural to define probability functions directly for sentences in the object language. We will refer to them, respectively, as *probability on sets* versus *probability on sentences* (see discussion in Section 8.6).

These two approaches are equivalent in classical logic, but such an equivalence is not immediate for probability based on other logics, since the algebraic relationship between the two approaches is not obvious for non-classical logics, or even not yet established. Another point is that probability functions in set-theoretical settings are required to satisfy countable additivity, but since propositional language is compact, for probability on sentences it suffices to require finite additivity (see [BSC16], section 5, for a quick survey of the equivalence in the classical case).

Following [BSC16], paraconsistent probabilities are defined here directly on sentences, emphasizing the role of the cautious logic **Ci** behind probabilistic reasoning, and its effects on an extended version of Bayes' rule.

Definition 8.9 *A probability function for the language $\mathcal{L}$ of* **Ci** *is a real-valued function $P : \mathcal{L} \mapsto \mathbb{R}$ satisfying the following conditions, where $\vdash_{\mathbf{Ci}}$ stands for the derivability relation of* **Ci***:*

1. *Non-negativity:* $0 \leq P(\varphi) \leq 1$ *for all* $\varphi \in \mathcal{L}$
2. *Tautologicity: If* $\vdash_{\mathbf{Ci}} \varphi$*, then* $P(\varphi) = 1$
3. *Anti-Tautologicity: If* $\varphi \vdash_{\mathbf{Ci}}$*, then* $P(\varphi) = 0$
4. *Comparison: If* $\psi \vdash_{\mathbf{Ci}} \varphi$*, then* $P(\psi) \leq P(\varphi)$
5. *Finite additivity:* $P(\varphi \vee \psi) = P(\varphi) + P(\psi) - P(\varphi \wedge \psi)$

This set of meta-axioms defines an arbitrary class of probability functions, just by taking $\vdash_L$ for an appropriate logic L, instead of $\vdash_{\mathbf{Ci}}$, including the classical and the intuitionistic case (the latter treated in [Wea03]). Clearly, the concept of probability can be regarded as entirely logic-dependent, and the choice of the underlying logic is a matter of interest, convenience or philosophical views.

Definition 8.10 *(Paraconsistent probability distribution)*
Let $\Sigma = \{\alpha_1, \cdots, \alpha_n, \cdots\}$ be a collection of sentences in the language $\mathcal{L}$ of **Ci**. *A (***Ci***)-paraconsistent probability distribution over Σ is an assignment of probability values $P*$ to the elements of Σ such that $P*$ can be extended to a full probability function $P : \mathcal{L} \mapsto \mathbb{R}$ in the sense of Definition 8.9.*

Two events, expressed by sentences α and β, are said to be *independent* if $P(\alpha \wedge \beta) = P(\alpha) \cdot P(\beta)$. Two events can be independent (see Section 8.8 and Section 8.10 for a discussion) relative to one probability distribution[3] and dependent relative to another. A question related to this specific point is proposed in Problem 8.25.

Two events expressed by sentences α and β are said to be *logically incompatible* if $\alpha, \beta \vdash \varphi$, for any φ (or equivalently, if $\alpha \wedge \beta$ act as a bottom particle).

Intuitively, two events are independent if the occurrence of one of them does not have any effect on the occurrence of the other event, and vice-versa. Differently, two events are logically incompatible if the occurrence of one of them prohibits the occurrence of the other. Logically incompatible events α and β with non-zero probabilities are always *dependent* since $0 = P(\alpha \wedge \beta) \neq P(\alpha) \cdot P(\beta)$.

In **Ci** there are different forms of bottom particle, and as a consequence of Definition 8.9 it follows that $P(\alpha \wedge \neg\alpha \wedge \circ\alpha) = 0$, $P(\circ\alpha \wedge \bullet\alpha) = 0$, $P(\circ\alpha \wedge \neg\circ\alpha) = 0$ and $P(\bullet\alpha \wedge \neg\bullet\alpha) = 0$, for any probability function P.

It is to be noted that in **Ci** the events α and $\sim\alpha$ are always logically incompatible (and hence dependent), while events α and $\neg\alpha$ are independent. In a certain sense, $P(\bullet\alpha) = P(\alpha \wedge \neg\alpha)$ measures the 'degree of independence' between α and $\neg\alpha$.

Some simple calculation rules that substantiate such intuitions follow:

Theorem 8.11

1. $P(\alpha \vee \beta) = P(\alpha) + P(\beta)$, *if α and β are logically incompatible.*

2. $$P(\alpha \vee \beta \vee \gamma) = P(\alpha) + P(\beta) + P(\gamma) - P(\alpha \wedge \beta) - P(\alpha \wedge \gamma) \\ -P(\beta \wedge \gamma) + P(\alpha \wedge \beta \wedge \gamma)$$

3. $P(\oplus\alpha) + P(\sim\alpha) + P(\bullet\alpha) = 1$

[3]Probability distributions are also referred to as probability measures.

4. $P(\circ\alpha) = 2 - (P(\alpha) + P(\neg\alpha))$
5. $P(\circ\alpha) = P(\alpha \wedge \circ\alpha) + P(\neg\alpha \wedge \circ\alpha)$
6. $P(\circ\alpha \wedge \beta) = P(\alpha \wedge \circ\alpha \wedge \beta) + P(\neg\alpha \wedge \circ\alpha \wedge \beta)$
7. $P(\alpha \wedge \neg\alpha) = P(\alpha) + P(\neg\alpha) - 1$
8. $P({\sim}\alpha) = P(\neg\alpha \wedge \circ\alpha) = 1 - P(\alpha)$
9. $P(\neg\alpha) = 1 - P(\alpha \wedge \circ\alpha)$
10. $P(\bullet\alpha) = P(\neg \circ \alpha) = 1 - P(\circ\alpha)$
11. $P(\oplus\alpha) = P(\circ\alpha) + P(\alpha) - 1$
12. $P({\sim}\alpha) = P(\circ\alpha) + P(\neg\alpha) - 1$
13. $P(\circ\alpha) = P(\oplus\alpha) + P({\sim}\alpha)$

Proof. Only items (10), (11), (12), and (13) will be proved (the rest is routine):

(10) follows from the fact that $\vdash_{\mathbf{Ci}} \circ\alpha \vee \neg \circ \alpha$, Tautologicity, Finite Additivity, and Theorem 8.7 item (4).

(11) follows from Tautologicity, Finite Additivity, and Theorem 8.7, item (9).

(12) follows from Tautologicity, Finite Additivity, and Theorem 8.7, item (10). It also follows by summing up equations (2), (8), and (10) above.

(13) follows by summing up equations (8) and (11) above.

□

An important property concerning the behavior of consistent and inconsistent events in extreme cases (i.e., when $P(\circ\alpha) = 1$ or $P(\bullet\alpha) = 1$) is that, in those cases, they are independent of any other distinct event, and any other distinct event such that $P(\beta) = 1$ is independent of them:

Lemma 8.12 *(Independence of consistent and inconsistent events in extreme cases)*

1. *If* $P(\circ\alpha) = 1$ *then* $P(\circ\alpha \wedge \beta) = P(\beta)$, *for* $\beta \neq \circ\alpha$
2. *If* $P(\bullet\alpha) = 1$ *then* $P(\bullet\alpha \wedge \beta) = P(\beta)$, *for* $\beta \neq \bullet\alpha$
3. *If* $P(\beta) = 1$ *then* $P(\circ\alpha \wedge \beta) = P(\circ\alpha)$, *for* $\beta \neq \circ\alpha$
4. *If* $P(\beta) = 1$ *then* $P(\bullet\alpha \wedge \beta) = P(\bullet\alpha)$, *for* $\beta \neq \bullet\alpha$

Proof. For item (1): If $P(\circ\alpha) = 1$ then $P(\circ\alpha \vee \beta) = 1$ from **Ax6** and Tautologicity, since $\circ\alpha \vdash_{\mathbf{Ci}} \circ\alpha \vee \beta$. By Finite Additivity $1 = P(\circ\alpha \vee \beta) = P(\circ\alpha) + P(\beta) - P(\circ\alpha \wedge \beta)$. As $P(\circ\alpha) = 1$, it follows that $P(\circ\alpha \wedge \beta) = P(\beta)$. The argument for items (2), (3) and (4) is similar. □

The restriction in the clauses of the above lemma are intended to avoid the problematic cases of 'self-independence' of extreme events discussed in Section 8.8.

Probabilities are sometimes seen as generalized truth-values. The so-called *probabilistic semantics* in this way replace the valuations $v : L \mapsto \{0, 1\}$ of classical propositional logic with the probability functions ranging on the real unit interval $[0, 1]$, and valuations can be seen as degenerate probability functions. In this sense, classical logic is to be regarded as a special case of probability logic. We show below that an analogous property holds for **Ci** where paraconsistent valuations can be seen as degenerate probability functions. To this end, define $\Vdash_P$ as a probabilistic semantic relation whose meaning is $\Gamma \Vdash_P \varphi$ if and only if for every probability function P, if $P(\psi) = 1$ for every $\psi \in \Gamma$ then $P(\varphi) = 1$. It can be shown that **Ci** is (strongly) sound and complete with respect to such probabilistic semantics, that is: $\Gamma \vdash \varphi$ if and only if $\Gamma \Vdash_P \varphi$ (see [BSC16] for details).

8.5 Paraconsistent conditionalization and Bayes' rule

The notion of conditional probability of α given β is defined as usual, for $P(\beta) \neq 0$, as follows (but see Section 8.8 for alternatives):

$$P(\alpha/\beta) = \frac{P(\alpha \wedge \beta)}{P(\beta)}$$

Some useful theorems on conditional paraconsistent probability are the following, with the *caveat* that $P(\beta) \neq 0$ in all cases where $P(\alpha/\beta)$ is mentioned:

Theorem 8.13

1. $P(\alpha/\beta \wedge \gamma) = \frac{P(\alpha/\gamma).P(\beta/\alpha\wedge\gamma)}{P(\beta/\gamma)}$.
2. $P(\alpha \wedge \beta/\gamma) = P(\alpha/\gamma).P(\beta/\alpha \wedge \gamma) = P(\beta/\gamma).P(\alpha/\beta \wedge \gamma)$.
3. $P(\alpha/\beta) + P(\neg\alpha/\beta) - P(\bullet\alpha/\beta) = 1$.
4. $P(\alpha/\beta) + P(\neg\alpha/\beta) = 1$ *if* $P(\circ\alpha) = 1$.

5. $P(\alpha \vee \beta/\gamma) = P(\alpha/\gamma) + P(\beta/\gamma) - P(\alpha \wedge \beta/\gamma)$.

6. $P(\alpha \vee \beta/\gamma) = P(\alpha/\gamma) + P(\beta/\gamma)$ *if* α *and* β *are logically incompatible.*

7. $P(\beta/\circ\beta) + P(\neg\beta/\circ\beta) = 1$.

8. *(Total Probability Theorem, paraconsistent form)*
$P(\alpha) = P(\alpha/\beta \wedge \circ\beta).P(\beta \wedge \circ\beta) + P(\alpha/\neg\beta \wedge \circ\beta).P(\neg\beta \wedge \circ\beta) + P(\alpha/\bullet\beta).P(\bullet\beta)$.

9. $P(\alpha/\circ\alpha \wedge \beta) + P(\neg\alpha/\circ\alpha \wedge \beta) = 1$

10. *(Chain Rule)*

$$P(\alpha_1 \wedge \cdots \wedge \alpha_n) = P(\alpha_1/\alpha_2 \wedge \cdots \wedge \alpha_n) \cdots P(\alpha_{n-1}/\alpha_n)P(\alpha_n)$$

Proof. Proofs follow from Definition 8.9, Remark 8.5, Theorem 8.7 and Theorem 8.11. □

The standard Bayes' theorem proves that, for $P(\beta) \neq 0$:

$$P(\alpha/\beta) = \frac{P(\beta/\alpha) \cdot P(\alpha)}{P(\beta)}$$

Here $P(\alpha)$ denotes the prior probability, i.e, is the probability of α before β has been observed. $P(\alpha/\beta)$ denotes the posterior probability, i.e., the probability of α after β is observed. $P(\beta/\alpha)$ is the likelihood, or the probability of observing β given α, and $P(\beta)$ is called the marginal likelihood or "model evidence".

A first paraconsistent version of Bayes' theorem is the following:

Theorem 8.14 *([Car09]) Paraconsistent Bayes' Conditionalization Rule, first form (PBCR1):*

$$P(\alpha/\beta) = \frac{P(\beta/\alpha) \cdot P(\alpha)}{P(\beta/\alpha) \cdot P(\alpha) + P(\beta/\neg\alpha) \cdot P(\neg\alpha) - P(\beta/\bullet\alpha) \cdot P(\bullet\alpha)}$$

for $P(\beta) \neq 0$ *and* $P(\bullet\alpha) = P(\alpha \wedge \neg\alpha) \neq 0$.

Proof. Suppose we have two contradictory hypothesis, α and $\neg\alpha$, and wish to compute the probability of α based on evidence β. Since the definition of conditional probability gives, since $P(\beta) \neq 0$:

$$P(\alpha/\beta) = \frac{P(\beta/\alpha) \cdot P(\alpha)}{P(\beta)}$$

it remains to compute the marginal likelihood $P(\beta)$, which here will depend on $P(\alpha \wedge \neg\alpha)$.

By Theorem 8.7, item 12, $P(\beta) = P(\beta \wedge (\alpha \vee \neg\alpha)) = P((\beta \wedge \alpha) \vee (\beta \wedge \neg\alpha)) =$

$$= P(\beta \wedge \alpha) + P(\beta \wedge \neg\alpha) - P(\beta \wedge \alpha \wedge \neg\alpha) =$$
$$= P(\beta/\alpha) \cdot P(\alpha) + P(\beta/\neg\alpha) \cdot P(\neg\alpha) - P(\beta/\alpha \wedge \neg\alpha) \cdot P(\alpha \wedge \neg\alpha)$$

and the rule follows, since $P(\bullet\alpha) = P(\alpha \wedge \neg\alpha) \neq 0$. □

Another form of Paraconsistent Bayes' Conditionalization Rule is the following,

Theorem 8.15 *Paraconsistent Bayes' Conditionalization Rule, second form (PBCR2):*

$$P(\alpha/\beta) = \frac{P(\beta/\alpha) \cdot P(\alpha)}{P(\beta/\alpha \wedge \circ\alpha).P(\alpha \wedge \circ\alpha) + P(\beta/\neg\alpha \wedge \circ\alpha).P(\neg\alpha \wedge \circ\alpha) + P(\beta/\bullet\alpha).P(\bullet\alpha)}$$

for $P(\beta) \neq 0$ *and* $0 < P(\bullet\alpha) = P(\alpha \wedge \neg\alpha) < 1$.

Rephrasing this statement in terms of assertion and denial one obtains:

$$P(\alpha/\beta) = \frac{P(\beta/\alpha) \cdot P(\alpha)}{P(\beta/\oplus\alpha).P(\oplus\alpha) + P(\beta/{\sim}\alpha).P({\sim}\alpha) + P(\beta/\bullet\alpha).P(\bullet\alpha)}$$

for $P(\beta) \neq 0$ *and* $0 < P(\bullet\alpha) = P(\alpha \wedge \neg\alpha) < 1$.

Proof. The proof, similar to the proof of Theorem 8.14, uses item 8 of Theorem 8.13. □

Since probability values cannot be greater than 1, Theorem 8.14 only makes sense if the following restrictions apply for computing $P(\alpha/\beta)$ (recall by Remark 8.5 that $\alpha \wedge \neg\alpha$ and $\bullet\alpha$ are equivalent in **Ci**): $P(\beta/\alpha) \cdot P(\alpha) + P(\beta/\neg\alpha) \cdot P(\neg\alpha) \geq P(\beta/\bullet\alpha) \cdot P(\bullet\alpha)$

Analogous restrictions may apply for Theorem 8.15 depending on values determined by applications.

It is possible to argue that, when approaching extreme cases, that is, $P(\circ\alpha)$ or $P(\bullet\alpha)$ approach 1, (or even $P(\circ\alpha)$ or $P(\bullet\alpha)$ approach 0, see Theorem 8.11, item 10) Theorem 8.15 tends to the standard case, as expected. This can be proved by an appropriate notion of convergence, defining the limit $P(\circ\alpha) \longrightarrow 1$ if there is a sequence of probability paraconsistent functions $\langle P_1, \cdots, P_n, \cdots \rangle$ that increases on $\circ\alpha$ as $n \longrightarrow \infty$, that is, $P_1(\circ\alpha) < \cdots < P_n(\circ\alpha) < \cdots$.

Of course, if $P(\circ\alpha) \longrightarrow 1$ then $P(\bullet\alpha) \longrightarrow 0$, and $[P(\beta/\oplus\alpha).P(\oplus\alpha) + P(\beta/{\sim}\alpha).P({\sim}\alpha) + P(\beta/\bullet\alpha).P(\bullet\alpha)] \longrightarrow [P(\beta/\alpha).P(\alpha) + P(\beta/\neg\alpha).P(\neg\alpha)]$.

Therefore PBCR2 in the form of Theorem 8.15 tends to the standard case.

An analogous reasoning shows that PBCR1 in the form of Theorem 8.14 also tends to the standard case. A problem we leave in this section is to formalize this informal argument:

Problem 8.16 *Define precisely an appropriate notion of convergence that can be used to show rigorously that both paraconsistent Bayes' rules converge to the standard Bayes' rule as* $P(\circ\alpha) \longrightarrow 1$.

Example 8.17 *Suppose that a test for an illegal drug is such that it is 95% sensitive (that is, 95% accurate in the case of a regular user of that drug or in the case that the tested individual sometimes uses the drug) and 85% specific (that is, 85% accurate in the case of a non-user of the drug).*

Suppose, additionally, that independent tests gave the following results: (i) 10% of the entire population often or sometimes uses this drug, (ii) that 95% of the entire population of all athletes rarely use the drug, or has never used it. Moreover, for the contradictory (inconsistent) region the test is 92% accurate.

Let's use mnemonically the following abbreviations:

D: the event that the drug test has declared "doping" (positive) for an individual; $\neg D$: the event that the drug test has declared "no doping" (negative) for an individual; A: the event that the person tested often or sometimes uses the drug; $\neg A$: the event that the person tested rarely or never uses the drug.

This gives the following information:

(1) $P(D/A) = 0.95$, $P(\neg D/\neg A) = 0.85$.

In this case there are contradictory probabilities with respect to the events A and $\neg A$, as they are not mutually exclusive. So we should compute the probability that a user is in the inconsistent region $A \wedge \neg A$: by Finite Additivity, $P(A \vee \neg A) = 1 = (P(A) + P(\neg A)) - P(A \wedge \neg A)$, and thus:

(2) $P(A \wedge \neg A) = (P(A) + P(\neg A)) - 1 = 0.05$

(3) The probability that the test is critical, *i.e.,* $P(D/A \wedge \neg A) = 0.92$.

The results of the test itself have no paraconsistent character, since the events D ('doping') and $\neg D$ ('no doping') are consistent, that is, exclude each other. Thus, by Theorem 8.13, item 4:

(4) $P(D/\neg A) = 1 - P(\neg D/\neg A) = 0.15$ *and* $P(\neg D/A) = 1 - P(D/A) = 0.05$.

Suppose someone has been tested, and the test is positive ('doping'). What is the probability that the tested individual uses (respectively, does not use) regularly, or sometimes, this illegal drug given the test accused the presence of the illegal drugs, that is, what is $P(A/D)$ *(respectively,* $P(\neg A/D)$*)?*

By applying Paraconsistent Bayes Rule, first form:

$$P(A/D) = \frac{P(D/A) \cdot P(A)}{P(D/A) \cdot P(A) + P(D/\neg A) \cdot P(\neg A) - P(D/A \wedge \neg A) \cdot P(A \wedge \neg A)}$$

since $P(A \wedge \neg A) \neq 0$.

Therefore, plugging the values one obtains:

$$P(A/D) = 0.4960 \text{ and } P(\neg A/D) = 0.7441$$

Such values are clearly distinct from the classical case in any way the contradictory critical region $P(D/A \wedge \neg A)$ *could be eliminated by any consistency restoring mechanism, since* $P(A/D) + P(\neg A/D) > 1$.

This suggests that Paraconsistent Bayes' Conditionalization Rule is significantly different than its classical counterpart, and might be more informative than traditional conditionalization in certain cases.

The analysis of the Example 1 given in [BSC16], section 4, is not correct. The example was too ambitious (and contained some typos), and should not be considered conclusive. It just indicates that there is much to be learned from the point of view of applications of paraconsistent probabilities.

As a comparison, let us apply the same values of Example 8.17 to PBCR2 (Theorem 8.15):

$$P(A/D) = \frac{P(D/A) \cdot P(A)}{P(D/\oplus A).P(\oplus A) + P(D/{\sim}A).P({\sim}A) + P(D/\bullet A).P(\bullet A)}$$

In this case, we need to add the extra information $P(D/\oplus A)$ and $P(D/{\sim}A)$ (the other being the same as in the previous example). Taking, for instance $P(D/\oplus A) = 0.98$ and $P(D/{\sim}A) = 0.9$ we obtain via PBCR2 that

$$P(A/D) = 0.5135 \text{ and } P(\neg A/D) = 0.7702$$

distinct from Example 8.17 using PBCR1, and also distinct from the standard case.

This shows that both forms of paraconsistent Bayesian conditionalization are meaningful and non-coincident; they offer us ampliative reasoning tools, whose usefulness seems to be relevant.

8.6 Paraconsistent Probability Spaces

The notion of Paraconsistent Probability Spaces was introduced in [BSC16] as the mathematical counterpart for the axiomatic concept of paraconsistent probability. Although for the classical case the approaches to probability on sentences and probability on sets are inherently equivalent this is not granted in the paraconsistent paradigm, as pointed out at Section 8.4.

Definition 8.18
A Paraconsistent Probability Space is a structure $\langle \Omega, \Sigma, \Pi, P_\mu \rangle$ *where:*

1. Ω *is sample space composed by all possible outcomes*
2. $\Sigma \subseteq \wp(\Omega)$ *is a set of events such that* Σ *is a* $\sigma^\star$*-algebra, i.e.:*
   (a) $\emptyset \in \Sigma$, $\Omega \in \Sigma$ *and* $\Pi \in \Sigma$;
   (b) Σ *is closed under* $\cup$, $\cap$ *and countable unions;*
   (c) Σ *is closed under the following two binary operations:*
      i. $X^\star : \Sigma \mapsto \Sigma$, *such that* $\bar{X} \subseteq X^\star$, *where* $\bar{X}$ *is the usual complement.*
      ii. $\bigcirc X : \Sigma \mapsto \Sigma$, *such that* $\bigcirc X \cap X^\star = \bar{X}$
   (d) Π *is the set of all consistent outcomes;*
3. *The map* $P_\mu : \Sigma \mapsto [0, 1]$ *is a probability measure satisfying the following conditions:*
   (a) $P_\mu(\Omega) = 1$ *and* $P_\mu(\emptyset) = 0$
   (b) *If* $S_1, \cdots, S_n \in \Sigma$ *are pairwise disjoint, then* $P_\mu(\bigcup_{i=1}^{\infty} S_i) = \sum_{i=1}^{\infty} P_\mu(S_i)$

A (classical) probability space is a particular case of a paraconsistent probability space $\langle \Omega, \Sigma, P \rangle$ where $\Pi = \Sigma$ and the operation $\bigcirc X : \Sigma \mapsto \Sigma$ is the identity on Σ (and consequently $X^\star = \bar{X}$). In the cases where Σ is closed under $\cup$, $\cap$ and finite unions a $\sigma^\star$-algebra is referred to as a σ-algebra.

A concrete example of a paraconsistent probability space is given by the structure:

Example 8.19 $\langle \Omega, \Sigma, \Pi, P_\mu \rangle$ *where:*

1. Ω *is any set (representing all possible outcomes)*
2. $\Sigma \subseteq \wp(\Omega)$ *is the set of all events such that* Σ *is a* σ*-algebra, i.e.:*

(a) $\emptyset \in \Sigma$, $\Omega \in \Sigma$ *and* $\Pi \in \Sigma$;

(b) Σ *is closed under* $\cup$ *and* $\cap$;

(c) Σ *is closed under the binary operations, for every* $X \in \Sigma$:

  i. $X^\star = \bar{X} \cup \bar{\Pi}$

  ii. $\bigcirc X = \Pi$

Just to be clear, Π represents the consistent events (from the point of view of the logic **Ci**). The structure $\langle \Omega, \Sigma, \Pi \rangle$ is a particular case of a *paraconsistent algebra of sets*, as investigated in [CdA84], and the results therein (particularly theorems 4 and 5) can be adapted to give a precise connection between the concepts of paraconsistent probability on events and probability on sentences in the **Ci** logic (again, in the cases of finite additivity).

A challenging problem we leave in this part is the following, intended to show that also in the paraconsistent case the notion of an *event* can be regarded either as a proposition, or as an element of the Paraconsistent Probability Space:

Problem 8.20 *Show that each Paraconsistent Probability Space induces a paraconsistent probability measure, and vice-versa, so that the two models express the same stochastic situations. For a guide in the classical case see [Hai96], section 4.7, pp, 210-212.*

8.7 Paraconsistent Bayesian Networks

This section is inspired by Chapter 8 of [CCG+07], but with the bonus of offering a generalization of the concept of standard Bayesian networks.

A *paraconsistent Bayesian network* is an annotated acyclic graph (a set of direct edges between variables) that represents a joint (paraconsistent) probability distribution over a finite set of random variables $V = \{V_1, \ldots, V_n\}$. Following the praxis, we suppose that each variable will have only a finite number of possible values (though this is not a mandatory restriction – numeric or continuous variables that take values from a set of continuous numbers can also be used).

For such discrete random variables, conditional probabilities are usually represented by a table containing the probability that a child node takes on each of the values, taking into account the combination of values of its parents, that is, to each variable V_i with parents $\{B_1, \ldots, B_{n_i}\}$ there is attached a conditional probability table relating V_i to its parents (regarded as the "causes" of V_i).

Following [CCG+07] we stick to two-valued variables, denoting the assignment of V_i to its values by v_i and $\neg v_i$ respectively, for $i = 1, \ldots, n$.

An *assignment* u to a subset $U \subseteq V$ of variables is considered to be a conjunction of assignments to each of the variables in U. For instance, $v_1 \wedge \neg v_2 \wedge \neg v_5$ is an assignment to $\{V_1, V_2, V_5\}$.

The network can be thought as a pair composed by a directed acyclic graph whose nodes represents random variables, and whose edges represent the direct dependencies between these variables. The graph is supposed to encode independence assumptions, so that each variable is independent of its non-descendants. The directed acyclic graph structure is considered to be the "qualitative" part of the model, whereas the "quantitative" part is given by the conditional probability distribution at each node. Paraconsistent Bayesian networks extend this "quantitative" part to paraconsistent conditional probability distributions, following Definition 8.10. More formally:

Definition 8.21 *A paraconsistent Bayesian network b on V consists of two components:*

- *A directed acyclic graph G with nodes from V, representing the causal relations amongst the variables;*
- *A paraconsistent probability distribution S, specifying, for each $V_i \in V$, the probability distribution of V_i conditional on its parents (expressing direct causes in G), where S consists of statements of the form*

$$(P(v_i/Par_i) = x_{i,Par_i})$$

for each $i = 1, \ldots, n$, such that Par_i is an assignment of values to the parents of V_i and $x_{i,Par_i} \in \{0, 1\}$.

For a given Bayesian network with probability specification S, if the value of a variable V_i is known then V_i is said to be *instantiated* to that value and the corresponding probability specifiers $P(v_i|Par_i)$ are 1 or 0 according to whether v_i or $\neg v_i$ is the instantiated value.

The graph and probability specification of a Bayesian network are governed by a fundamental assumption known as the *causal Markov condition*, meaning that the conditional probability distribution at each node depends only on its parents, written

$$V_i \perp\!\!\!\perp ND_i \mid Par_i$$

where ND_i and Par_i are respectively the sets of non-descendants and parents of V_i.

Because of this independence condition, and using the Chain Rule, a (paraconsistent) Bayesian network determines a joint paraconsistent probability distribution over its nodes, since, for each assignment v on V,

$$P(v) = \prod_{i=1}^{n} P(v_i/Par_i)$$

where v_i is the assignment v gives to V_i, and Par_i is the assignment v gives to the parents Par_i of V_i.

As much as in the case of Bayes conditionalization, probability distributions taking into account events such as α, $\oplus\alpha$, $\neg\alpha$, and $\sim\alpha$ will be fully informative.

Although orthodox Bayesian networks are a powerful tool for reasoning probabilistically, difficult problems arise when contradictory probability values are assigned: there is no efficient contradiction handling method that can identify the set of contradictory assignments and propose the retraction of one (or some) of them. This question is addressed in [RR94] where the notion of 'ignorant belief networks' is introduced, a mechanism that in principle can detect and handle contradictions.

Paraconsistent Bayesian networks, notably when combined with paraconsistent belief revision (as in [TCR16]), and with belief maintenance systems can lead to a new approach to detecting and handling contradictions, and producing explanations for its conclusions. This is naturally relevant, for instance, in medical diagnosis, natural language understanding, forensic sciences and other areas where evidence interpretation is an important issue.

It is not difficult to be convinced that paraconsistent Bayesian networks may be useful and stimulating in a series of circumstances where contradictions are around, although we do not treat any concrete example here. A problem for this part is:

Problem 8.22 *Establish some main connections among paraconsistent Bayesian networks, (paraconsistent) Belief Revision, and Belief Maintenance Systems, investigating properties of paraconsistent Bayesian networks as studied in [CCG+07].*

8.8 The troubles with conditional probability

The standard account of conditional probability is problematic, as it is based on the ratio $P(\alpha/\beta) = P(\alpha \wedge \beta)/P(\beta)$, and thus requires that $P(\beta) \neq 0$. There are several situations, however, when $P(\beta) = 0$ while β is a meaningful event that may affect α particularly in the continuous case (a famous example was suggested by E. Borel and known as the Borel-Klmogorov paradox, see [JB03], section 15.7, pp. 467-470).

Moreover, in Kolmogorovian terms, two events α and β are considered to be independent, as mentioned before, if $P(\alpha \wedge \beta) = P(\alpha) \cdot P(\beta)$, for some probability distribution P. This leads to a puzzling situation: extreme events α where $P(\alpha) = 0$ or $P(\alpha) = 1$ are such that $P(\alpha) = P(\alpha \wedge \alpha) = P(\alpha) \cdot P(\alpha)$ in both cases. Therefore, extreme events can be regarded as independent of themselves.

The troubles with the orthodox account of conditional probability as based on a ratio also affects our paraconsistent theory of probability as the examples may suggest. Several axiomatizations of conditional probability (as primitive) have been proposed by R. Carnap (1950, 1952), K. Popper (1959), A. Kolmogorov (1950), and A. Renyi (1970) (see [FH14] for more details).

Popper's axiomatization of conditional probability as a two-place relation $Pr(\alpha, \beta)$ is one of the most general accounts and generalizes conventional probability theory in such a way that unconditional probability can be regarded as probability conditional on a logical truth. As argued in [FH14], independence should be rethought: a new theory of probabilistic independence emerges when conditional probabilities are taken as primitive and defined as total functions, fixing some of the above-mentioned inconveniences. There are some surprising connection between Popper's philosophy and paraconsistency: D. Miller in the chapter 13 of his [Mil06] emphatically suggests that the best logic to meet Popper's account of falsificationism is paraconsistent (in particular, the Browerian system that Popper himself have considered, and later on dismissed as 'too weak'). An altogether natural suggestion, on our own grounds, is to propose a further extension of Popper's axiomatization of conditional probability by taking the logic **Ci** as its underlying logic, which leads to the following problem:

Problem 8.23 *Modify K. Popper's axiomatization of conditional probability as presented in [RL99] (see also [FH14]) just extending axiom P3 in the following way: For every* α *and* β, $Pr(\alpha, \beta) + Pr(\neg\alpha, \beta) - Pr(\bullet\alpha, \beta) = 1$. *Show that this modification leads to a satisfying theory of paraconsistent conditional probability (which naturally reduces to the Popperian axiomatization when* $P(\circ\alpha) = 1$*).*

8.9 From paraconsistent probability to paraconsistent possibility

Possibility theory is a natural generalization of probability measures, a kind of imprecise probability theory regarded as a theory of uncertainty used in several areas to express uncertain knowledge in scenarios of incomplete information, sometimes as a rival of probability (cf. e.g. [DP01]). Possibility theory uses a pair of dual set functions (possibility and necessity measures) instead of only one measure, in this way differing from probability.

Probability, possibility and other credal calculi are alternative formalisms, such as the Dempster-Shafer theory of evidence (cf. [DP88]), usually starting from different motivations. Possibility theory can be easily and naturally extended over LFIs, by defining possibility and necessity functions (as e.g. in [BL94] and especially in [DG92], where the notion of degrees of support are investigated as a generalizations of degrees of belief) simulating our paraconsistent probability measures over LFIs. A deeper study on the meaning and applications of paraconsistent possibility theory, as well as its inter-relations to paraconsistent probability, will be left to further work. An immediate, almost obvious task, is:

Problem 8.24 *Develop the main properties of paraconsistent possibility theory and compare it, conceptually and technically, with paraconsistent probability theory.*

8.10 Comments and conclusions

Paraconsistent probabilities can be regarded as degrees of belief that a rational agent attaches to events, even if such degrees of belief may be contradictory. Thus it is not impossible for an agent to believe in the proposition α and $\neg\alpha$ and to be rational, if this belief is justified by evidence, as defended in [CR16]. In the literature, George Orwell in his [Orw61] even praised the power of holding two contradictory beliefs in one's mind simultaneously, and accepting both of them, as *doublethink*. Under such a view, contradictory beliefs and probability measures upon them, having as underlying logics the Logics of Formal Inconsistency as studied in [CCM07] and [CC16], can be tolerated and used profitably as new tools for rationality. This adjusts quite well with the view on probability as extending logic, as much as a theory of uncertain reasoning will extend the theory of certain reasoning.

This new probability theory based on the logic **Ci** poses new and challenging problems, as some of the questions we have left in this paper. It is important to make explicitly clear that other paraconsistent theories, based on other LFIs, are plainly possible and equally exciting. We finish with three additional difficult problems that may help to comprehend the new areas of non-standard probabilities.

D. Mundici proved in [Mun15] that the notion of de Finetti coherence is preserved under taking products of coherent books on two sets of independent events, showing that this justifies the 'product rule': the probability of the conjunction of two independent events as the product of the probability of the events. Mundici's proof relies strongly on mathematical concepts and results that are unlikely to be reproducible in non-standard terrain, as represented by de Finetti Dutch Book theorem, independent Boolean subalgebras, Carathéodory extension theorem and Stone duality.

Even if J. B. Paris in [Par01] has shown that a generalization of the standard Dutch Book theorem (his Theorem 5) applies to several non-classical logics, among them to paraconsistent logics, the rest of the mathematical background seems elusive, although, despite the problems raised in Section 8.8 the product rule for paraconsistent probability seems intuitive and philosophically reasonable.

Problem 8.25 *Find and argument independent from probability theory that justifies the product rule for paraconsistent probability theory and that could replace the arguments in [Mun15].*

Now, towards our two final problems:

The logical consistency of a collection of assertions about events can be viewed as a particular case of coherent probability assessments in the sense of de Finetti. A probability assignment to the events in a collection D is said to be *coherent* (in the sense of de Finetti) if, when prices (or odds) for bets are based on such assignment, a gambler cannot construct a Dutch book, that is, a set of odds and bets which guarantee a profit, regardless of the outcome of the gamble. On the other hand, the events in D regarded as logical assertions are said to be *belief-consistent* if, by assigning truth-values $v(A) = 1$ if A is taken to be certain and $v(A) = 0$ if A is taken to be impossible for each $A \in D$, the collection D does not entail any contradiction, in the sense that logically equivalent propositions would receive different truth-values on the basis of this assignment through the following rules:

$v(A \vee B) = max\{v(A), v(B)\}$
$v(A \wedge B) = min\{v(A), v(B)\}$

$v(\neg) = 1 - v(A)$

As proved in [DES09], the belief-consistency of a collection of assertions about events can be viewed as a special case of coherent probability assessments in the sense of de Finetti.

Problem 8.26 *Find a relationship among the notions of consistency in the sense of the LFIs expressed by the operator* $\circ$ *([CCM07], [CC16]), and belief-consistency and coherence in the sense of de Finetti.*

Paraconsistent probability theory leads to a natural quantum analogue of total probability rule that may replace quantum probability based on the structure of Hilbert spaces. Paraconsistent and quantum phenomena are naturally related: in [AC10], for instance, a model of paraconsistent Turing machines is shown to simulate superposed states of quantum computing, solving (under certain restrictions) the well-known Deutsch's and Deutsch-Jozsa problems. As shown in [FAS13] and discussed in [SJFD14], quantum total probability rule is equivalent to Max Born's rule (the rule that estimates the probability that a measurement on a quantum system will yield a given result). All the discussion in [SJFD14] (see also [Roy16]) can be profitably reframed into **Ci**-probability theory.

Problem 8.27 *Show that the paraconsistent Bayesian probability theory based on the logic* **Ci** *is directly applicable to the quantum total probability rule of Quantum mechanics.*

Acknowledgements

We thank all the participants of the IST for their comments and remarks, and especially the dear Amílcar for his sharp criticisms to some of these ideas in seminars held in Lisbon in 2015. None of them is, of course, responsible for any deficiencies in this essay.

Both authors acknowledge support from FAPESP Thematic Project LogCons 2010/51038-0, Brazil, and the first author thanks a research grant from the National Council for Scientific and Technological Development (CNPq), Brazil.

Bibliography

[AC10] J. C. Agudelo and W. A. Carnielli. Paraconsistent machines and their relation to quantum computing. *Journal of Logic and Computation*, 20:573–595, 2010.

[BL94] P. Besnard and J. Lang. Possibility and necessity functions over non-classical logics. In *UAI'94 Proceedings of the Tenth International Conference on Uncertainty in Artificial Intelligence*, pages 69–76. Morgan Kaufmann Publishers Inc., 1994.

[BSC15] J. Bueno-Soler and W. A. Carnielli. Experimenting with consistency. In V. Markin and D. Zaitsev, editors, *The Logical Legacy of Nikolai Vasiliev and Modern Logic*, page In print. Synthese Library - Studies in Epistemology, Logic, Methodology, and Philosophy of Science-Springer, 2015. Pre-print available at: `http://www.cle.unicamp.br/e-prints/vol_14,n_3,2014.html`.

[BSC16] J. Bueno-Soler and W. A. Carnielli. Paraconsistent probabilities: Consistency, contradictions and Bayes theorem. In J. Stern, editor, *Special Issue Statistical Significance and the Logic of Hypothesis Testing*, page On line. Entropy 18(9), 2016. Open access `http://www.mdpi.com/1099-4300/18/9/325/htm`.

[Car50] R. Carnap. *Logical Foundations of Probability*. University of Chicago Press, 1950.

[Car09] W. A. Carnielli. Uma lógica da modalidade econômica? *Revista Brasileira de Filosofia*, pages 01–18, 2009.

[Car11] W. A. Carnielli. The single-minded pursuit of consistency and its weakness. *Studia Logica*, 97(1):81–100, 2011.

[CC16] W. A. Carnielli and M. E. Coniglio. *Paraconsistent Logic: Consistency, Contradiction and Negation*. Logic, Epistemology, and the Unity of Science. Springer International Publishing, 2016.

[CCG+07] W. A. Carnielli, M. E. Coniglio, D. Gabbay, P. Gouveia, and C. Sernadas. *Analysis and Synthesis of Logics*. Sringer, 2007.

[CCM07] W. A. Carnielli, M. E. Coniglio, and J. Marcos. Logics of formal inconsistency. In D. Gabbay and F. Guenthner, editors, *Handbook of Philosophical Logic*, volume 14, pages 1–93, Amsterdam, 2007. Springer-Verlag.

[CCPR15] W. Carnielli, M.E. Coniglio, R. Podiacki, and T. Rodrigues. On the way to a wider model theory: Completeness theorems for first-order logics of formal inconsistency. *Review of Symbolic Logic*, 3:548–578, 2015.

[CCR] W. A. Carnielli, M. E. Coniglio, and A. Rodrigues. On formal aspects of the epistemic approach to paraconsistency. Submitted.

[CdA84] W. A. Carnielli and L. P. de Alcantara. Paraconsistent algebras. *Studia Logica*, 43:79–88, 1984.

[CM02] W. A. Carnielli and J. Marcos. A taxonomy of **C**-systems. In W. A. Carnielli, M. E. Coniglio, and I. M. L. D'Ottaviano, editors, *Paraconsistency - the Logical Way to the Inconsistent*, volume 228 of *Lecture Notes in Pure and Applied Mathematics*, pages 1–94, Nova Iorque, 2002. Marcel Dekker.

[CR15] W. A. Carnielli and A Rodrigues. Contradictions, inconsistencies and other oximora. In E. H. Haeusler, W. de C. Sanz, and B. Lopes, editors, *Why is this a proof? Festschrift for Luiz Carlos Pereira*, pages 197–223. College Publications, 2015.

[CR16] W. Carnielli and A. Rodrigues. An epistemic approach to paraconsistency: a logic of evidence and truth. Submitted, 2016.

[DES09] J. M. Dickey, M. L. Eaton, and W. D. Sudderth. De finetti coherence and logical consistency. *Notre Dame Journal of Formal Logic*, 50(2):133–139, 2009.

[dF37] B. de Finetti. La preévision : ses lois logiques, ses sources subjectives. *Annales de l'Institut Henri Poincaré*, 7(1):1–68, 1937.

[dF72] B. de Finetti. *Probability, Induction and Statistics. The Art of Guessing.* John Wiley & Sons, 1972. Wiley Series in Probability and Mathematical Statistics.

[dF74] B. de Finetti. *Theory of Probability: A critical introductory treatment.* John Wiley & Sons, Ltd, 1974.

[DG92] A. Darwiche and M. L. Ginsberg. A symbolic generalization of probability theory. In *AAAI-92 10th Conference of the American Association for Artificial Intelligence*, pages 622–627, 1992.

[DP88] D. Dubois and H. Prade. Representation and combination of uncertainty with belief functions and possibility measures. *Computational Intelligence*, 4:244–264, 1988.

[DP01] D. Dubois and H. Prade. Possibility theory, probability theory and multiple-valued logics: A clarification. *Annals of Mathematics and Artificial Intelligence*, 32:35–66, 2001.

[FAS13] C. Fuchs, A. Christopher A., and R. Schack. Quantum-Bayesian coherence. *Rev. Mod. Phys.*, 85:1693–1715, Dec 2013.

[FH14] B. Fitelson and A. Hájek. Declarations of independence. *Synthese*, on line:DOI 10.1007/s11229–014–0559–2, 2014.

[Hai96] T. Hailperin. *Sentential Probability Logic: Origins, Development, Current Status, and Technical Applications.* Bethlehm: Lehigh University Press, 1996.

[JB03] E. T. Jaynes and G. L Bretthorst, editors. *Probability Theory: The Logic of Science.* Cambridge University Press, 2003.

[Jef92] R. Jeffrey. *Probability and the Art of Judgement.* Cambridge University Press, Cambridge, 1992.

[Key21] J. M. Keynes. *A Treatise on Probability.* Macmillan, 1921.

[Kol56] A. N. Kolmogorov. *Foundations of the Theory of Probability.* New York:Chelsea, 1956. Translation of Grundbegriffe der Wahrscheinlichkeitsrechnung. Springer, Berlin, 1933.

[Mar97] E. Mares. Paraconsistent probability theory and paraconsistent Bayesianism. *Logique et Analyse*, 160:375–384, 1997.

[Mil06] D. Miller. *Out of Error: Further Essays on Critical Rationalism.* Routledge, 2006.

[Mun06] D. Mundici. Bookmaking over infinite-valued events. *International Journal of Approximate Reasoning*, 43:223–240, 2006.

[Mun15] D. Mundici. Coherence of de Finetti coherence. *Synthese*, on line:DOI 10.1007/s11229–016–1126–9, 2015.

[Orw61] G. Orwell. *1984.* Signet Classic, 1961.

[Par01] J. B. Paris. A note on the Dutch Book method. In G. De Cooman, T. Fine, and T. Seidenfeld, editors, *ISIPTA'01-Proceedings of the Second International Symposium on Imprecise Probabilities and Their Applications, Ithaca, NY, USA*, pages 301–306. Shaker Publishing, 2001.

[Pri87] G. Priest. *In Contradiction: A Study of the Transconsistent.* Martinus Nijhoff, Amsterdam, 1987.

[Ram26] F. P. Ramsey. Truth and probability. In H. E. Kyburg and H. E. Smokler, editors, *Studies in Subjective Probability, 2nd ed. 2002*, pages 23–52. Huntington, New York: Robert E. Krieger Publishing Co, 1926.

[RL99] P. Roeper and H. Leblanc. *Probability Theory and Probability Logic.* University of Toronto Press, 1999.

[Rob00] L. Roberts. Maybe, maybe not: Probabilistic semantics for two paraconsistent logic. In D. Batens, C. Mortensen, G. Priest, and J. P van Bendegem, editors, *Frontiers of Paraconsistent Logic: Proceedings of the I World Congress on Paraconsistency*, Logic and Computation Series, pages 233–254. Baldock: Research Studies Press, King's College Publications, 2000.

[Roy16] S. Roy. Quantum probability, paraconsistent logic and decision making in cognitive domain. `http://www.academia.edu/23116795/Quantum_Probability_Paraconsistent_Logic_and_Decision_Making_in_Cognitive_Domain`, 2016.

[RR94] M. Ramoni and A. Riva. Belief maintenance in Bayesian networks. In *Proceedings of the Tenth International Conference on Uncertainty in Artificial Intelligence*, UAI'94, pages 498–505, San Francisco, CA, USA, 1994. Morgan Kaufmann Publishers Inc.

[RSS16] J. Rasga, C. Sernadas, and A. Sernadas. On probability and logic. *ArXiv e-prints*, 2016. 1601.02915.

[SJFD14] R. Salazar, C. Jara-Figueroa, and A. Delgado. Analogue of the quantum total probability rule from paraconsistent Bayesian probability theory. arXiv:1408.5308 v 1, 2014.

[TCR16] R. R. Testa, M. E. Coniglio, and M. M. Ribeiro. AGM-like paraconsistent belief change. To appear, 2016.

[Wea03] B. Weatherson. From classical to intuitionistic probability. *Notre Dame Journal of Formal Logic*, 44:111–123, 2003.

[Wil81] J. N. Williams. Inconsistency and contradiction. *Mind*, 90:600– 602, 1981.

[Wil14] J. R. G. Williams. Non-classical logic and probability. In A. Hajek and C. Hitchcocki, editors, *Oxford Handbook to Probability and Philosophy, forthcoming.* Oxford University Press, 2014. `http://www.personal.leeds.ac.uk/~phljrgw/wip/nonclassicalprobabilitypostrevisions.pdf`.

9

Analysis in weak systems

António M. Fernandes[1], Fernando Ferreira[2] and Gilda Ferreira[2]

[1]Dep. de Matemática, Instituto Superior Técnico, Universidade de Lisboa, Portugal
amfernandes@netcabo.pt

[2]Dep. de Matemática, Faculdade de Ciências, Universidade de Lisboa, Portugal
{fjferreira,gmferreira}@fc.ul.pt

Abstract

The authors survey and comment their work on weak analysis. They describe the basic set-up of analysis in a feasible second-order theory and consider the impact of adding to it various forms of weak König's lemma. A brief discussion of the Baire categoricity theorem follows. It is then considered a strengthening of feasibility obtained (fundamentally) by the addition of a counting axiom and showed how it is possible to develop Riemann integration in the stronger system. The paper finishes with three questions in weak analysis.

9.1 Introduction

This paper is the third and last part of a triptych whose two other parts consist of "Techniques in weak analysis for conservation result" [13] and "Interpretability in Robinson's Q" [20]. This triptych of papers mostly surveys the work of the authors in weak analysis. The present paper is representative of our work in this field and it is the way that the authors have found to jointly render homage to Amílcar Sernadas on the occasion of his 64th birthday (Amílcar was born on April 20, 1952). The influence and scientific leadership of Amílcar on *his side* of logic has been enormous at Instituto Superior Técnico and, indeed, in Portugal. One should also mention his international collaborations, specially with Brazilian logicians. Amílcar's influence was also felt in Portugal on the *other side*, namely in important matters of academic support and institutional collaboration. The authors of this paper are grateful for the chance to present here the final paper of their triptych. They accepted with pleasure the kind invitation of Francisco Dionísio and the other organizers of the *Festschrift* to participate in the conference (via a presentation of Fernando Ferreira) and to make a contribution to this volume. We must also thank them for giving us the opportunity to render a public homage to Amílcar. Unfortunately, less than a year after the conference, Amílcar passed away (on February 7, 2017). This paper is dedicated to his memory.

Both this paper and the *Techniques* paper (also written by the three of us) survey our work on the formalization of analysis in weak sub-exponential systems of second-order arithmetic (the theme of this paper) and on the metamathematical properties of these systems (the theme of the *Techniques* paper). This line of research was inaugurated by F. Ferreira in his doctoral dissertation [14], written in 1988 in *The Pennsylvania State University* under the supervision of Stephen Simpson. Three years before, Samuel Buss had submitted to Princeton University a doctoral dissertation [5] studying weak systems of arithmetic connected to well-known classes of computational complexity (polytime and polyspace computability, for instance). Simpson asked F. Ferreira to look into it and see if a conservation result regarding weak König's lemma could be proved in the weak setting. Of course, the project was motivated by similar results in reverse mathematics according to which the second-order system WKL_0 is Π^1_1-conservative over the base system RCA_0 (see [26] for this result and for reverse mathematics in general). It was also connected with the following challenge of Wilfried Sieg: Find a mathematically significant subsystem of analysis whose class of provably recursive functions consists only of the computationally "feasible" ones (see the end of [25]).

In his dissertation, F. Ferreira re-worked Buss's arithmetical theory S_2^1 in terms of a system of binary words (finite 0-1 strings), thereby obtaining the first-order system Σ_1^b-NIA. The recast was done not only because the work with binary strings is more congenial with computational complexity, but also because it made the formulation of weak König's lemma very transparent. The dissertation proceeded model-theoretically and it was soon realized that, as opposed to the usual setting of reverse mathematics, weak König's lemma has various interesting formulations in the weak setting. It can be formulated by saying that every infinite binary set-tree (i.e., a tree in the range of the second-order variables of the weak system) has an infinite path, but this most natural form turns out to be insufficient for basic reversals (e.g., Theorem 9.7). It can also be formulated thus: For each bounded formula of arithmetic which defines an infinite binary tree (not necessarily a set in the weak system) there is an infinite path through the tree. The latter formulation of weak König's lemma is denoted by Σ_∞^b-WKL. Intermediate formulations do exist and are important for weak analysis (see Section 9.3). Stronger forms of weak König's lemma can also be envisaged, as it is the case with the principle of strict Π_1^1-reflection discussed in Section 9.6. Note that all of these different formulations of weak König's lemma collapse over RCA_0. F. Ferreira was able to prove that the theory BTFA together with Σ_∞^b-WKL is Π_2^0-conservative over Σ_1^b-NIA. The corresponding first-order conservation result is false because BTFA + Σ_∞^b-WKL proves bounded collection, and this form of collection is independent from Σ_1^b-NIA. The matter was finally clarified in [16], where it was shown that the first-order consequences of BTFA + Σ_∞^b-WKL are exactly the consequences of bounded collection over Σ_1^b-NIA. It was actually shown that BTFA + Σ_∞^b-WKL is first-order conservative over the theory Σ_1^b-NIA + $\mathsf{B}\Sigma_\infty^b$, where $\mathsf{B}\Sigma_\infty^b$ is the bounded collection scheme. Unsurprisingly, the proof of this result used Harrington's forcing argument (N.B. It is sometimes thought that Harrington's conservation result can only be proved using a forcing technique whereas, in fact, it can also be proved by pure proof-theoretic means, as it was shown in [18] via a cut-elimination argument). Most of these matters are discussed in our *Techniques* paper, but in the above we wanted to convey to the reader how the results first appeared.

In F. Ferreira's doctoral dissertation there is little of analysis (nevertheless, e.g., the dissertation considers and studies the Heine/Borel theorem for the Cantor space). The formalization of the real number system and of continuous real functions was only worked out systematically in [11], together with António Fernandes. Section 9.2 below outlines this systematization. The section finishes with the proof of the intermediate value theorem in BTFA (Theorem 9.4). This result entails that Tarski's theory of real closed ordered fields RCOF is interpretable in BTFA. The interest of this remark is that F. Ferreira showed in the last section of [11] that BTFA is interpretable in Raphael Robinson's theory of arithmetic Q. We have, therefore, the following interesting relationship between the very weak arithmetical theory Q and the geometric theory RCOF: The first is not interpretable in the second (because Q is essentially undecidable and RCOF is a decidable theory by Tarski's theorem of quantifier-elimination), whereas the second is interpretable in the first. The theories Q and RCOF could not be more different from the metamathematical point of view, and this issue was the theme of the talk of F. Ferreira in the *Festschrift* (the title of the talk was "Arithmetic and geometry from the formal point of view: some notes, some lessons"). The fact that BTFA is interpretable in Q is *prima facie* quite amazing. It rests on the work of people like Edward Nelson, Alex Wilkie, Robert Solovay, Petr Hájek and others on interpretability in Q. This work was surveyed by F. Ferreira and G. Ferreira in the triptych paper "Interpretability in Robinson's Q," where an old unpublished argument of Solovay appeared for the first time. From this body of work it transpires that the interpretability of BTFA in Q is far from being a strange and isolated phenomena. It is a practical application of the rule of thumb according to which a bounded theory of arithmetic, short of proving the totality of exponentiation (actually, for the knowledgeable, short of $\mathsf{I}\Delta_0 + \Omega_\infty$), must be interpretable in Q. (Alex Wilkie proved that $\mathsf{I}\Delta_0 + \mathsf{exp}$ is *not* interpretable in Q, a result discussed in [20], but as far as we know it is still an open question whether $\mathsf{I}\Delta_0 + \Omega_\infty$ is interpretable in Q.) For instance, the *Interpretability* paper shows that a weak theory of analysis associated with polyspace computability (which contains the counting theory of Section 9.5, where Riemann integration can be developed) is interpretable in Q.

We have already disclosed a bit of the structure of this paper. In the next section, we show how to develop the basic notions of analysis in a weak base second-order theory. Section 9.3 considers various formulations of weak König's lemma and their relationships with basic theorems of analysis. In Section 9.4 we briefly discuss the Baire category theorem in the feasible setting. We answer a question of [9] and remark that Cohen's forcing does not preserve bounded collection in the absence of the totality of exponentiation. The following section strengthens the base theory to a theory in which counting is possible. We sketch how Riemann integration can be developed in this strengthened theory. The material of this section comes mainly from the doctoral dissertation of Gilda Ferreira [21] and the paper [19]. It is perhaps worth remarking at this point that recently Stephen Cook and Akitoshi Kawamura considered a polytime version of Weihrauch reducibility with the view of classifying the computational complexity of problems in analysis (cf. [7]). Even more recently, in the executive summary [3] of a conference on Weihrauch reducibility, the following passage can be read: "one could expect relations between weak complexity theoretic versions of arithmetic as studied by Fernando Ferreira *et al.*, on the one hand, and the polynomial-analogue of Weihrauch reducibility studied by Cook, Kawamura *et al.*, on the other hand." The relationships are certainly there, but they are perhaps more subtle than the known relationships between plain Weihrauch reducibility and ordinary reverse mathematics. We close the paper with a section where we pose three questions in weak analysis.

9.2 Groundwork for weak analysis

The theory BTFA (an acronym for 'Base Theory for Feasible Analysis') is a second-order theory of 0-1 strings. Its language directly describes finite binary words. The intended *standard* model has first-order domain $^{<\omega}2$ (the set of finite sequences of zeros and ones). Let $\mathcal{L}$ be the first-order language with three constant symbols ϵ (for the empty word), 0 and 1, two binary function symbols $\frown$ (for concatenation, usually omitted) and $\times$ (the intended interpretation of $x \times y$ is that of the word x concatenated with itself length of y times) and two binary relation symbols $=$ and $\subseteq$ (for *equality* and *initial subwordness*, respectively).

$\mathcal{L}_2$, the second-order language of BTFA, extends the language $\mathcal{L}$ with:

- second-order set variables X, Y, Z, ... (we reserve lower-case roman variables x, y, z, ... for first-order variables).
- a binary relation symbol $\in$ which infixes between a term t of $\mathcal{L}$ and a second-order variable.

A structure for $\mathcal{L}_2$ has domain (M, S) with M the domain of a structure for $\mathcal{L}$ and $S \subseteq \mathcal{P}(M)$. The first-order variables range over M and the second-order variables range over S. Note that we are allowing Henkin models, i.e., S need not be $\mathcal{P}(M)$. As it is well-known, second-order logic with Henkin semantics is essentially first-order. The *full* standard model for $\mathcal{L}_2$ has domain $(^{<\omega}2, \mathcal{P}(^{<\omega}2))$.

We denote by $x \preceq y$ (respectively, $x \equiv y$) the formula $1 \times x \subseteq 1 \times y$ (respectively, $1 \times x = 1 \times y$). In the standard model $x \preceq y$ (respectively, $x \equiv y$) expresses that the length of x is less than or equal (respectively, equal) to the length of y. By $l(x)$ we denote $1 \times x$ (the tally length of x). Quantifications of the form $\forall x\,(x \preceq t \to \ldots)$ and $\exists x\,(x \preceq t \wedge \ldots)$, usually abbreviated by $\forall x \preceq t\,(\ldots)$ and $\exists x \preceq t\,(\ldots)$, are called *bounded* *quantifications.* A *subword quantification formula* is a formula where all quantifications appear in the form $\forall x\,(x \subseteq^* t \to A)$ or in the form $\exists x\,(x \subseteq^* t \wedge A)$, where $\subseteq^*$ stands for *subwordness*, i.e., $x \subseteq^* t$ abbreviates $\exists z(z\frown x \subseteq t)$. A subword quantification can be seen as a very particular type of bounded quantification. Note also that a bounded quantification, over elements x such that $x \preceq t$, ranges over exponential many elements in the length of t, whereas a subword quantification over elements x such that $x \subseteq^* t$, ranges over polynomial many elements in the length of t.

Let us introduce some important classes of formulas. The Σ_1^b*-formulas* (respectively Π_1^b*-formulas*) are the formulas of the form $\exists y \preceq t\, A$, (respectively $\forall y \preceq t\, A$) where A is a subword quantification formula, possibly with parameters, and t is a term in which y does not occur. The *bounded formulas* (also known as the class of Σ_∞^b*-formulas*) are the formulas where all the quantifications are bounded (i.e., there are no second-order quantifications and all first-order quantifications are bounded). It is well-known that in the first-order language, the Σ_1^b-formulas define exactly the NP-predicates in the standard model of domain ${}^{<\omega}2$ (a detailed proof of this fact can be found in [14]); that, dually, the Π_1^b-formulas define the co-NP predicates; and that the Σ_∞^b-formulas define the predicates in the polytime hierarchy.

Definition 9.1 BTFA *is the second-order theory, in the language $\mathcal{L}_2$, which has the following axioms:*

- Basic axioms

  $x\epsilon = x$, $x(y0) = (xy)0$ and $x(y1) = (xy)1$;

  $x \times \epsilon = \epsilon$, $x \times y0 = (x \times y)x$ and $x \times y1 = (x \times y)x$;

  $x \subseteq \epsilon \leftrightarrow x = \epsilon$, $x \subseteq y0 \leftrightarrow x \subseteq y \vee x = y0$ and $x \subseteq y1 \leftrightarrow x \subseteq y \vee x = y1$;

  $x0 = y0 \to x = y$ and $x1 = y1 \to x = y$;

  $x0 \neq y1$, $x0 \neq \epsilon$ and $x1 \neq \epsilon$;

- Induction on notation scheme

  $$A(\epsilon) \wedge \forall x\,(A(x) \to A(x0) \wedge A(x1)) \to \forall x A(x),$$

  where A is a Σ_1^b-formula (possibly with first and second-order parameters);

- Bounded collection scheme ($\mathsf{B}\Sigma_\infty^b$)

  $$\forall x \preceq w \exists y A(x,y) \to \exists z \forall x \preceq w \exists y \preceq z A(x,y),$$

where A is a bounded formula (possibly with first and second-order parameters) and z is a new variable;

- Comprehension scheme

$$\forall x\,(\exists y A(x,y) \leftrightarrow \forall z B(x,z)) \rightarrow \exists X \forall x\,(x \in X \leftrightarrow \exists y A(x,y))$$

where A is a Σ_1^b-formula and B is a Π_1^b-formula (possibly with first and second-order parameters) and X does not occur in A nor in B.

If, instead of $\mathcal{L}_2$, we take the language $\mathcal{L}$, the basic axioms together with the above scheme of induction on notation form the first-order theory Σ_1^b-NIA. It is well-known that Σ_1^b-NIA is equivalent to Buss's theory S_2^1 defined in his doctoral dissertation (see [23] for a formal interpretation of S_2^1 in Σ_1^b-NIA). Thus, by the celebrated witnessing theorem of Buss in his dissertation, the provably total functions of Σ_1^b-NIA (with Σ_1^b-graphs) are the polytime computable functions. The theory BTFA is Π_2^0-conservative over Σ_1^b-NIA and, therefore, these two theories are of the same proof-theoretic strength. It also follows that whenever $\mathsf{BTFA} \vdash \forall x \exists y A(x,y)$, where A is a Σ_1^b-formula, then there exists a polytime computable function $f : {}^{<\omega}2 \rightarrow {}^{<\omega}2$ such that $A(x, f(x))$, for all $x \in {}^{<\omega}2$. The reader can consult our *Techniques* paper for these and related results and for pointers to the original papers.

The structure $({}^{<\omega}2, \textit{Recursive Sets})$ is the smallest model of BTFA with the first-order part ${}^{<\omega}2$, and this fact may give the impression that the comprehension scheme is stronger than in fact is. The comprehension scheme states that recursive sets exist *under the condition* that their recursiveness is shown. In a weak theory like BTFA, this condition may be impossible to show. In any case, the *unbounded* quantifications in the antecedent of the comprehension scheme are extremely convenient for the development of analysis within BTFA. Since in this weak base system functions are formalized as *sets* of codes of ordered pairs, to show the existence of the composition of two functions $f : X \rightarrow Y$ and $g : Y \rightarrow Z$ we just need to write $(g \circ f)(x) = z$ in the following two $\exists/\forall$ forms: $(g \circ f)(x) = z$ iff $x \in X \wedge \exists y\,((x,y) \in f \wedge (y,z) \in g)$ iff $x \in X \wedge \forall y\,((x,y) \in f \rightarrow (y,z) \in g)$. A similar observation ensures the existence of the inverse image: $f(x) \in Z$ can be stated by $x \in X \wedge \exists y\,((x,y) \in f \wedge y \in Z)$ or by $x \in X \wedge \forall y\,((x,y) \in f \rightarrow y \in Z)$.

In the remainder of this section we formalize the basics of real analysis (e.g. the real number system, the notion of continuous function on the reals, etc.) in BTFA. This material is taken from [11], where more details can be found. We start by considering two sorts of natural numbers in BTFA: the tally numbers and the dyadic natural numbers.

The *tally numbers*, denoted by $\mathbb{N}_1$, are the binary words satisfying $x = 1 \times x$. These numbers are concatenations of 1s (tallies). We can define $0_{\mathbb{N}_1}$, $\leq_{\mathbb{N}_1}$, $+_{\mathbb{N}_1}$ and $\cdot_{\mathbb{N}_1}$ as ϵ, $\subseteq$, $\hat{}$ and $\times$ respectively. BTFA proves that $\mathbb{N}_1$ is an ordered semi-ring.

The *dyadic natural numbers*, denoted by $\mathbb{N}_2$, are the binary words satisfying $x = \epsilon \vee x = 1\hat{\ }y$ (with y a word). The idea is that the dyadic number $1x_1x_2\cdots x_{n-1}$, where each x_i is 0 or 1, represents the number $2^{n-1} + \sum_{i=1}^{n-1} x_i 2^{n-i-1}$. The empty string ϵ represents the number zero. We can define $0_{\mathbb{N}_2}$, $\leq_{\mathbb{N}_2}$, $+_{\mathbb{N}_2}$ and $\cdot_{\mathbb{N}_2}$ in order to reproduce the usual operations of the natural numbers and show in BTFA that $\mathbb{N}_2$ is an ordered semi-ring.

The *dyadic rational numbers*, denoted by $\mathbb{D}$, are the triples (i, x, y) (coded as strings in a smooth way), with $i = 0$ or $i = 1$, $x \in \mathbb{N}_2$ and $y = \epsilon \vee y = z\hat{\ }1$ (with z a word). The idea is that the triple $(s, x_0 \ldots x_{n-1}, y_0 \ldots y_{m-1})$ represents the rational number

$$(-1)^s\left(\sum_{i=0}^{n-1} x_i 2^{n-i-1} + \sum_{j=0}^{m-1} \frac{y_j}{2^{j+1}}\right).$$

We denote such dyadic rational number by $\pm x_0x_1 \ldots x_{n-1} \cdot y_0 \ldots y_{m-1}$. By x^* we denote the binary word x with its rightmost zeros chopped off. Thus $\cdot x^*$ is a (positive) dyadic rational number. We define $0_{\mathbb{D}}$, $\leq_{\mathbb{D}}$, $+_{\mathbb{D}}$ and $\cdot_{\mathbb{D}}$ extending, to the dyadic rational numbers, the operations already mentioned in the dyadic natural numbers. Such operations reproduce the usual operations in the rational numbers and turn $\mathbb{D}$ in an ordered ring. The dyadic rational numbers do not form a field, but divisions by 2 are possible. We can also introduce in $\mathbb{D}$ the operations $-_{\mathbb{D}}$ and $|.|_{\mathbb{D}}$ with the expected meaning of subtraction and absolute value function, respectively.

We use the notations 2^n and 2^{-n}, with n a tally number, to stand for the dyadic rational numbers

$$+1\underbrace{00\ldots0}_{n \text{ zeros}}\cdot\epsilon \quad \text{and} \quad +\epsilon\cdot\underbrace{00\ldots0}_{n-1 \text{ zeros}}1$$

respectively. Note that this exponential notation makes sense, even though BTFA does not prove the totality of exponentiation.

Definition 9.2 *A function* $\alpha : \mathbb{N}_1 \to \mathbb{D}$ *is a* real number if $|\alpha(n) - \alpha(m)| \leq 2^{-n}$ for all tallies n and m with $n \leq m$. Two real numbers α and β are said to be *equal* (written $\alpha = \beta$) if $\forall n \in \mathbb{N}_1\, |\alpha(n) - \beta(n)| \leq 2^{-n+1}$.

The above definition resembles the definition of real number in RCA_0. The point in need of attention is that in the feasible context the domain of the function α is $\mathbb{N}_1$. In systems in which the totality of exponentiation is provable, there is no difference between $\mathbb{N}_1$ and $\mathbb{N}_2$. More precisely, in BTFA it is possible to define (in a natural way) an embedding of $\mathbb{N}_1$ into $\mathbb{N}_2$, and the totality of exponentiation may be taken to affirm that this embedding is surjective. In short, in a system where the totality of exponentiation is not provable (like BTFA), $\mathbb{N}_1$ and $\mathbb{N}_2$ are essentially different entities.

A particular real numbers is a *set* (of ordered pairs) whose existence needs to be shown by the comprehension available in BTFA. The relation of real equality is defined by a formula of the form $\forall x A$, with A a Π^b_1-formula (it is a $\forall\Pi^b_1$-formula). A *dyadic real number* is a triple (i, x, X) with $i = 0$ or $i = 1$, $x \in \mathbb{N}_2$ and X an infinite binary sequence. Informally, the idea is that the triple $(s, x_0 \ldots x_{n-1}, X)$ gives the real number $(-1)^s(\sum_{i=0}^{n-1} x_i 2^{n-i-1} + \sum_{i=0}^{\infty} \frac{X(i)}{2^{i+1}})$, where $X(i)$ is the (i+1)-th bit of X ($i \in \mathbb{N}_1$). We denote dyadic real numbers by $\pm x_0 x_1 \ldots x_{n-1} \cdot X$. Dyadic real numbers are a natural generalization of the dyadic rational numbers. It is shown in [12] that, over BTFA, the two alternative definitions of real numbers give the same numbers (this is far from straightforward, and in [11] it is deduced as a consequence of the definition of the value of a continuous function at a point of its domain: see a later comment on this issue). More precisely, to a dyadic real number $\pm x \cdot X$ we can associate the real number $\alpha_X : \mathbb{N}_1 \to \mathbb{D}$ given by $\alpha_X(n) := \pm x \cdot X[n]^*$ ($X[n]$ denotes the first n bits of X), and BTFA is able to prove that every real number α (according to Definition 9.2) is equal, as a real number, to a dyadic real number α_X (this is the hard part).

The arithmetical operations on the real numbers can be defined as follows:

- $\alpha + \beta$ is the real number $n \rightsquigarrow \alpha(n+1) + \beta(n+1)$
- $\alpha - \beta$ is the real number $n \rightsquigarrow \alpha(n+1) - \beta(n+1)$
- $\alpha \cdot \beta$ is the real number $n \rightsquigarrow \alpha(n+k) \cdot \beta(n+k)$, where k is the least *tally* such that $|\alpha(0)| + |\beta(0)| + 2 \leq 2^k$ (the symbol $\cdot$ is usually omitted)
- $\alpha \leq \beta$ is defined by the $\forall\Pi^b_1$-formula $\forall n(\alpha(n) \leq \beta(n) + 2^{-n+1})$
- $\alpha < \beta$ is defined by the formula $\alpha \leq \beta \wedge \alpha \neq \beta$ (which is equivalent to a $\exists\Sigma^b_1$-formula)
- $|\alpha|$ is the real number $n \rightsquigarrow |\alpha(n)|$,

and it is possible to prove (in BTFA) that the real numbers form an ordered field.

By $\forall\alpha \in \mathbb{R}\,(\ldots)$ or $\alpha \in [\beta, \gamma]$ we abbreviate $\forall\alpha$ (*if* α *is a real number then* ...) or α *is a real number and* $\beta \leq \alpha \leq \gamma$, respectively. Note that the language of BTFA does not allow for the formation of sets of sets, and so $\mathbb{R}$ does not make literal sense in BTFA. With this proviso, there plainly exists a natural embedding of $\mathbb{D}$ into $\mathbb{R}$, by identifying each dyadic rational number x with the real number α_x defined by the constant function $\alpha_x(n) = x$, for all $n \in \mathbb{N}_1$.

In the following definition $(x, n)\Phi(y, k)$ can informally be seen as stating that the elements in the interval $]x - 2^{-n}, x + 2^{-n}[$ are applied under Φ into elements of the interval $[y - 2^{-k}, y + 2^{-k}]$.

Definition 9.3 *A* continuous partial function *from* $\mathbb{R}$ *to* $\mathbb{R}$ *is a set* Φ *of codes of quintuples (denoted by* $\langle w, x, n, y, k\rangle$*) satisfying:*

- *if* $\langle w, x, n, y, k\rangle \in \Phi$ *then* w *is a first-order element,* $x, y \in \mathbb{D}, n, k \in \mathbb{N}_1$
- *if* $(x, n)\Phi(y, k)$ *and* $(x, n)\Phi(y', k')$ *then* $|y - y'| \leq 2^{-k} + 2^{-k'}$
- *if* $(x, n)\Phi(y, k)$ *and* $(x', n') < (x, n)$ *then* $(x', n')\Phi(y, k)$
- *if* $(x, n)\Phi(y, k)$ *and* $(y, k) < (y', k')$ *then* $(x, n)\Phi(y', k')$,

where $(x, n)\Phi(y, k)$ *stands for* $\exists w\, \langle w, x, n, y, k\rangle \in \Phi$ *and* $(x', n') < (x, n)$ *abbreviates* $|x - x'| + 2^{-n'} < 2^{-n}$.

The above definition follows closely Simpson's definition of continuous function in the context of reverse mathematics. Simple examples of continuous functions (as sets of quintuples) can be given, like the constant functions, the identity function, the modulus of a function, and the sum and product of two functions (cf. [11] for details). Now we present some standard definitions:

- Let Φ be a continuous partial function from $\mathbb{R}$ to $\mathbb{R}$. A real number α *is in the domain of* Φ, denoted by $\alpha \in \mathrm{dom}(\Phi)$, if
$$\forall k \in \mathbb{N}_1 \exists n \in \mathbb{N}_1 \exists x, y \in \mathbb{D}\,(|\alpha - x| < 2^{-n} \wedge (x, n)\Phi(y, k)).$$
- Let Φ be a continuous partial function from $\mathbb{R}$ to $\mathbb{R}$, and let α be a real number in the domain of Φ. We say that a real number β *is the value of* α *under the function* Φ, denoted by $\Phi(\alpha) = \beta$, if
$$\forall x, y \in \mathbb{D}\, \forall n, k \in \mathbb{N}_1((x, n)\Phi(y, k) \wedge |\alpha - x| < \frac{1}{2^n} \rightarrow |\beta - y| \leq \frac{1}{2^k}).$$

In BTFA it is possible to prove that if Φ is a continuous partial function from $\mathbb{R}$ into $\mathbb{R}$ and $\alpha \in \mathrm{dom}(\Phi)$, then there is a unique real number β satisfying $\Phi(\alpha) = \beta$ (the uniqueness condition is easy to check and, of course, it refers to uniqueness with respect to equality of reals). The proof of this fact (see [11] pages 569-572) is very detailed and sensitive, making a strong use of classical logic. It is worth remarking that the real-number theoretic relations $\Phi(\alpha) = \beta$ and $\Phi(\alpha) \leq \beta$ are given by $\forall\Pi_1^b$-formulas, while the relation $\Phi(\alpha) < \beta$ is given by a $\exists\Sigma_1^b$-formula.

Theorem 9.4 *Let Φ be a continuous function total in the interval $[0,1]$ such that $\Phi(0) < 0 < \Phi(1)$. Then there is a real number $\alpha \in [0,1]$ such that $\Phi(\alpha) = 0$.*

Proof. If there is a dyadic rational number $x \in [0,1]$ such that $\Phi(\alpha_x) = 0$, the proof is done. Suppose this is not the case. Thus, the value of a dyadic rational number in the interval $[0,1]$ under Φ is strictly positive or strictly negative. With the comprehension available in BTFA it is possible to form the sets

$$X_1 = \{x : x \in \mathbb{D} \cap [0,1] \wedge \Phi(\alpha_x) < 0\}$$

and

$$X_2 = \{x : x \in \mathbb{D} \cap [0,1] \wedge \Phi(\alpha_x) > 0\}$$

Note that the formulas defining the sets above, say $A_1(x)$ and $A_2(x)$ respectively, are both $\exists\Sigma_1^b$-formulas and are such that $\neg A_1(x)$ is equivalent to $A_2(x) \vee x \notin \mathbb{D} \cap [0,1]$ and $\neg A_2(x)$ is equivalent to $A_1(x) \vee x \notin \mathbb{D} \cap [0,1]$.

We now use a divide and conquer argument to construct a real number α such that $\Phi(\alpha) = 0$. Let

$$\begin{array}{rccc} f: & \mathbb{N}_1 & \to & \mathbb{D} \times \mathbb{D} \\ & n & \mapsto & \langle f_0(n), f_1(n) \rangle \end{array}$$

be the function, defined by bounded recursion along the tally part, by $f(0) = \langle 0, 1 \rangle$ and:

$$f(n+1) = \begin{cases} \langle (f_0(n) + f_1(n))/2, f_1(n) \rangle, & \text{if } (f_0(n) + f_1(n))/2 \in X_1; \\ \langle f_0(n), (f_0(n) + f_1(n))/2 \rangle, & \text{otherwise.} \end{cases}$$

By the induction on notation (for Σ_1^b-formulas) available in BTFA, it is possible to prove that, for all tally n, $f_0(n) \in X_1$, $f_1(n) \in X_2$, $f_0(n) \leq f_0(n+1)$, $f_1(n) \geq f_1(n+1)$, $f_0(n) < f_1(n)$ and $f_1(n) - f_0(n) = 2^{-n}$. But, then, f_0 and f_1 are real numbers such that $f_0 = f_1$. With $\alpha = f_0 = f_1$, we have that $\Phi(\alpha) = 0$. □

Within BTFA, we can speak of polynomials of *tally* degree. Given $d \in \mathbb{N}_1$, a sequence $(\gamma)_{i \leq d}$ of real numbers of length $d+1$ is a function

$$F : \{i \in \mathbb{N}_1 : i \leq d\} \times \mathbb{N}_1 \to \mathbb{D}$$

such that, for every $i \leq d$, the function γ_i defined by $\gamma_i(n) = F(i, n)$ is a real number. A real polynomial $P(X)$ of (tally) degree d is just such a sequence with the proviso that $\gamma_d \neq 0$. As usual, we write $P(X) = \gamma_d X^d + \cdots + \gamma_1 X + \gamma_0$. It is not difficult to define smoothly $P(\alpha)$, for each real number α. (Note that if d was not tally then a^d, for $a \in \mathbb{N}_2$, would not make sense in general because BTFA does not prove that exponentiation is total. On the other hand a^d, with $a \in \mathbb{N}_2$ and $d \in \mathbb{N}_1$, is always a well-defined dyadic number.) It is even possible to show in BTFA that to each tally polynomial P as described, we can associate a total continuous function Φ_P such that, for each real number α, $\Phi_P(\alpha) = P(\alpha)$. As a *very particular* case of this discussion, it makes sense to speak in BTFA of polynomials of *standard* degree and to see that they are given by continuous functions in the sense of Definition 9.3. It should now be clear, using the intermediate value theorem, that – as observed in the introduction – Tarski's theory of real closed ordered fields RCOF is interpretable in BTFA.

9.3 The role of weak König's lemma

Given A a formula of $\mathcal{L}_2$ and x a first-order variable, we denote by $Tree(A_x)$ the formula:

$$\forall x \forall y\, (A(x) \wedge y \subseteq x \to A(y)) \wedge \forall n \in \mathbb{N}_1 \exists x\, (l(x) = n \wedge A(x)).$$

The first condition expresses that initial subwords of words satisfying A still satisfy A, and the second condition says that there are binary sequences that satisfy A of arbitrarily large tally length (the tree given by the formula $A(x)$ is infinite). Given X a second-order variable, we denote by $Path(X)$ the formula:

$$Tree((x \in X)_x) \wedge \forall x \forall y\, (x \in X \wedge y \in X \to x \subseteq y \vee y \subseteq x).$$

Weak König's lemma for trees defined by bounded formulas, denoted by Σ^b_∞-WKL, is the following scheme:

$$Tree(A_x) \to \exists X\, (Path(X) \wedge \forall x\, (x \in X \to A(x))),$$

where A is a bounded formula and X is a fresh variable. Informally it says that every infinite binary tree (defined by a bounded formula) has an infinite path. Note that, while the tree needs not to be a set in the system, the path is a set.

Theorem 9.5 BTFA $+ \Sigma^b_\infty$-WKL *is* Π^1_1*-conservative over* BTFA.

A consequence of this theorem is that the class of provably total functions (with Σ^b_1-graphs) of BTFA $+ \Sigma^b_\infty$-WKL is still the class of poly-time computable functions. Even though BTFA$+\Sigma^b_\infty$-WKL has the same proof-theoretic strength as BTFA, weak König's lemma is a useful form of a compactness principle that increases the demonstrative power of BTFA. In what follows, we denote by Π^b_1-WKL the Σ^b_∞-WKL scheme restricted to Π^b_1-formulas. By WKL we denote the Σ^b_∞-WKL scheme above restricted to sets i.e., to formulas $A(x)$ of the form $x \in X$.

Definition 9.6 (BTFA) *An* open set *of* $\mathbb{R}$ *is a set* U *of codes of triples of the form* $\langle w, z, n\rangle$ *such that* w *is a first-order element,* $z \in \mathbb{D}$ *and* $n \in \mathbb{N}_1$. *We say that a real number* α *is an element of* U*, and we write* $\alpha \in U$*, if*

$$\exists z \in \mathbb{D}\, \exists n \in \mathbb{N}_1\, (|\alpha - z| < \frac{1}{2^n} \wedge \exists w\, \langle w, z, n\rangle \in U).$$

The formulation of open set may appear unfamiliar at first sight, but it merely says that the open sets of $\mathbb{R}$ are given by countable unions of the form

$$\bigcup_{w} \bigcup_{\substack{z,n \\ \langle w,z,n\rangle \in U}}]z - \frac{1}{2^n}, z + \frac{1}{2^n}[.$$

The Heine/Borel theorem for the closed unit interval says that if U is an open set such that $[0, 1] \subseteq U$ (i.e., every real number in the closed unit interval is an element of U), then there is $k \in \mathbb{N}_1$ such that: For all $\alpha \in [0, 1]$, there are w, $z \in \mathbb{D}$, $n \in \mathbb{N}_1$, all of length less than k, such that $|\alpha - z| < \frac{1}{2^n}$ and $\langle w, z, n\rangle \in U$.

Theorem 9.7 (BTFA) *The Heine/Borel theorem for* $[0, 1]$ *is equivalent to* Π^b_1-WKL.

We are not going to prove this result here. The proof adapts well-known arguments of reverse mathematics and the details can be found in [12]. Instead, we will try to convey to the reader the need of Π^b_1-WKL instead of just plain WKL. How does the argument from the right side to the left side goes? It hinges on considering a tree given by a formula $T(x)$ of the form $\forall y\, (\langle x, y\rangle \in X)$, for a certain *set* X. The tree is proven to be bounded by contradiction. If it were infinite, it would have an infinite path and this gives rise to absurdity. Now, and this is the critical point, how can we conclude from the infinitude of the tree that it has an infinite path? It is well-known that we can associate to the above tree $T(x)$ the tree $T'(x) :\equiv \forall w \subseteq x \forall y \preceq x (\langle w, y \rangle \in X)$. Moreover, T' is infinite if T is, and any path through T' is also a path through T. We apply weak König's lemma to T'. This tree is defined by a Π_1^b-formula, and this is why plain WKL is not enough in the feasible setting. Over RCA_0 the problem does not arise because the formulas in the comprehension scheme of RCA_0 are closed under bounded quantifications.

The above result shows the need for fine-tuned versions of weak König's lemma when doing reverse mathematics over a feasible base theory.

Definition 9.8 *Let $\Phi : [0,1] \to \mathbb{R}$ be a continuous total function. We say that Φ is* uniformly continuous *if*

$$\forall k \in \mathbb{N}_1 \exists m \in \mathbb{N}_1 \forall \alpha, \beta \in [0,1](|\alpha - \beta| \leq \tfrac{1}{2^m} \to |\Phi(\alpha) - \Phi(\beta)| < \tfrac{1}{2^k}).$$

Proposition 9.9 (BTFA) *Let $\Phi : [0,1] \mapsto \mathbb{R}$ be a uniformly continuous function. Then there is $n \in \mathbb{N}_1$ such that, for all $\alpha \in [0,1]$, $|\Phi(\alpha)| \leq 2^n$.*

Proof. Take $m \in \mathbb{N}_1$ such that if $|\alpha - \beta| \leq 2^{-m}$ then $|\Phi(\alpha) - \Phi(\beta)| < 1$, for $\alpha, \beta \in [0,1]$. It is easy to see, by bounded collection, that there is $r \in \mathbb{N}_1$ such that $\forall x\, (\ell(x) = m \to |\Phi(.x^*)| < 2^r)$. Since every real in the closed unit interval is within 2^{-m} of a certain $.x^*$ for x of length m, it is clear that $n = r + 1$ does the job. □

In the next result there is a gap that we are unable to fill.

Theorem 9.10 (BTFA) *The principle that every total real valued continuous function defined on $[0,1]$ is uniformly continuous implies* WKL *and is implied by* Π_1^b-WKL.

Proof. To prove the first assertion, suppose that WKL is not valid. Take T an infinite binary (set) tree which has no infinite paths. In BTFA it is possible to prove that there is a total continuous function Φ, defined on $[0,1]$ such that, for all end nodes x of T, we have $\Phi(\cdot x^*) = 2^{l(x)}$ (see [12], pages 5-6 for a proof of a more general result). Since T has nodes of arbitrarily large length, Φ is unbounded. This contradicts Proposition 9.9.

To prove the second assertion we reason within BTFA+Π_1^b-WKL. Let Φ be a (total) real valued continuous function on [0,1] and fix $k \in \mathbb{N}_1$. Let U be the open set defined by $\{\langle \langle w, y \rangle, x, n+1 \rangle : \langle w, x, n, y, k+2 \rangle \in \Phi\}$.

Since Φ is a total function, it can be proved that $[0,1] \subseteq U$. By Theorem 9.7, we have the Heine/Borel theorem. Thus, there is $m \in \mathbb{N}_1$ such that: For all $\alpha \in [0,1]$, there are $x, y \in \mathbb{D}$, $n \in \mathbb{N}_1$, all of length less than m, such that $|\alpha - x| < \frac{1}{2^{n+1}}$ and $(x,n)\Phi(y, k+2)$. We claim that, for all $\alpha, \beta \in [0,1]$, if we have $|\alpha - \beta| \leq \frac{1}{2^m}$ then $|\Phi(\alpha) - \Phi(\beta)| < \frac{1}{2^k}$. Take α, β as in the claim. Take $x, y \in \mathbb{D}$ and $n \in \mathbb{N}_1$ with $n <_{\mathbb{N}_1} m$, $|\alpha - x| < \frac{1}{2^{n+1}}$ and $(x,n)\Phi(y, k+2)$. By definition of continuous function, we have $|\Phi(\alpha) - y| \leq \frac{1}{2^{k+2}}$. Now, $|\beta - x| \leq |\beta - \alpha| + |\alpha - x| < \frac{1}{2^m} + \frac{1}{2^{n+1}} \leq \frac{1}{2^n}$. Hence, by definition of continuous function again, $|\Phi(\beta) - y| \leq \frac{1}{2^{k+2}}$. We conclude that $|\Phi(\alpha) - \Phi(\beta)| \leq \frac{1}{2^{k+1}} < \frac{1}{2^k}$. □

Before we finish this section with a last result, we need to make some brief considerations concerning induction. The induction available in BTFA is induction *on notation* for Σ_1^b-formulas. In BTFA we can introduce, in a natural way, a *successor function* S defined by: $S(\epsilon) = 0$, $S(x0) = x1$ and $S(x1) = S(x)0$, i.e., we order the binary words according to length and, within the same length, lexicographically, and S is the successor function induced by this order (usually denoted by $\leq_l$, l for lexicographic order).

The scheme of "slow" induction is $A(\epsilon) \wedge \forall x\,(A(x) \rightarrow A(S(x))) \rightarrow \forall x A(x)$, where A is a Σ_1^b-formula. It corresponds to the usual $+1$ scheme of induction in ordinary theories of arithmetic (it gives Buss's theory T_2^1). This scheme does not seem to be available in BTFA (in [24] it is shown that if this is the case – actually if Buss's theories S_2^1 and T_2^1 are the same – then the polytime computable hierarchy collapses).

Theorem 9.11 *Over* BTFA + Σ_∞^b-WKL, *the following are equivalent:*

(a) Every continuous real valued function defined on $[0,1]$ *has a maximum.*

(b) Every continuous real valued function defined on $[0,1]$ *has a supremum.*

(c) Slow induction for Σ_1^b*-formulas.*

See [12] for a detailed proof. Here we just sketch the strategy to prove that (b) implies (c). We consider (c) in the following (BTFA equivalent) formulation: every non-empty set of binary words of equal length has a lexicographically greatest element. It is easy to construct a piecewise linear continuous real valued function Φ defined on [0,1] such that, if $x \in X$ then $\Phi(\cdot x^*) = \cdot x^*$, and if $x \notin X$ then $\Phi(\cdot x^*) = 0$. By hypothesis (b), Φ has a supremum, from which we can extract the desired maximum.

The proof that (c) implies (a) is much more involved and requires (first) the formation of an infinite tree defined by a Π^0_1-formula so that an infinite path Y through it satisfies $\forall\alpha \in [0,1](\Phi(\alpha) \leq \cdot Y)$ and (second) to prove, using Σ^b_∞-WKL again, that there is $\alpha \in [0,1]$ such that $\Phi(\alpha) = \cdot Y$.

9.4 A brief digression on the Baire category theorem

Let us consider in BTFA the following formulation of the Baire category theorem for the Cantor space:

$$\forall z \forall a \exists x (a \subseteq x \wedge A(x,z)) \rightarrow \exists X (Path(X) \wedge \forall z \exists x\, (x \in X \wedge A(x,z)),$$

where A is an arithmetical formula (possibly with first and second-order parameters). The above scheme says that if an arithmetical formula $A(x,z)$ defines a countable family $D^A_z := \{x : A(x,z)\}$ of dense open sets in the Cantor space, then there is a path X intersecting all these sets. Let us call Π^0_∞-BCT this form of the Baire category theorem. It is well-known that $\mathsf{RCA}_0 + \Pi^0_\infty$-BCT is Π^1_1-conservative over RCA_0. This was shown by Simpson and Douglas Brown in [4] using Cohen forcing. Fernandes asked in [9] whether the theory BTFA+Π^0_∞-BCT is Π^1_1-conservative over BTFA. The answer is negative by a wide margin:

Theorem 9.12 *The theory* BTFA $+ \Pi^0_\infty$-BCT *proves the totality of exponentiation.*

Remark 9.13 *We will use the fact that in* BTFA *the totality of exponentiation can be formulated in the following manner:* $\forall a \exists c \forall z \preceq a\, (z \subseteq^* c)$. *See [15].*

Proof. We reason within BTFA $+ \Pi^0_\infty$-BCT. Clearly, $\forall z \forall a \exists x\, (a \subseteq x \wedge z \subseteq^* x)$. By Π^0_∞-BCT, there is an infinite path X such that $\forall z \exists x\, (x \in X \wedge z \subseteq^* x)$. Let us now show that exponentiation is total. Consider an arbitrary element a. Since we have $\forall z \preceq a\, \exists x\, (x \in X \wedge z \subseteq^* x)$, by bounded collection we know that there is b such that $\forall z \preceq a\, \exists x \preceq b\, (x \in X \wedge z \subseteq^* x)$. Now, take c such that $c \equiv b$ (c has the same length as b) and $c \in X$. It is clear that $\forall z \preceq a\, (z \subseteq^* c)$. We are done. □

The above proof is implicit in the final section of [9], but it is brought here into the open for the first time. The use of bounded collection in the above argument is crucial. The situation seems to be different if bounded collection is not included. In fact, Takeshi Yamazaki in [28] and Fernandes in [9] studied Π^0_∞-BCT over a feasible theory weaker than BTFA. The weaker second-order theory that they considered is Σ^b_1-NIA+ ∇^b_1-CA, where ∇^b_1-CA is the following comprehension scheme:

$$\forall x\,(A(x) \leftrightarrow B(x)) \to \exists X \forall x\,(x \in X \leftrightarrow A(x)),$$

where A is a Σ^b_1-formula and B is a Π^b_1-formula (possibly with first and second-order parameters) and X is a fresh variable. This theory allows only the formation of NP $\cap$ co-NP sets and does *not* include bounded collection. Yamazaki showed that Σ^b_1-NIA + ∇^b_1-CA + Π^0_∞-BCT is Π^1_1-conservative over Σ^b_1-NIA+∇^b_1-CA and Fernandes improved this conservation result to include uniqueness statements of the form $\forall X \exists! Y A(X,Y)$, for A arithmetical. Their proofs also use Cohen forcing. Brown and Simpson showed that Cohen forcing preserves Σ^0_1-induction but, as follows from Theorem 9.12, Cohen forcing does not preserve bounded collection (at least in the absence of the totality of exponentiation). On the other hand, Harrington's forcing preserves both bounded collection and Σ^0_1-induction.

9.5 Riemann integration and the theory TCA^2

How far can we go in the formalization of analysis in feasible systems? In [17] it is shown that if we are able to do a minimum of integration in BTFA, then it follows that we can count. This looks unsurprising. Given $X \subseteq \mathbb{N}_2$, we can associate (within BTFA) a continuous total function $\Phi_X : [0, \infty[\to \mathbb{R}$, with a modulus of uniform continuity (see Definition 9.21), in such a way that, given $w \in \mathbb{N}_2$, $\int_0^w \Phi_X(x)dx$ is the number of elements of X up to w. For instance, if $X = \{0, 2, 3\}$ then Φ_X is the function:

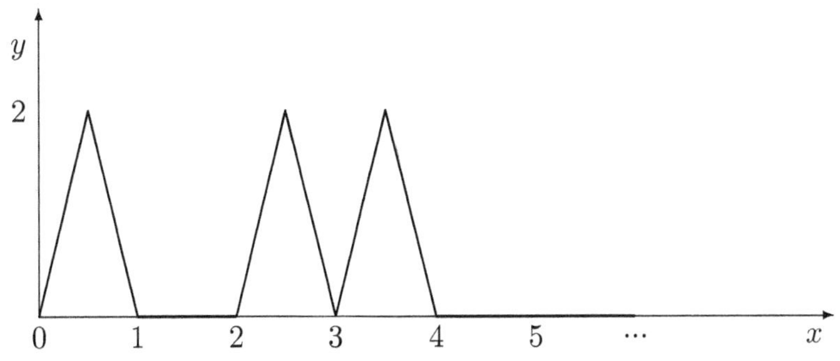

There are, however, some technical difficulties in pulling the above argument through in BTFA. The difficulties can, nevertheless, be met: see [17].

As mentioned in the introduction, in her doctoral dissertation (in Portuguese) G. Ferreira showed in detail how Riemann integration can be developed in a theory whose class of provably total functions is the *hierarchy of counting functions* (a computational complexity class between PTIME and PSPACE). The theme is also the subject of the paper [19]. In the rest of this section we are going to sketch why/how such formalization is possible, focusing on the logical problems of the formalization.

By $\exists X \preceq t\ A$ we abbreviate the formula $\exists X(X \preceq t \wedge A)$, where $X \preceq t$ means that every word in X has length less than or equal to the length of t. The abbreviation $\forall X \preceq t\ A$ has the corresponding dual meaning. These are the *bounded second-order quantifiers.* The $\Sigma_1^{1,b}$*-formulas* (respectively $\Pi_1^{1,b}$*-formulas*) are the formulas of the form $\exists X \preceq t\, A$, (respectively $\forall X \preceq t\, A$) where A is bounded formula. A $\Sigma_\infty^{1,b}$*-formulas* is a formula of $\mathcal{L}_2$ where all the first and second-order quantifications are bounded. This class of formulas constitutes a natural generalization of the bounded formulas.

Before presenting the *counting axiom*, which is crucial in our theory for integration, let us motivate it. Informally, given $X \preceq w$ and $y \preceq w$ we want to be able to count the number of elements $\leq_l y$ which are in X. Let f be such that $f(\epsilon) = \epsilon$, if $\epsilon \notin X$, $f(\epsilon) = 0$, if $\epsilon \in X$, and

$$f(S(x)) = \begin{cases} f(x) & \text{if } S(x) \notin X \\ S(f(x)) & \text{if } S(x) \in X \end{cases}$$

It is clear that $f(y)$ gives the result (in the $\leq_l$-order) of the above counting. Formally, the *counting axiom* has the form:

$$\forall X \preceq w\ \exists C \preceq q(w)\ Count(C, X),$$

where $q(w)$ is a certain term which depends on the variable w (for the record, the term $q(w)$ can be taken to be $wwww1111$, cf. [21]), and $Count(C, X)$ is a bounded formula (we omit it) which says the following: given $y \preceq w$, $\langle y, j\rangle \in C$ if and only if $f(y) = j$. For the formulation of the axiom, with the exact expression for $Count$, see [19, 21]. It is easy to see that this counting axiom also permits to do counting with the result given in $\mathbb{N}_2$. This is the usual counting and it is this counting that we will use from now on.

Definition 9.14 TCA^2 (*acronym for* Theory for Counting Arithmetic) *is the second-order theory in the language* $\mathcal{L}_2$ *which has the following axioms:*

- Basic axioms (*the previous 14 basic axioms of Definition 9.1*);
- Induction on notation for bounded formulas

$$A(\epsilon) \wedge \forall x\,(A(x) \to A(x0) \wedge A(x1)) \to \forall x A(x),$$

  with A *a bounded formula (first and second-order parameters are permitted);*

- Substitution for bounded formulas (*This is a technical axiom scheme that permits a kind of "permutation" between a bounded first-order universal quantification and a bounded second-order existential quantification. See [19] for its formulation. This axiom is instrumental in showing that* $\Sigma_1^{1,b}$*-formulas are provably closed under bounded first-order quantifications. A dual property holds for* $\Pi_1^{1,b}$*-formulas. We use these properties without much ado.*);
- Counting axiom

$$\forall X \preceq w \exists C \preceq q(w)\ Count(C, X)$$

- Extended bounded collection scheme ($\mathsf{B}^1\Sigma_\infty^{1,b}$)

$$\forall X \preceq w \exists y A(y, X) \to \exists z \forall X \preceq w \exists y \preceq z A(y, X),$$

  where A *is a* $\Sigma_\infty^{1,b}$*-formula (possibly with first and second-order parameters) and* z *is a new variable;*

- Recursive comprehension scheme

$$\forall x\,(\exists y A(x, y) \leftrightarrow \forall z B(x, z)) \to \exists X \forall x\,(x \in X \leftrightarrow \exists y A(x, y))$$

  where A *is a* $\Sigma_1^{1,b}$*-formula and* B *is a* $\Pi_1^{1,b}$*-formula (possibly with first and second-order parameters) and* X *is a fresh variable.*

The class of provably total functions of TCA^2 with $\Sigma_1^{1,b}$-graphs is exactly the computational complexity class FCH. For a proof of this result see [21, 22]. For a general blueprint of how to construct theories for weak analysis related to concrete computational complexity classes see [13]. (The FCH class consists of the *hierarchy of counting functions*. It is a computational complexity class which lies between PTIME and

PSPACE introduced by Klaus Wagner [27] in 1986. More precisely, FCH is $\bigcup_{i\geq 0} i\#\mathsf{P}$, where $0\#\mathsf{P} = \mathsf{P}$ and $(i+1)\#\mathsf{P} = \#\mathsf{P}^{i\#\mathsf{P}}$, for $i \geq 0$, i.e. $(i+1)\#\mathsf{P}$ is the class of functions that "count" the number of accepting computations in a polynomial time nondeterministic Turing Machine, permitting a function in $i\#\mathsf{P}$ as an oracle.)

Before we start developing Riemann integration, we need to lay out a proposition indicating the forms of induction and minimization that are available in TCA^2. We will use them at will.

Proposition 9.15 *The following is provable in* TCA^2 *(cf. [21]):*

- *induction on notation for* $\Delta_1^{1,b}$*-formulas,*
- *plain* $(+1)$ *induction on* $\mathbb{N}_2$ *for* $\Delta_1^{1,b}$*-formulas,*
- *the minimization scheme* $\exists x\, A(x) \rightarrow \exists x\, (A(x) \wedge \forall y <_l x \neg A(y))$*, for* $\Delta_1^{1,b}$*-formulas* A*.*

A crucial step towards integration is the ability to sum. It is not difficult to see that the ability of counting is sufficient to perform sums. Informally, we want to show that given $f : X \times \mathbb{N}_2 \rightarrow \mathbb{D}$ there is (in TCA^2) a function $\sum_f : X \times \mathbb{N}_2 \rightarrow \mathbb{D}$ such that $\sum_f(x,n) = f(x,0) + \ldots + f(x,n)$. $\sum_f(x,n)$ will be denoted by $\sum_{i=0}^{n} f(x,i)$. We start with a preliminary lemma.

Lemma 9.16 *Let* f *be a function from* $X \times \mathbb{N}_2$ *to* $\mathbb{N}_2$*. Then there is a function* $g : X \times \mathbb{N}_2 \rightarrow \mathbb{N}_2$ *such that* $\forall x \in X \forall n \in \mathbb{N}_2 \forall i \leq n\, (f(x,i) \leq g(x,n))$*.*

Proof. Let us fix $x \in X$ and $n \in \mathbb{N}_2$. By bounded collection it is easy to see that $\exists r \forall i \leq n\, (f(x,i) \leq r)$. Let ϕ be the bounded formula $\forall i \leq n\, (f(x,i) \leq r)$. Since $\forall x \in X \forall n \in \mathbb{N}_2 \exists r \in \mathbb{N}_2\, \phi(x,n,r)$, we can apply minimization and have

$$\forall x \in X \forall n \in \mathbb{N}_2 \exists r\, (\phi(x,n,r) \wedge \forall r' < r \neg\phi(x,n,r')).$$

Thus $g := \{\langle\langle x,n\rangle, r\rangle : x \in X \wedge n \in \mathbb{N}_2 \wedge r \in \mathbb{N}_2 \wedge \phi(x,n,r) \wedge \forall r' < r \neg\phi(x,n,r')\}$ is a function from $X \times \mathbb{N}_2$ to $\mathbb{N}_2$ satisfying the desired condition. □

Theorem 9.17 *Given* $f : X \times \mathbb{N}_2 \rightarrow \mathbb{N}_2$*, there is a function* $\Sigma_f : X \times \mathbb{N}_2 \rightarrow \mathbb{N}_2$ *such that* $\forall x \in X \forall n \in \mathbb{N}_2\, [\Sigma_f(x,0) = f(x,0) \wedge \Sigma_f(x,n+1) = \Sigma_f(x,n) + f(x,n+1)]$*.*

Proof. Given $x \in X$ and $n \in \mathbb{N}_2$, let Z be the set

$$\{u : \exists i, r \preceq u\, (u = \langle r, i\rangle \wedge i \leq n \wedge r < f(x,i))\},$$

where we suppose that the pairing function (in $\mathbb{N}_2$) is monotone (in the sense of $\preceq$) in both arguments and is such that $r, i \preceq \langle r, i\rangle$. Informally, the idea is that $\Sigma_f(x,n)$ is the number of elements in Z. Note that from Lemma 9.16, there is a function g such that $u \in Z \to u \preceq \langle g(x,n), n\rangle$, i.e., Z is a bounded set. Let $w := \langle g(x,n), n\rangle$. By the counting axiom in $\mathbb{N}_2$, for a concrete term $q(w)$ there is $C \preceq q(w)$ such that $\langle u, j\rangle \in C$ iff there are j elements in $\mathbb{N}_2$ less than or equal to u in Z. Let $P(Z)$ abbreviate: $\forall u \preceq w\, (u \in Z \leftrightarrow \exists i, r \preceq u\, (u = \langle r, i\rangle \wedge i \leq n \wedge r < f(x,i)))$. Take Σ_f as

$$\{\langle\langle x, n\rangle, s\rangle : x \in X \wedge n \in \mathbb{N}_2 \wedge \exists Z \preceq w \exists C \preceq q(w) \\ (P(Z) \wedge Count(C, Z) \wedge \langle w, s\rangle \in C)\}.$$

It can be seen that the set Σ_f exists in TCA^2. This uses the counting axiom and the recursion comprehension scheme. Note that the latter condition above is equivalent to $\forall Z \preceq w \forall C \preceq q(w)\, (P(Z) \wedge Count(C,Z) \to \langle w, s\rangle \in C)$. Clearly, Σ_f defines a function satisfying the desired conditions. □

It is not difficult to see that the above result can be extended and the following proposition proved:

Proposition 9.18 *Given $f : X \times \mathbb{N}_2 \to \mathbb{D}$, there is a function $\Sigma_f : X \times \mathbb{N}_2 \to \mathbb{D}$ s.t. $\Sigma_f(x,0) = f(x,0)$ and $\Sigma_f(x,n+1) = \Sigma_f(x,n) + f(x,n+1)$, $\forall x \in X, \forall n \in \mathbb{N}_2$.*

Note that the relation $z = \sum_{i=0}^{n} f(x,i)$ is $\Delta_1^{1,b}$-definable. The usual properties of summations can be proved by plain induction on $n \in \mathbb{N}_2$. We will not make a list of them and assume that the new notations that will occur in the computations below can be given a straightforward sense in TCA^2. All the details can be found in G. Ferreira's doctoral dissertation [21].

Given α a real number and $n \in \mathbb{N}_1$, the dyadic rational number $\alpha(n)$ is well determined. Note that the formula $\alpha(n) =_{\mathbb{D}} d$ is a bounded formula, since it abbreviates $\langle n, d\rangle \in \alpha$. However, given Φ a continuous partial function from $\mathbb{R}$ to $\mathbb{R}$ and $\alpha \in \mathrm{dom}(\Phi)$, although the expression $\Phi(\alpha)$ is well-defined, the expression $\Phi(\alpha)(n)$ is ambiguous. This is because $\Phi(\alpha)$ is only defined modulo equality of the reals. In order to control the complexity of the formula which defines the integral, it is necessary to introduce the expression $\Phi(\alpha, n)$ that gives a canonical representative of $\Phi(\alpha)$. This is possible to do in TCA^2 but apparently not in BTFA (because of lack of minimization).

Let Φ be a continuous partial function from $\mathbb{R}$ to $\mathbb{R}$ and α a real number in the domain of Φ. Consider $\varphi(n, r)$ the formula

$$\exists w \exists k\, [\langle w, \alpha(k+1), k, r, n+1\rangle \in \Phi \wedge$$

$$\forall \langle r', w', k'\rangle < \langle r, w, k\rangle\, (\langle w', \alpha(k'+1), k', r', n+1\rangle \notin \Phi)].$$

It can be proved that, given $n \in \mathbb{N}_1$, there exists a unique $r \in \mathbb{D}$ such that $\varphi(n, r)$. The proof of this result (see [19], pages 926–927) uses minimization.

Definition 9.19 *Given Φ a continuous partial function from $\mathbb{R}$ to $\mathbb{R}$ and α a real number in the domain of Φ, we define $\Phi(\alpha, n) = r :\leftrightarrow \varphi(n, r)$.*

Note that $\Phi(\alpha, n)$ is the unique dyadic rational number such that $\varphi(n, \Phi(\alpha, n))$ and that $\{\langle n, r\rangle : \Phi(\alpha, n) = r\}$ is a set in TCA^2.

Proposition 9.20 *Let Φ be a continuous partial function from $\mathbb{R}$ to $\mathbb{R}$ and let $\alpha \in \mathrm{dom}(\Phi)$. The function $\lambda : \mathbb{N}_1 \to \mathbb{D}$ defined by $\lambda(n) = \Phi(\alpha, n)$ is a real number. Moreover, for every real number β such that $\Phi(\alpha) = \beta$, we have $|\Phi(\alpha, n) - \beta| \leq \frac{1}{2^n}$. In particular, $\Phi(\alpha) = \lambda$.*

The theory TCA^2 can formalize Riemann integration for functions with a modulus of uniform continuity.

Definition 9.21 *Let $\Phi : [0,1] \to \mathbb{R}$ be a continuous total function. A* modulus of uniform continuity (m.u.c.) *for Φ is a strictly increasing function h from $\mathbb{N}_1$ to $\mathbb{N}_1$ such that for all $n \in \mathbb{N}_1$ and for all $\alpha, \beta \in [0,1]$, if $|\alpha - \beta| < 2^{-h(n)}$ then $|\Phi(\alpha) - \Phi(\beta)| < 2^{-n}$.*

In order to simplify notation, we start by introducing the notion of Riemann integral for functions restricted to the interval $[0, 1]$.

Definition 9.22 *Take Φ a continuous total function in the interval $[0,1]$, with a modulus of uniform continuity h in that interval. We define* the definite integral *between 0 and 1 of Φ, denoted by $\int_0^1 \Phi(t)\, dt$, in the following way:*

$$\int_0^1 \Phi(t)\, dt :=_{\mathbb{R}} \lim S_n$$

where, for all $n \in \mathbb{N}_1$, $S_n = \sum_{i=0}^{2^{h(n)}-1} \frac{1}{2^{h(n)}} \Phi(\frac{i}{2^{h(n)}}, n)$.

Note that, $f : \mathbb{N}_1 \times \mathbb{N}_2 \to \mathbb{D}$, defined by $f(n,i) = \frac{1}{2^{h(n)}}\Phi(\frac{i}{2^{h(n)}}, n)$ is a function in TCA^2. Also observe that it is possible to consider sums of the form $\sum_{i=0}^{2^{h(n)}-1} f(n,i)$, for f a function from $\mathbb{N}_1 \times \mathbb{N}_2$ to $\mathbb{D}$. In fact, $\sum_{i=0}^{2^{h(n)}-1} f(n,i) = \Sigma_f(n, 2^{h(n)} - 1)$.

Of course, we need to make sense of the limit above. One way to do this is to develop a theory of limits within TCA^2 and prove that Cauchy sequences, with a modulus of Cauchy convergence, have limits. This is done in [21]. In here we rely on the fact (also proved in [21]) that α defined by $\alpha(n) = S_{n+5}$ is a real number. We just let the above limit to be α. It can be proved (see [21], pages 122–123) that the value of the integral (as a real number) does not depend on the function chosen as a modulus of uniform continuity.

Some of the usual properties of the integral can be established in TCA^2 (see [21], pages 123–126). For instance:

Proposition 9.23 *Let Φ and Ψ be continuous total functions in the interval $[0,1]$ with a modulus of uniform continuity in that interval and $\gamma \in \mathbb{R}$:*

a) $\int_0^1 \gamma\, dt = \gamma$

b) $\int_0^1 t\, dt = \frac{1}{2}$

c) $\int_0^1 (\Phi + \Psi)(t)\, dt = \int_0^1 \Phi(t)\, dt + \int_0^1 \Psi(t)\, dt$

d) $|\int_0^1 \Phi(t) dt| \leq \int_0^1 |\Phi|(t) dt$

e) If $\Phi(t) = \Psi(t)$ *for all* $t \in [0,1]$ *then* $\int_0^1 \Phi(t)\, dt = \int_0^1 \Psi(t)\, dt$

f) If $\Phi(t) \leq \Psi(t)$ *for all* $t \in [0,1]$ *then* $\int_0^1 \Phi(t)\, dt \leq \int_0^1 \Psi(t)\, dt$

g) $\int_0^1 \gamma\Phi(t)\, dt = \gamma \int_0^1 \Phi(t)\, dt.$

We can (as we sketch next) introduce the notion of Riemann integral with arbitrary dyadic rational limits in an analogous way.

Definition 9.24 *Take $x, y \in \mathbb{D}$ such that $x < y$ and Φ a continuous total function in the interval $[x,y]$, with a modulus of uniform continuity h in that interval. We define the* integral between x and y *of Φ, we denote by $\int_x^y \Phi(t)\, dt$, in the following way:*

$$\int_x^y \Phi(t)\, dt :=_{\mathbb{R}} \lim S_n$$

where, for all $n \in \mathbb{N}_1$, $S_n = \sum_{i=0}^{2^{h(n)}-1} \frac{y-x}{2^{h(n)}}\Phi(x + \frac{(y-x)i}{2^{h(n)}}, n)$.

With this definition, we have similar properties to those of Proposition 9.23. Note that S_n is obtained by taking a partition of $[x, y]$ with diameter $\frac{y-x}{2^{h(n)}}$. Such partitions are designated by *standard* partitions. The definition of integral is robust in the following (expected) sense: it does not depend upon the choice of points in the subintervals of the partitions nor on the adjunction of new points to the *standard* partitions. Such is crucial in proving the following property (see [21], pages 128–129):

Proposition 9.25 *Take z a dyadic rational number such that $x < z < y$ and Φ a continuous total function in $[x, y]$ with a modulus of uniform continuity in that interval, then*

$$\int_x^z \Phi(t)\, dt + \int_z^y \Phi(t)\, dt = \int_x^y \Phi(t)\, dt.$$

Given Φ a continuous total function in $[0, 1]$ with a modulus of uniform continuity h, we will define Ψ, a continuous total function in $[0, 1]$, such that $\Psi(x) = \int_0^x \Phi(t)\, dt$ for all dyadic rational number $x \in [0, 1]$. By Proposition 9.9, take $m \in \mathbb{N}_1$ such that $\forall \alpha \in [0, 1], |\Phi|(\alpha) \leq 2^m$. Consider $d : \mathbb{D} \to \mathbb{D}$ the function defined by $d(x) = 0$, for $x < 0$; $d(x) = x$, for $0 \leq x \leq 1$; and $d(x) = 1$, for $x > 1$. We define $(x, n)\Psi(y, k)$ as:

$$x, y \in \mathbb{D} \wedge n, k \in \mathbb{N}_1 \wedge \left| \int_0^{d(x)} \Phi(t)\, dt - y \right| < \frac{1}{2^k} - \frac{1}{2^{n-m-1}}.$$

The formula above is equivalent to a $\exists\Sigma^b_\infty$-formula of the form

$$\exists w \theta'(w, x, n, y, k),$$

with θ' bounded. The set $\{\langle w, x, n, y, k\rangle : \theta'(w, x, n, y, k)\}$ is officially the function Ψ.

We can prove (see [21], pages 130–131) the following result:

Proposition 9.26 *Let Ψ be the set of quintuples, as defined above. This set is a partial continuous function from $\mathbb{R}$ to $\mathbb{R}$. Moreover, if $\alpha \in [0, 1]$ then $\alpha \in \mathrm{dom}(\Psi)$ and, for all dyadic rational number r in $[0, 1]$, $\Psi(\alpha_r) = \int_0^r \Phi(t)\, dt$.*

Given Φ a continuous total function in $[0, 1]$ with a modulus of uniform continuity in that interval, Ψ is the *indefinite integral of* Φ. Note that, although the function Ψ, as a set, depends on the *tally* m chosen, the images of the reals in $[0, 1]$ under Ψ do not depend, as reals, on such a choice. Therefore, although rigorously the indefinite integral of Φ may be given by different sets of quintuples, we can indeed speak of the *indefinite integral function.*

The previous proposition permits to give a meaning to $\int_0^\alpha \Phi(t)dt$ also for real numbers $\alpha \in [0,1]$. We can easily define $\int_\alpha^\beta \Phi(t)dt$, for $\alpha, \beta \in [0,1]$ by taking an appropriate difference (the definite integral with real upper and lower limits can, more generally, be defined by approximations, as it is shown in [21]). Propositions 9.23 and 9.25 extend to integrals with real limits.

Definition 9.27 *Let Φ be a continuous total function in $[0,1]$, $\alpha \in [0,1]$ and $\beta \in \mathbb{R}$. β* is the derivative of Φ at α, *written $\Phi'(\alpha) = \beta$, if $\forall n \in \mathbb{N}_1 \exists m \in \mathbb{N}_1$ such that*

$$\forall h \neq 0\,(0 \leq \alpha + h \leq 1 \wedge |h| < \frac{1}{2^m} \rightarrow |\frac{\Phi(\alpha+h) - \Phi(\alpha)}{h} - \beta| \leq \frac{1}{2^n}).$$

Definition 9.28 *Let Φ and Ψ be continuous total functions in $[0,1]$. We say that Φ* is the derivative of Ψ *if $\Phi(\alpha) = \Psi'(\alpha)$, $\forall \alpha \in [0,1]$.*

Theorem 9.29 (The fundamental theorem of calculus) *If Φ is a continuous total function in $[0,1]$ with a m.u.c. and Ψ is such that $\Psi(\alpha) = \int_0^\alpha \Phi(t)\,dt, \forall \alpha \in [0,1]$, then Φ is the derivative of Ψ.*

Proof. The usual proof of the theorem goes through in TCA^2. Take $\alpha \in [0,1]$. Let us prove that, if Φ is a continuous total function in $[0,1]$ with a m.u.c. and Ψ is a continuous total function in $[0,1]$ such that $\Psi(\alpha) = \int_0^\alpha \Phi(t)\,dt, \forall \alpha \in [0,1]$, then $\Phi(\alpha) = \Psi'(\alpha)$, i.e., given $n \in \mathbb{N}_1$ there is $m \in \mathbb{N}_1$ such that

$$\forall h \neq 0\,\big(0 \leq \alpha + h \leq 1 \wedge |h| < \frac{1}{2^m} \rightarrow |\frac{\Psi(\alpha+h) - \Psi(\alpha)}{h} - \Phi(\alpha)| \leq \frac{1}{2^n}\big).$$

Consider p a m.u.c. for Φ. Given $n \in \mathbb{N}_1$, take $m := p(n)$. Let $h \neq 0$ be such that $0 \leq \alpha + h \leq 1 \wedge |h| < \frac{1}{2^m}$. Since p is a m.u.c. for Φ, we have $|\Phi(\alpha) - \Phi(\alpha+k)| < \frac{1}{2^n}$ for all k such that $|k| \leq |h|$.

If $0 < h$, we have $h(\Phi(\alpha) - \frac{1}{2^n}) \leq \int_0^{\alpha+h} \Phi(t)\,dt - \int_0^\alpha \Phi(t)\,dt \leq h(\Phi(\alpha) + \frac{1}{2^n})$ and if $h < 0$ we know that $h(\Phi(\alpha) + \frac{1}{2^n}) \leq \int_0^{\alpha+h} \Phi(t)\,dt - \int_0^\alpha \Phi(t)\,dt \leq h(\Phi(\alpha) - \frac{1}{2^n})$. Therefore, in each case, $\Phi(\alpha) - \frac{1}{2^n} \leq \frac{\int_0^{\alpha+h} \Phi(t)\,dt - \int_0^\alpha \Phi(t)\,dt}{h} \leq \Phi(\alpha) + \frac{1}{2^n}$. We proved that $|\frac{\int_0^{\alpha+h} \Phi(t)\,dt - \int_0^\alpha \Phi(t)\,dt}{h} - \Phi(\alpha)| \leq \frac{1}{2^n}$, i.e., $|\frac{\Psi(\alpha+h) - \Psi(\alpha)}{h} - \Phi(\alpha)| \leq \frac{1}{2^n}$. □

9.6 Three questions in weak analysis

9.6.1 Integration in BTFA

In weak systems of second-order arithmetic the *representation* of analytic notions is of great importance. The notion of continuous function given in Definition 9.3 is based on the definition which appears in ordinary studies of reverse mathematics. We saw in Section 9.5 that the theory TCA^2 is able to develop a decent theory of Riemann integration for continuous functions with a modulus of uniform continuity. As explained in the beginning of that section, the availability of counting is also a necessary condition for this development. However, continuous functions can be represented in other ways. These different ways coincide in set theory, but in weak systems they need not be equivalent. For instance, Takeshi Yamazaki presents in [29] an alternative definition. Essentially, a continuous function (on a closed bounded interval) for Yamazaki is defined as the uniform limit (given by a modulus of uniform convergence) of piecewise-linear continuous functions (N.B. With this definition, he proves in the referred paper some theorems of reverse mathematics). Another alternative would be to replace in Yamazaki's definition the piecewise-linear functions by polynomials (see the end of Section 9.2). It seems to us that with this new definition one could develop a theory of integration in BTFA that encompasses a good class of functions, including the most important transcendental functions (like the sine and cosine functions). Since integration of polynomials can be done via primitivations, the idea is that integration lifts to functions (suitably) approximated by polynomials. The project is clear but its implementation may face some technical difficulties. So, the problem is the following:

Problem: *To develop in* BTFA *a good theory of integration for a sufficiently robust class of continuous functions.*

9.6.2 Weierstrass approximation theorem

In the previous problem, we suggested that continuous functions (on a closed bounded interval) could be defined as uniform approximations of polynomials. Weierstrass approximation theorem ensures that we indeed obtain all the continuous functions (in ordinary set-theoretic mathematics). Is this also the case in BTFA? In other words, can we prove Weierstrass approximation theorem in BTFA? We do not offer a precise statement of the question, but we insist that by a continuous function we mean a total continuous function, as given by Definition 9.3, with a modulus of uniform continuity. The answer should be negative because, otherwise, this would entail that we could count in BTFA (if the previous problem has a positive solution). We actually have a stronger conjecture:

Conjecture: *Over* BTFA, *Weierstrass approximation theorem is equivalent to the totality of the exponential function.*

The right-to-left direction should be doable by formalizing a suitable proof of Weierstrass approximation theorem in BTFA + exp. Note that BTFA + exp is equivalent to *Elementary Arithmetic* (cf. [1]). The left-to-right conjecture is based on the following informal considerations. For each tally number n, we can consider in BTFA the (very oscillating) continuous function (in the sense of Definition 9.3) defined on the interval [0,1] as follows: It is the piecewise-linear (continuous) function Φ_n that, on the numbers of the form $\frac{x}{2^{n+1}}$ (with $0 \leq x \leq 2^{n+1}$), takes the value 0 if $x \equiv 0 \pmod 4$, takes the value 1 if $x \equiv 1 \pmod 4$ and takes the value -1 if $x \equiv 3 \pmod 4$. A polynomial $P_n(X)$ sufficiently close to Φ_n must have at least 2^n roots and, hence, must be of degree at least 2^n. Now, our polynomials are of tally degree. The statement that for each tally number n, the number 2^n is also tally is equivalent to the totality of exponentiation.

9.6.3 Cantini's conjecture

The scheme of *strict* Π^1_1*-reflection*, denoted by sΠ^1_1-ref is

$$\forall X \exists x\, A(X,x) \to \exists w \forall X \exists x \preceq w\, A(X,x),$$

where $A(X,x)$ is a bounded formula, possibly with first and second-order parameters. This scheme is a set-theoretic truth and is closely related with weak König's lemma. The beginning of chapter VIII of Jon Barwise's book [2] is a good introduction to these matters and, from the arguments in there, one can extract the following fact: over RCA_0, the principles WKL and sΠ^1_1-ref are equivalent (the equivalence even holds in BTFA + exp). In [6], Andrea Cantini considered the principle of strict Π^1_1-reflection over the base theory BTFA. He showed that BTFA+sΠ^1_1-ref is a Π^0_2-conservative extension of Σ^b_1-NIA (Fernandes has extended this result in [10]). Therefore, the class of provably total functions (with Σ^b_1-graphs) of BTFA + sΠ^1_1-ref is still the class of polytime computable functions. Cantini also showed that, over BTFA, the principle sΠ^1_1-ref implies Σ^b_∞-WKL and conjectured the following:

Conjecture: *Over* BTFA, Σ^b_∞-WKL *implies* sΠ^1_1-ref.

Our formulation of strict Π^1_1-reflection is different from the one in Cantini but, in fact, it is an equivalent formulation in the presence of bounded collection $\mathsf{B}\Sigma^b_\infty$. Still, Cantini's conjecture as presented in [6] is not exactly the one given above. Cantini's original formulation is seemingly a bit stronger because it has the theory $\Sigma^b_1\text{-}\mathsf{NIA} + \nabla^b_1\text{-}\mathsf{CA}$ in place of BTFA (see Section 9.4 for the nabla principle). Do notice, however, that $\Sigma^b_1\text{-}\mathsf{NIA} + \nabla^b_1\text{-}\mathsf{CA} + \Sigma^b_\infty\text{-}\mathsf{WKL}$ proves $\mathsf{B}\Sigma^b_\infty$ (cf. [16]). In fact, there is a close relationship between forms of weak König's lemma and forms of bounded collection.

Let the principle of *extended strict* Π^1_1*-reflection* be the generalization of the scheme of $\mathsf{s}\Pi^1_1\text{-}\mathsf{ref}$ by allowing $\Sigma^{1,b}_\infty$-formulas in the schematic position A (see Section 9.5 for this class of formulas). This is the scheme considered in section 7 of our tryptic paper [13], where it is shown that BTFA together with extended strict Π^1_1-reflection proves $\mathsf{B}^1\Sigma^{1,b}_\infty$ (see Section 9.5 for this form of bounded collection).

Theorem 9.30 *The theory* BTFA *with extended strict* Π^1_1*-reflection is conservative over* $\mathsf{BTFA}+\mathsf{B}^1\Sigma^{1,b}_\infty$ *with respect to formulas without second-order unbounded quantifiers.*

A sketch of the proof of this result appears in [13]. It uses a Harrington forcing argument in which the forcing conditions are binary trees, taken in a certain generalized sense. This kind of forcing appeared originally in the doctoral dissertation of Fernandes [8].

The above discussion shows that there is a tight connection between extended strict Π^1_1-reflection and $\mathsf{B}^1\Sigma^{1,b}_\infty$. It is the same sort of connection that exists between $\Sigma^b_\infty\text{-}\mathsf{WKL}$ and $\mathsf{B}\Sigma^b_\infty$. The principle $\mathsf{s}\Pi^1_1\text{-}\mathsf{ref}$ falls between $\Sigma^b_\infty\text{-}\mathsf{WKL}$ and extended strict Π^1_1-reflection. In [10], Fernandes was able to isolate a principle (implied by $\mathsf{B}^1\Sigma^{1,b}_\infty$) that must be provable in BTFA if Cantini's conjecture is true. This principle, as opposed to $\mathsf{s}\Pi^1_1\text{-}\mathsf{ref}$, does not have unbounded second-order quantifiers. The authors believe that Cantini's conjecture is false.

Acknowledgements

The three authors acknowledge Centro de Matemática, Aplicações Fundamentais e Investigação Operacional (Universidade de Lisboa). The second author also acknowledge the support of Fundação para a Ciência e a Tecnologia (FCT) [UID/MAT/04561/2013]. The third author is also grateful to FCT [UID/MAT/04561/2013, UID/CEC/00408/2013 and grant SFRH/BPD/93278/2013] and to Large-Scale Informatics Systems Laboratory (Universidade de Lisboa).

Bibliography

[1] J. Avigad. Number theory and elementary arithmetic. *Philosophia Mathematica*, 11(3):257–284, 2003.

[2] J. Barwise. *Admissible Set Theory and Structures: An Approach to Definability Theory*. Perspectives in Mathematical Logic. Springer-Verlag, Berlin, 1975.

[3] V. Brattka, A. Kawamura, A. Marcone, and A. Pauly. Executive Summary of Measuring the Complexity of Computational Content. In *Dagstuhl Reports*, volume 5, pages 77–78, 9 2015. Dagstuhl Seminar 15392.

[4] D. Brown and S. Simpson. The Baire category theorem in weak subsystems of second-order arithmetic. *The Journal of Symbolic Logic*, 58(2):557–578, 1993.

[5] S. R. Buss. *Bounded arithmetic*. PhD thesis, Princeton University, 1985. A revision of this thesis was published by Bibliopolis (Naples) in 1986.

[6] A. Cantini. Asymmetric interpretations for bounded theories. *Mathematical Logic Quarterly*, 42(1):270–288, 1996.

[7] S. Cook and A. Kawamura. Complexity theory for operators in analysis. *A.C.M. Transactions on Computation Theory*, 4(2):5:1–5:24, 2012.

[8] A. M. Fernandes. *Investigações em sistemas de análise exequível* (in Portuguese). PhD thesis, Universidade de Lisboa, 2001.

[9] A. M. Fernandes. The Baire category theorem over a feasible base theory. In Stephen Simpson, editor, *Reverse Mathematics 2001*, volume 21 of *Lecture Notes in Logic*, pages 164–174. A K Peters, Massachusetts, 2005.

[10] A. M. Fernandes. Strict Π^1_1-reflection in bounded arithmetic. *Archive for Mathematical Logic*, 49:17–34, 2010.

[11] A. M. Fernandes and F. Ferreira. Groundwork for weak analysis. *The Journal of Symbolic Logic*, 67(2):557–578, 2002.

[12] A. M. Fernandes and F. Ferreira. Basic applications of weak König's lemma in feasible analysis. In Stephen Simpson, editor, *Reverse Mathematics 2001*, volume 21 of *Lecture Notes in Logic*, pages 175–188. A K Peters, Massachusetts, 2005.

[13] A. M. Fernandes, F. Ferreira, and G. Ferreira. Techniques in weak analysis for conservation results. In P. Cégielski, C. Cornaros, and C. Dimitracopoulos, editors, *New Studies in Weak Arithmetics*, volume 211 of *CSLI Lecture Notes*, pages 115–147. CSLI Publications, 2013.

[14] F. Ferreira. *Polynomial time computable arithmetic and conservative extensions.* PhD thesis, The Pennsylvania State University, USA, 1988.

[15] F. Ferreira. Binary models generated by their tally part. *Archive for Mathematical Logic*, 33:283–289, 1994.

[16] F. Ferreira. A feasible theory for analysis. *The Journal of Symbolic Logic*, 59(3):1001–1011, 1994.

[17] F. Ferreira and G. Ferreira. Counting as integration in feasible analysis. *Mathematical Logic Quarterly*, 52(3):315–320, 2006.

[18] F. Ferreira and G. Ferreira. Harrington's conservation theorem redone. *Archive for Mathematical Logic*, 47(2):91–100, 2008.

[19] F. Ferreira and G. Ferreira. The Riemann integral in weak systems of analysis. *Journal of Universal Computer Science*, 14(6):908–937, 2008.

[20] F. Ferreira and G. Ferreira. Interpretability in Robinson's Q. *The Bulletin of Symbolic Logic*, 19:289–317, 2013.

[21] G. Ferreira. *Sistemas de Análise Fraca para a Integração (in Portuguese).* PhD thesis, Universidade de Lisboa, 2006.

[22] G. Ferreira. The counting hierarchy in binary notation. *Portugaliae Mathematica*, 66:81–94, 2009.

[23] G. Ferreira and I. Oitavem. An interpretation of S^1_2 in Σ^b_1-NIA. *Portugaliae Mathematica*, 63(4):427–450, 2006.

[24] J. Krajíček, P. Pudlák, and G. Takeuti. Bounded arithmetic and the polynomial hierarchy. *Annals of Pure and Applied Logic*, 52(1):143 – 153, 1991.

[25] W. Sieg. Hilbert's program sixty years later. *The Journal of Symbolic Logic*, 53:338–348, 1988.

[26] S. G. Simpson. *Subsystems of Second Order Arithmetic.* Perspectives in Mathematical Logic. Springer-Verlag, Berlin, 1999.

[27] K. W. Wagner. Some observations on the connection between counting and recursion. *Theoretical Computer Science*, 47(2):131–147, 1986.

[28] T. Yamazaki. Some more conservation results on the Baire category theorem. *Mathematical Logic Quarterly*, 46(1):105–110, 2000.

[29] T. Yamazaki. Reverse mathematics and weak systems of 0-1 strings for feasible analysis. In Stephen Simpson, editor, *Reverse Mathematics 2001*, volume 21 of *Lecture Notes in Logic*, pages 394–401. A K Peters, Massachusetts, 2005.

10

EPISTEMIC NATURE OF QUANTUM REASONING

Alfredo B. Henriques[1] and Amílcar Sernadas[2]

[1]Centro Multidisciplinar de Astrofísica and Dep. de Física,
Instituto Superior Técnico,
Universidade de Lisboa, Portugal

[2]Dep. de Matemática, Instituto Superior Técnico, Universidade de Lisboa, Portugal

{alfredo.henriques,amilcar.sernadas}@tecnico.ulisboa.pt

Abstract

Doubts are raised concerning the usual interpretation of the alleged failure, by quantum mechanics, of the distributive law of classical logic. The difficulty raised by incompatible sets of observables is overcome within an epistemic enrichment of classical logic that provides the means for distinguishing between the value of a variable and its observation while retaining the classical connectives.

Since the seminal work of von Neumann and Birkhoff, in the 1930's (see [2]), it is commonly assumed (see Putnam [10]) that the experimental propositions of quantum mechanics can break the distributive law of classical logic. Due to the importance of this statement to the construction and rise of quantum logic, we re-examine a simple, but rather typical, example borrowed from [11]. Consider a particle moving along a line, taken as x-axis, and write the following propositions (we use the so called natural units, taking the velocity of light $c = 1$ and $\hbar = 1$):

p ="the particle has linear momentum p_x in the interval $[0, 1/6]$";

q = "the particle is in the interval on the x-axis $[-1, 1]$";

r = "the particle is in the interval on the x-axis $[1, 3]$".

In this system of units, the uncertainty principle is expressed as

$$\Delta p_x \Delta x \geq \frac{1}{2}.$$

The distributive law of propositional logic is given by the expression

$$p \wedge (q \vee r) \equiv (p \wedge q) \vee (p \wedge r).$$

Consider the usual reasons given for the failure of this law. The left hand side (lhs) is read as telling us that the particle, with momentum in the interval $[0, 1/6]$, has position in the interval $[-1, 3]$, compatible with the uncertainty principle: $\Delta p_x = 1/6$ and $\Delta x = 4$, their product being $2/3 > 1/2$. Thus, it has been argued that $p \wedge (q \vee r)$ is true. Rigorously, one should say that $p \wedge (q \vee r)$ can be true (depending on the truth values of p, q and r). As for the right-hand-side (rhs), we immediately see that $p \wedge q$ cannot be true, as the product of the uncertainties is $\Delta p_x \Delta x = 1/3 < 1/2$. The same applies to the second term, where, again, we have $\Delta p_x \Delta x = 1/3 < 1/2$. The usual conclusion is that the rhs $(p \wedge q) \vee (p \wedge r)$ is false. Rigorously, one should say that $(p \wedge q) \vee (p \wedge r)$ must always be false. Hence, the distributive law, valid in classical logic, can fail when dealing with quantum mechanics.

We now look more carefully at what has been argued.

With the intervals defined as above, we immediately see that, when considering propositions $p \wedge q$ and $p \wedge r$, we are, in fact, considering propositions which violate the uncertainty principle, that is, they break basic rules of quantum mechanics, with the consequence that they cannot express *experimental/testable* propositions. We cannot perform experiments allowing us to make simultaneous measurements of the values of the linear momentum and the position within those intervals; they correspond to mutually exclusive experimental situations. If we took them as being made simultaneously, we would be making the sin of a *counterfactual reasoning*: we would be assuming as being observed something which in fact has not been observed since it cannot be observed at all.

To show that quantum theory breaks some established rule of classical logic, like the distributive property, are we not supposed to argue in a way consistent with the theory itself and should we not avoid using counterfactual reasoning?

Is it appropriate, in any reasonable interpretation of the expressions of classical logic, to take cases and situations - perfectly possible in classical physics, but not within quantum mechanics - and use them to claim that quantum mechanics breaks some rule of classical logic and, then, using this to claim that logic is empirical?

Measurements *actually performed* would certainly give us values in agreement with the uncertainty principle. For example, if the intervals in the propositions q and r were slightly larger, as would result from actual measurements, then, either $p \wedge q$ or $p \wedge r$ could be true and the distributive law would be valid.

The same problem can be formulated when we use the more sophisticated language of the Hilbert spaces, for instance in Bacciagaluppi (see [1] also [10]). The equivalent to the counterfactual error above is made, although in a more rigorous setting. Other examples could be found, like those based on the Young interference observations, where, again, the same kind of incorrect use is made of incompatible observations.

To advance a little more with our analysis, we go back to our example and construct the truth table corresponding to the propositions p, q, r. We have the possibilities depicted in Figure 10.1.

	$p\ q\ r$	$p \wedge (q \vee r)$	$(p \wedge q) \vee (p \wedge r)$
	0 0 0	0	0
	0 0 1	0	0
	0 1 0	0	0
	0 1 1	0	0
	1 0 0	0	0
*	1 0 1	×	×
*	1 1 0	×	×
*	1 1 1	×	×

Figure 10.1: Truth table for $p \wedge (q \vee r)$ and $(p \wedge q) \vee (p \wedge r)$.

Sign (*) marks those table entries that correspond to incompatible values of the propositions, corresponding to the mutually exclusive experimental situations mentioned above; these incompatibilities mean that we cannot even complete the truth table for the distributive law. Can we, then, use the connectives of classical logic, without the truth table being complete? Are we then allowed to talk about the distributive law? Definitely not, at least according to [4].

There are, however, a few additional critical points that we want to address. When we write p, q or r, we take it as meaning that those measurements have been made, and that we know the corresponding results. Once again, remember these measurements are assumed to be made simultaneously, meaning that $p \wedge (q \vee r)$ is not a legitimate expression, as we have in fact two determinations, two measurements, of the position, each one of them incompatible with the measurement of the momentum, given the intervals previously defined. If it is to be a valid expression, we must refer to a single measurement of position in the interval $[-1, 3]$. This suggests that there has been an initial bad choice of propositions. What we need is a proposition like

$s =$ "the particle was observed in the interval on the x-axis $[-1, 3]$"

which, according to quantum mechanics, is not the same thing as saying "the particle was observed in the interval $[-1, 1]$ *or* was observed in the interval $[1, 3]$".

Nevertheless, one may want to work with possibly incompatible propositions. In such a scenario, two possibilities seem to be open to us. Either we accept that the connectives may have different properties from those in classical logic, and we work with a different logic (adopting the stance first proposed in [2]), or we may instead try to retain the classical connectives, as they are, and adopt an epistemic approach as we proceed to explain.

In order to overcome the difficulties mentioned above, raised by incompatible observables, we propose to extend classical logic with the means for distinguishing between, for instance, "linear momentum is in $[0, 1/6]$" and "linear moment was measured to be in $[0, 1/6]$". To this end, if we introduce the epistemic operator K (see [6, 5, 9]), for knowing (after observing), we can express that the knowledge of p is incompatible with the knowledge of q and with the knowledge of r:

$$K(p) \supset \neg K(q)$$

and

$$K(p) \supset \neg K(r).$$

Hence, $K(p) \wedge (K(q) \vee K(r))$ must always be false, while $K(p \wedge s) \equiv K(p) \wedge K(s)$ may be true.

Observe that the law $K(a \wedge b) \equiv K(a) \wedge K(b)$ is characteristic of an epistemic operator (see, for instance, Chapter 1 of [9]). On the other hand, one should not expect $K(a \vee b) \equiv K(a) \vee K(b)$ to hold in general (ibidem). Hence, the epistemic K does reflect the quantum physics reality:

$$K(q \vee r) \not\equiv K(q) \vee K(r),$$

that is, observing that the position is in $[-1, 3]$ is not equivalent to observing that it is in $[-1, 1]$ or to observing that it is in $[1, 3]$.

Therefore, according to the properties of the epistemic operator K (known = quantum observed), there is no surprise at all in

$$K(p \wedge (q \vee r)) \not\equiv K(p \wedge q) \vee K(p \vee r).$$

Indeed, while

$$K(p \wedge (q \vee r)) \equiv K(p) \wedge K(q \vee r)$$

and

$$K(p \wedge q) \vee K(p \wedge r) \equiv (K(p) \wedge K(q)) \vee (K(p) \wedge K(r)) \equiv K(p) \wedge (K(q) \vee K(r))$$

because K distributes over conjunction, it does not distribute over disjunction and, so, $K(q \vee r)$ is not necessarily equivalent to $K(q) \vee K(r)$.

Recall that none of the difficulties we have discussed appear when we describe the state of the quantum system using what is called a complete set of compatible observables, whose operators commute among themselves, giving a complete description of the state of the system (see, for instance, Section 13 of Chapter II in [3] or Section 1 of Chapter 1 in [8]). These observables can, in principle, be precisely and simultaneously measured, and no problems arise with the distributive law. This is to be expected, as quantum mechanics is based on a 'mathematical formalism that is, itself, based firmly on a classical two-valued logic' (see [7]). In such a scenario, $K(a) = a$, since we do not need to distinguish between them, and the epistemic enrichment of classical logic would collapse back into classical logic.

After all that has been said, can we still claim that logic is empirical, by appealing to quantum mechanics? Quantum mechanics is very different from classical physics, but does not break classical logic. It just shows that classical logic is not rich enough. In our opinion, given the epistemological constraints imposed by quantum mechanics on what can be measured, the missing ingredient is necessarily of an epistemic nature.

Acknowledgements

This work was partially supported by the PQDR initiative of SQIG at Instituto de Telecomunicações, Fundação para a Ciência e a Tecnologia via projects UID/FIS/00099/2013 and UID/EEA/50008/2013, and the European Union through project GA 318287 LANDAUER.

Bibliography

[1] G. Bacciagaluppi. Is logic empirical?. In D. Gabbay, K. Engesser, and D. Lehmann, editors, *Handbook of Quantum Logic and Quantum Structures: Quantum Logic*, pages 49–78. Elsevier, 2009.

[2] G. Birkhoff and J. von Neumann. The logic of quantum mechanics. *Annals of Mathematics*, 37(1):823–843, 1936.

[3] P. A. M. Dirac. *The Principles of Quantum Mechanics.* Oxford University Press, 1958.

[4] M. Dummett. Is logic empirical? In H. D. Lewis, editor, *Contemporary British Philosophy, 4th Edition*, pages 45–68. Allen and Unwin, 1976.

[5] R. Fagin, J. Y. Halpern, and M. Y. Vardi. A nonstandard approach to the logical omnscience problem. *Artificial Intelligence*, 79(2):203–240, 1995.

[6] J. Hintikka. *Knowledge and Belief.* Cornell University Press, 1962.

[7] C. J. Isham. *Lectures on Quantum Theory.* Imperial College Press, 2001.

[8] L. Landau and E. Lifchitz. *Mécanique Quantique.* Éditions Mir, 1966.

[9] J.-J. Ch. Meyer and W. van der Hoek. *Epistemic Logic for AI and Computer Science.* Cambridge University Press, 1995.

[10] H. Putnam and M. Wartofsky. Is logic empirical? In R. Cohen, editor, *Boston Studies in the Philosophy of Science, vol. 5*, pages 216–241. Dordrecht: Reidel, 1968. Reprinted as “The logic of quantum mechanics” in H. Putnam, Mathematics, Matter and Method, Philosophical Papers, vol.1 (Cambridge University Press, 1975), pages 174-197.

[11] Wikipedia. Quantum logic.
http://en.wikipedia.org/wiki/Quantum_logic.

11

QUANTUM MACHINES WITH CLASSICAL CONTROL

Paulo Mateus[1], Daowen Qiu[2] and André Souto[3]

[1]Dep. Matemática, Instituto Superior Técnico, Universidade de Lisboa, Portugal
SQIG, Instituto de Telecomunicações

[2]Department of Computer Science, Sun Yat-sen University, Guangzhou, China

[3]LaSIGE, Dep. de Informática, Fac. de Ciências, Universidade de Lisboa, Portugal

Corresponding author contact: paulo.mateus@tecnico.ulisboa.pt

Abstract

Herein we survey the main results concerning quantum automata and machines with classical control. These machines were originally proposed by Sernadas *et al* in [37] during the FCT QuantLog project. First, we focus on the expressivity of quantum automata with both quantum and classical states. We revise the result obtained in [32] where it was proved that such automata are able to recognise, with exponentially less states than deterministic finite automata, a family of regular languages that cannot be recognised by other types of quantum automata.

Finally, we revise the concept of quantum Turing machine with classical control introduced in [25].

The novelty of these machines consists in the fact that their termination problem is completely deterministic, in opposition to other notions in the literature. Concretely, we revisit the result that such machines fulfil the *s-m-n* property, while keeping the expressivity of a quantum model for computation.

11.1 Introduction

Quantum based machines were thought of by Feynman [11] when it became clear that quantum systems were hard to emulate with classical computers. The first notion of quantum Turing machine was devised by Deutsch [10], and although it is sound and full working, it evolves using a quantum superposition of states. The main problem with Deutsch Turing machine is that checking its termination makes the machine evolution to collapse, interfering in this way with the quantum evolution itself. Due to this issue, it is not obvious how one can extend the classical computability results, such as the s-m-n property, for these quantum Turing machines.

To avoid these complications, the community adopted other models of computation, such as quantum circuits, where it is relatively easy to present quantum algorithms. Moreover, these models were closer to what one was expecting, at that time, to be a physically implementable quantum computer. However, it soon became clear that implementing a full-fledged quantum computer was a long-term goal. For this reason, the community looked into restricted models of quantum computation, that required only a finite amount of memory – quantum automata. Interestingly, although finite quantum automata (or measure once one-way quantum finite automata – MO-1QFA) were able to accept some regular languages with exponentially less states, MO-1QFA do not accept all regular languages.

One of the main goals of the FCT QuantLog project was to understand if by endowing quantum systems or devices with classical control one could, in one hand, avoid the termination problem of quantum Turing machines, and on the other hand, extend the expressiveness of quantum automata, while keeping the exponential conciseness in the number of states. Some of these problems were introduced in the seminal paper by Sernadas *et al* [37].

During the QuantLog project two main results were attained along this line of research. Firstly, an extension of classical logic was proposed that was able to deal with quantum systems – the exogenous quantum propositional logic (EQPL) [21, 22, 23, 9, 8]. A second result, was an algorithm to minimize quantum automata [20, 43], which was an essential step to fully understand the exponential conciseness of quantum automata. Interestingly, the method also allowed to minimize constructively probabilistic automata (a problem that was open for more than 30 years) and that was previously characterized in terms of category theory in [24, 36, 13, 14].

The previous results set the ground to show that quantum automata endowed with classical control recognise a family of regular languages that cannot be recognised by other types of quantum automata, and, moreover, with exponentially less states than deterministic finite automata [32]. Along the same line, by endowing classical Turing machines with a quantum tape, one can define a deterministic well behaved quantum Turing machine, where the usual classical theorems from computability can be derived [25]. Given the contribution of Amílcar Sernadas to these two elegant results, it is worthwhile to revisit them in this volume. The first set of results is presented in Section 2 and the second set of results is revised in Section 3.

11.2 Quantum automata

A Measure Only One-way Quantum Finite Automaton (MO-1QFA) is defined as a quintuple $\mathcal{A} = (Q, \Sigma, |\psi_0\rangle, \{U(\sigma)\}_{\sigma\in\Sigma}, Q_{acc})$, where Q is a set of finite states, $|\psi_0\rangle$ is the initial state that is a superposition of the states in Q, Σ is a finite input alphabet, $U(\sigma)$ is a unitary matrix for each $\sigma \in \Sigma$, and $Q_{acc} \subseteq Q$ is the set of accepting states.

As usual, we identify Q with an orthonormal base of a complex Euclidean space and every state $q \in Q$ is identified with a basis vector, denoted by Dirac symbol $|q\rangle$ (a column vector), and $\langle q|$ is the conjugate transpose of $|q\rangle$. We describe the computing process for any given input string $x = \sigma_1\sigma_2\cdots\sigma_m \in \Sigma^*$. At the beginning the machine $\mathcal{A}$ is in the initial state $|\psi_0\rangle$, and upon reading σ_1, the transformation $U(\sigma_1)$ acts on $|\psi_0\rangle$. After that, $U(\sigma_1)|\psi_0\rangle$ becomes the current state and the machine reads σ_2. The process continues until the machine has read σ_m ending in the state $|\psi_x\rangle = U(\sigma_m)U(\sigma_{m-1})\cdots U(\sigma_1)|\psi_0\rangle$. Finally, a measurement is performed on $|\psi_x\rangle$ and the accepting probability $p_a(x)$ is equal to

$$p_a(x) = \langle\psi_x|P_a|\psi_x\rangle = \||P_a|\psi_x\rangle\|^2$$

where $P_a = \sum_{q\in Q_{acc}} |q\rangle\langle q|$ is the projection onto the subspace spanned by $\{|q\rangle : q \in Q_{acc}\}$.

Now we further recall the definition of multi-letter QFA [4].

A k-letter 1QFA $\mathcal{A}$ is defined as a quintuple $\mathcal{A} = (Q, \Sigma, |\psi_0\rangle, \nu, Q_{acc})$ where Q, $|\psi_0\rangle$, Σ, $Q_{acc} \subseteq Q$, are the same as those in MO-1QFA above, and ν is a function that assigns a unitary transition matrix U_w on $\mathbb{C}^{|Q|}$ for each string $w \in (\{\Lambda\} \cup \Sigma)^k$, where $|Q|$ is the cardinality of Q.

The computation of a k-letter 1QFA $\mathcal{A}$ works in the same way as the computation of an MO-1QFA, except that it applies unitary transformations corresponding not only to the last letter but the last k letters received. When $k = 1$, it is exactly an MO-1QFA as defined before. According to [4, 33], the languages accepted by k-letter 1QFA are a proper subset of regular languages for any k.

A Measure Many One-way Quantum Finite Automaton (MM-1QFA) is defined as a 6-tuple $\mathcal{A} = (Q, \Sigma, |\psi_0\rangle, \{U(\sigma)\}_{\sigma \in \Sigma \cup \{\$\}}, Q_{acc}, Q_{rej})$, where $Q, Q_{acc} \subseteq Q, |\psi_0\rangle, \Sigma, \{U(\sigma)\}_{\sigma \in \Sigma \cup \{\$\}}$ are the same as those in an MO-1QFA defined above, $Q_{rej} \subseteq Q$ represents the set of rejecting states, and $\$ \notin \Sigma$ is a tape symbol denoting the right end-mark. For any input string $x = \sigma_1 \sigma_2 \cdots \sigma_m \in \Sigma^*$, the computing process is similar to that of MO-1QFAs except that after every transition, $\mathcal{A}$ measures its state with respect to the three subspaces that are spanned by the three subsets Q_{acc}, Q_{rej}, and Q_{non}, respectively, where $Q_{non} = Q \setminus (Q_{acc} \cup Q_{rej})$. In other words, the projection measurement consists of $\{P_a, P_r, P_n\}$ where $P_a = \sum_{q \in Q_{acc}} |q\rangle\langle q|$, $P_r = \sum_{q \in Q_{rej}} |q\rangle\langle q|$, $P_n = \sum_{q \in Q \setminus (Q_{acc} \cup Q_{rej})} |q\rangle\langle q|$. The machine stops after the right end-mark \$ has been read. Of course, the machine may also stop before reading \$ if the current state, after the machine reading some σ_i $(1 \leq i \leq m)$, does not contain the states of Q_{non}. Since the measurement is performed after each transition with the states of Q_{non} being preserved, the accepting probability $p_a(x)$ and the rejecting probability $p_r(x)$ are given as follows (for convenience, we denote $\$ = \sigma_{m+1}$):

$$p_a(x) = \sum_{k=1}^{m+1} \|P_a U(\sigma_k) \prod_{i=1}^{k-1} (P_n U(\sigma_i)) |\psi_0\rangle\|^2,$$

$$p_r(x) = \sum_{k=1}^{m+1} \|P_r U(\sigma_k) \prod_{i=1}^{k-1} (P_n U(\sigma_i)) |\psi_0\rangle\|^2.$$

Here we define $\prod_{i=1}^{n} A_i = A_n A_{n-1} \cdots A_1$.

Bertoni *et al* [5] introduced a 1QFA, called 1QFACL that allows a more general measurement than the previous models. Similar to the case in MM-1QFA, the state of this model can be observed at each step, but an observable $\mathcal{O}$ is considered with a fixed, but arbitrary, set of possible results $\mathcal{C} = \{c_1, \ldots, c_n\}$, without limit to $\{a, r, g\}$ as in MM-1QFA. The accepting behavior in this model is also different from that of the previous models. On any given input word x, the computation displays a sequence $y \in \mathcal{C}^*$ of results of $\mathcal{O}$ with a certain probability $p(y|x)$, and the computation is accepted if and only if y belongs to a fixed regular language $\mathcal{L} \subseteq \mathcal{C}^*$. Bertoni *et al* [5] called such a language $\mathcal{L}$ *control language*.

More formally, given an input alphabet Σ and the end-marker symbol $\$ \notin \Sigma$, a 1QFACL over the working alphabet $\Gamma = \Sigma \cup \{\$\}$ is a five-tuple

$\mathcal{M} = (Q, |\psi_0\rangle, \{U(\sigma)\}_{\sigma\in\Gamma}, \mathcal{O}, \mathcal{L})$, where

- Q, $|\psi_0\rangle$ and $U(\sigma)$ ($\sigma \in \Gamma$) are defined as in the case of MM-1QFA;
- $\mathcal{O}$ is an observable with the set of possible results $\mathcal{C} = \{c_1, \ldots, c_s\}$ and the projector set $\{P(c_i) : i = 1, \ldots, s\}$ of which $P(c_i)$ denotes the projector onto the eigenspace corresponding to c_i;
- $\mathcal{L} \subseteq \mathcal{C}^*$ is a regular language (control language).

The input word w to 1QFACL $\mathcal{M}$ is in the form: $w \in \Sigma^*\$$, with symbol \$ denoting the end of a word. Now, we define the behavior of $\mathcal{M}$ on word $x_1 \ldots x_n\$$. The computation starts in the state $|\psi_0\rangle$, and then the transformations associated with the symbols in the word $x_1 \ldots x_n\$$ are applied in succession. The transformation associated with any symbol $\sigma \in \Gamma$ consists of two steps:

1. First, $U(\sigma)$ is applied to the current state $|\phi\rangle$ of $\mathcal{M}$, yielding the new state $|\phi'\rangle = U(\sigma)|\phi\rangle$.
2. Second, the observable $\mathcal{O}$ is measured on $|\phi'\rangle$. According to quantum mechanics principle, this measurement yields result c_k with probability $p_k = ||P(c_k)|\phi'\rangle||^2$, and the state of $\mathcal{M}$ collapses to $P(c_k)|\phi'\rangle/\sqrt{p_k}$.

Thus, the computation on the word $x_1 \ldots x_n\$$ leads to a sequence $y_1 \ldots y_{n+1} \in \mathcal{C}^*$ with probability $p(y_1 \ldots y_{n+1}|x_1 \ldots x_n\$)$ given by

$$p(y_1 \ldots y_{n+1}|x_1 \ldots x_n\$) = \| \prod_{i=1}^{n+1} P(y_i)U(x_i)|\psi_0\rangle\|^2, \tag{11.1}$$

where we let $x_{n+1} = \$$ as stated before. A computation leading to the word $y \in \mathcal{C}^*$ is said to be accepted if $y \in \mathcal{L}$. Otherwise, it is rejected. Hence, the accepting probability of 1QFACL $\mathcal{M}$ is defined as:

$$\mathcal{P}_{\mathcal{M}}(x_1 \ldots x_n) = \sum_{y_1 \cdots y_{n+1} \in \mathcal{L}} p(y_1 \ldots y_{n+1}|x_1 \ldots x_n\$). \tag{11.2}$$

11.2.1 One-way quantum automata together with classical states

In the introduction we gave the motivation for introducing the new one-way quantum finite automata model, i.e., 1QFAC. We now define formally the model. To this end, we need the following notations. Given a finite set B, we denote by $\mathcal{H}(B)$ the Hilbert space freely generated by B. Furthermore, we denote by I and O the identity operator and zero operator on $\mathcal{H}(Q)$, respectively.

Definition 11.1 A 1QFAC $\mathcal{A}$ is defined by a 9-tuple

$$\mathcal{A} = (S, Q, \Sigma, \Gamma, s_0, |\psi_0\rangle, \delta, \mathbb{U}, \mathcal{M})$$

where:

- Σ is a finite set (the *input alphabet*);
- Γ is a finite set (the *output alphabet*);
- S is a finite set (the set of *classical states*);
- Q is a finite set (the *quantum state basis*);
- s_0 is an element of S (the *initial classical state*);
- $|\psi_0\rangle$ is a unit vector in the Hilbert space $\mathcal{H}(Q)$ (the *initial quantum state*);
- $\delta : S \times \Sigma \to S$ is a map (the *classical transition map*);
- $\mathbb{U} = \{U_{s\sigma}\}_{s\in S,\sigma\in\Sigma}$ where $U_{s\sigma} : \mathcal{H}(Q) \to \mathcal{H}(Q)$ is a unitary operator for each s and σ (the *quantum transition operator* at s and σ);
- $\mathcal{M} = \{\mathcal{M}_s\}_{s\in S}$ where each $\mathcal{M}_s$ is a projective measurement over $\mathcal{H}(Q)$ with outcomes in Γ (the *measurement operator at s*).

Hence, each $\mathcal{M}_s = \{P_{s,\gamma}\}_{\gamma\in\Gamma}$ such that $\sum_{\gamma\in\Gamma} P_{s,\gamma} = I$ and $P_{s,\gamma}P_{s,\gamma'} = \begin{cases} P_{s,\gamma}, & \gamma = \gamma', \\ O, & \gamma \neq \gamma'. \end{cases}$ Furthermore, if the machine is in classical state s and quantum state $|\psi\rangle$ after reading the input string, then $\|P_{s,\gamma}|\psi\rangle\|^2$ is the probability of the machine producing outcome γ on that input.

Note that the map δ can be extended to a map $\delta^* : \Sigma^* \to S$ as usual. That is, $\delta^*(s, \epsilon) = s$; for any string $x \in \Sigma^*$ and any $\sigma \in \Sigma$, $\delta^*(s, \sigma x) = \delta^*(\delta(s, \sigma), x)$.

A specially interesting case of the above definition is when $\Gamma = \{a, r\}$, where a denotes *accepting* and r denotes *rejecting*. Then, $\mathcal{M} = \{\{P_{s,a}, P_{s,r}\} : s \in S\}$ and, for each $s \in S$, $P_{s,a}$ and $P_{s,r}$ are two projectors such that $P_{s,a} + P_{s,r} = I$ and $P_{s,a}P_{s,r} = O$. In this case, $\mathcal{A}$ is an acceptor of languages over Σ.

For the sake of convenience, we denote the map $\mu : \Sigma^* \to S$, induced by δ, as $\mu(x) = \delta^*(s_0, x)$ for any string $x \in \Sigma^*$.

We further describe the computing process of

$$\mathcal{A} = (S, Q, \Sigma, s_0, |\psi_0\rangle, \delta, \mathbb{U}, \mathcal{M})$$

for input string $x = \sigma_1\sigma_2\cdots\sigma_m$ where $\sigma_i \in \Sigma$ for $i = 1, 2, \cdots, m$.

The machine $\mathcal{A}$ starts at the initial classical state s_0 and initial quantum state $|\psi_0\rangle$. On reading the first symbol σ_1 of the input string, the states of the machine change as follows: the classical state becomes $\mu(\sigma_1)$; the quantum state becomes $U_{s_0\sigma_1}|\psi_0\rangle$. Afterward, on reading σ_2, the machine changes its classical state to $\mu(\sigma_1\sigma_2)$ and its quantum state to the result of applying $U_{\mu(\sigma_1)\sigma_2}$ to $U_{s_0\sigma_1}|\psi_0\rangle$.

The process continues similarly by reading $\sigma_3, \sigma_4, \cdots, \sigma_m$ in succession. Therefore, after reading σ_m, the classical state becomes $\mu(x)$ and the quantum state is as follows:

$$U_{\mu(\sigma_1\cdots\sigma_{m-2}\sigma_{m-1})\sigma_m}U_{\mu(\sigma_1\cdots\sigma_{m-3}\sigma_{m-2})\sigma_{m-1}}\cdots U_{\mu(\sigma_1)\sigma_2}U_{s_0\sigma_1}|\psi_0\rangle. \quad (11.3)$$

Let $\mathcal{U}(Q)$ be the set of unitary operators on Hilbert space $\mathcal{H}(Q)$. For the sake of convenience, we denote the map $v : \Sigma^* \to \mathcal{U}(Q)$ as: $v(\epsilon) = I$ and

$$v(x) = U_{\mu(\sigma_1\cdots\sigma_{m-2}\sigma_{m-1})\sigma_m}U_{\mu(\sigma_1\cdots\sigma_{m-3}\sigma_{m-2})\sigma_{m-1}}\cdots U_{\mu(\sigma_1)\sigma_2}U_{s_0\sigma_1} \quad (11.4)$$

for $x = \sigma_1\sigma_2\cdots\sigma_m$ where $\sigma_i \in \Sigma$ for $i = 1, 2, \cdots, m$, and I denotes the identity operator on $\mathcal{H}(Q)$, indicated as before.

By means of the denotations μ and v, for any input string $x \in \Sigma^*$, after $\mathcal{A}$ reading x, the classical state is $\mu(x)$ and the quantum states $v(x)|\psi_0\rangle$.

Finally, the probability $\mathsf{Prob}_{\mathcal{A},\gamma}(x)$ of machine $\mathcal{A}$ producing result γ on input x is as follows:

$$\mathsf{Prob}_{\mathcal{A},\gamma}(x) = ||P_{\mu(x),\gamma}v(x)|\psi_0\rangle||^2. \quad (11.5)$$

In particular, when $\mathcal{A}$ is thought of as an acceptor of languages over Σ ($\Gamma = \{a, r\}$), we obtain the probability $\mathsf{Prob}_{\mathcal{A},a}(x)$ for accepting x:

$$\mathsf{Prob}_{\mathcal{A},a}(x) = ||P_{\mu(x),a}v(x)|\psi_0\rangle||^2. \quad (11.6)$$

If a 1QFAC $\mathcal{A}$ has only one classical state, then $\mathcal{A}$ reduces to an MO-1QFA [26]. Therefore, the set of languages accepted by 1QFAC with only one classical state is a proper subset of regular languages, the languages whose syntactic monoid is a group [6]. However, we revisit here the result obtained in [32] that 1QFAC can accept all regular languages with no error.

Proposition 11.1 Let Σ be a finite set. Then each regular language over Σ that is accepted by a minimal DFA of k states is also accepted by some 1QFAC with no error and with 1 quantum basis state and k classical states.

Proof. Let $L \subseteq \Sigma^*$ be a regular language. Then there exists a DFA $M = (S, \Sigma, \delta, s_0, F)$ accepting L, where, as usual, S is a finite set of states, $s_0 \in S$ is an initial state, $F \subseteq Q$ is a set of accepting states, and $\delta : Q \times \Sigma \to Q$ is the transition function. We construct a 1QFAC $\mathcal{A} = (S, Q, \Sigma, \Gamma, s_0, |\psi_0\rangle, \delta, \mathbb{U}, \mathcal{M})$ accepting L without error, where S, Σ, s_0, and δ are the same as those in M, and, in addition, $\Gamma = \{a, r\}$, $Q = \{0\}$, $|\psi_0\rangle = |0\rangle$, $\mathbb{U} = \{U_{s\sigma} : s \in S, \sigma \in \Sigma\}$ with $U_{s\sigma} = I$ for all $s \in S$ and $\sigma \in \Sigma$, $\mathcal{M} = \{\{P_{s,a}, P_{s,r}\} : s \in S\}$ assigned as: if $s \in F$, then $P_{s,a} = |0\rangle\langle 0|$ and $P_{s,r} = O$ where O denotes the zero operator as before; otherwise, $P_{s,a} = O$ and $P_{s,r} = |0\rangle\langle 0|$.

By the above definition of 1QFAC $\mathcal{A}$, it is easy to check that the language accepted by $\mathcal{A}$ with no error is exactly L. □

Observe that for any regular language L over $\{0, 1\}$ accepted by a k state DFA, it was proved that there exists a 1QFACL accepting L with no error and with $3k$ classical states ($3k$ is the number of states of its minimal DFA accepting the control language) and 3 quantum basis states [27]. Here, for 1QFAC, we require only k classical states and 1 quantum basis states. Therefore, in this case, 1QFAC have better state complexity than 1QFACL.

On the other hand, any language accepted by a 1QFAC is regular. We can prove this result in detail, based on a well-know idea for one-way probabilistic automata by Rabin [35], that was already applied for MM-1QFA by Kondacs and Watrous [17] as well as for MO-1QFA by Brodsky and Pippenger [6]. However, the process is much longer and further results are needed, since both classical and quantum states are involved in 1QFAC. Another possible approach is based on topological automata [3, 16]. However, in next section we obtain this result while studying the state complexity of 1QFAC and so we postpone the proof of regularity to the next section.

11.2.2 State complexity of 1QFAC

State complexity of classical finite automata has been a hot research subject with important practical applications [40]. In this section, we consider this problem for 1QFAC. First, we prove a lower bound on the state complexity of 1QFAC which states that 1QFAC are at most exponentially more concise than DFA. Second, we show that our bound is tight by giving some languages that witness the exponential advantage of 1QFAC over DFA. Particularly, these languages can not be accepted by any MO-1QFA, MM-1QFA or multi-letter 1QFA.

Here we prove a lower bound for the state complexity of 1QFAC which states that 1QFAC are at most exponentially more concise than DFA.

Also, we show that the languages accepted by 1QFAC with bounded error are regular. Some examples given in the next subsection shows that our lower bound is tight.

Given a 1QFAC $\mathcal{A} = (S, Q, \Sigma, \Gamma, s_0, |\psi_0\rangle, \delta, \mathbb{U}, \mathcal{M})$, we shall consider the triple

$$(\mathcal{H}, |\phi_0\rangle, \{M(\sigma) : \sigma \in \Sigma\}, \{P_\gamma : \gamma \in \Gamma\})$$

where

- $\mathcal{H} = \mathcal{H}(S) \otimes \mathcal{H}(Q)$;
- $|\phi_0\rangle = |s_0\rangle|\psi_0\rangle$;
- $M(\sigma) = \sum_{s\in S} |\delta(s,\sigma)\rangle\langle s| \otimes U_{s\sigma}$ for $\sigma \in \Sigma$;
- $P_\gamma = \sum_{s\in S} |s\rangle\langle s| \otimes P_{s\gamma}$ for each $\gamma \in \Gamma$.

It is easy to verify that

$$\mathsf{Prob}_{\mathcal{A},\gamma}(x) = \|P_\gamma M(x)|\phi_0\rangle\|^2 \tag{11.7}$$

for each $\gamma \in \Gamma$ and $x \in \Sigma^*$, where $M(x_1 \cdots x_n) = M(x_n) \cdots M(x_1)$. Furthermore, we let

$$\mathcal{V} = \{|\phi_x\rangle : |\phi_x\rangle = M(x)|\phi_0\rangle, x \in \Sigma^*\}. \tag{11.8}$$

Then we have the following result.

Lemma 11.2 It holds that

(i) each $|\phi\rangle \in \mathcal{V}$ has the form $|\phi\rangle = |s\rangle|\psi\rangle$ where $s \in S$ and $|\psi\rangle \in \mathcal{H}(Q)$;

(ii) $\||\phi\rangle\|^2 = 1$ for all $|\phi\rangle \in \mathcal{V}$;

(iii) $\|M(x)|\phi_1\rangle - M(x)|\phi_2\rangle\| \leq \sqrt{2}\||\phi_1\rangle - |\phi_2\rangle\|$ for all $x \in \Sigma^*$.

Proof. Items (i) and (ii) are easy to be verified. In the following, we prove item (iii). Let $|\phi_i\rangle = |s_i\rangle|\psi_i\rangle$ and $|\phi'_i\rangle = M(x)|\phi_i\rangle = |s'_i\rangle|\psi'_i\rangle$ for $i = 1, 2$ and $x \in \Sigma^*$, where $s_i, s'_i \in S$ and $|\psi_i\rangle, |\psi'_i\rangle \in \mathcal{H}(Q)$. The discussion is divided into two cases.

Case (a): $|s_1\rangle = |s_2\rangle$. In this case it necessarily holds that $|s'_1\rangle = |s'_2\rangle$ and furthermore we have

$$\||\phi'_1\rangle - \phi'_2\rangle\| = \||\psi'_1\rangle - |\psi'_2\rangle\| = \||\psi_1\rangle - |\psi_2\rangle\| = \||\phi_1\rangle - |\phi_2\rangle\|, \tag{11.9}$$

where the first and third equations hold because of $\||\alpha\rangle|\beta\rangle\| = \||\alpha\rangle\|.\||\beta\rangle\|$ and the second holds since $|\psi_1'\rangle$ and $|\psi_2'\rangle$ are obtained by performing the same unitary operation on $|\psi_1\rangle$ and $|\psi_2\rangle$, respectively.
Case (b): $|s_1\rangle \neq |s_2\rangle$. First it holds that $\||\phi_1\rangle - |\phi_2\rangle\| = \sqrt{2}$. Indeed, let $|\psi_1\rangle = \sum_i \alpha_i|i\rangle$ and $|\psi_2\rangle = \sum_i \beta_i|i\rangle$. Then, we have

$$\begin{aligned}
\||\phi_1\rangle - |\phi_2\rangle\| &= \||s_1\rangle|\psi_1\rangle - |s_2\rangle|\psi_2\rangle\| && (11.10)\\
&= \left\|\sum_i \alpha_i|s_1\rangle|i\rangle + \sum_i(-\beta_i)|s_2\rangle|i\rangle\right\| && (11.11)\\
&= \left(\sum_i |\alpha_i|^2 + \sum_i |\beta_i|^2\right)^{\frac{1}{2}} && (11.12)\\
&= \sqrt{\||\psi_1\rangle\|^2 + \||\psi_1\rangle\|^2} && (11.13)\\
&= \sqrt{2}. && (11.14)
\end{aligned}$$

Therefore,

$$\begin{aligned}
\||\phi_1'\rangle - \phi_2'\rangle\| &= \||s_1'\rangle|\psi_1'\rangle - |s_2'\rangle|\psi_2'\rangle\| && (11.15)\\
&= \begin{cases} \||\psi_1'\rangle - |\psi_2'\rangle\|, & \text{if } s_1' = s_2'; \\ \sqrt{2}, & \text{else.} \end{cases} && (11.16)
\end{aligned}$$

Note that $\||\psi_1'\rangle - |\psi_2'\rangle\| \leq 2 = \sqrt{2}\||\phi_1\rangle - |\phi_2\rangle\|$.

In summary, item (iii) holds in any case. □

Next we present another lemma which is critical for obtaining the lower bound on 1QFAC.

Lemma 11.3 Let $\mathcal{V}_\theta \subseteq \mathbb{C}^n$ such that $\||\phi_1\rangle - |\phi_2\rangle\| \geq \theta$ for any two elements $|\phi_1\rangle, |\phi_2\rangle \in \mathcal{V}_\theta$. Then $\mathcal{V}_\theta$ is a finite set containing $k(\theta)$ elements where $k(\theta) \leq (1 + \frac{2}{\theta})^{2n}$.

Proof. Arbitrarily choose an element $|\phi\rangle \in \mathcal{V}_\theta$. Let $U(|\phi\rangle, \frac{\theta}{2}) = \{|\chi\rangle : \||\chi\rangle - |\phi\rangle\| \leq \frac{\theta}{2}\}$, i.e., a sphere centered at $|\phi\rangle$ with the radius $\frac{\theta}{2}$. Then all these spheres do not intersect pairwise except for their surface, and all of them are contained in a large sphere centered at $(0, 0, \cdots, 0)$ with the radius $1 + \frac{\theta}{2}$. The volume of a sphere of a radius r in $\mathbb{C}^n$ is cr^{2n} where c is a constant. Note that $\mathbb{C}^n$ is an n-dimensional complex space and each element from it can be represented by an element of $\mathbb{R}^{2n}$. Therefore, it holds that

$$k(\theta) \leq \frac{c(1 + \frac{\theta}{2})^{2n}}{c(\frac{\theta}{2})^{2n}} = (1 + \frac{2}{\theta})^{2n}. \qquad (11.17)$$

□

Below we recall a result that will be used later on (c.f. Lemma 8 in [41] for a complete proof).

Lemma 11.4 For any two elements $|\phi\rangle, |\varphi\rangle \in \mathbb{C}^n$ with $\||\phi\rangle\| \leq c$ and $\||\varphi\rangle\| \leq c$, it holds that $\left|\|P|\phi\rangle\|^2 - \|P|\varphi\rangle\|^2\right| \leq c\||\phi\rangle - |\varphi\rangle\|$ where P is a projective operator on $\mathbb{C}^n$.

Given a language $L \subseteq \Sigma^*$, define an equivalence relation "$\equiv_L$" as: for any $x, y \in \Sigma^*$, $x \equiv_L y$ if for any $z \in \Sigma^*$, $xz \in L$ iff $yz \in L$. If x, y do not satisfy the equivalence relation, we denote it by $x \not\equiv_L y$. Then the set Σ^* is partitioned into some equivalence classes by the equivalence relation "$\equiv_L$". In the following we recall a well-known result that will be used in the sequel.

Lemma 11.5 (Myhill-Nerode theorem [15]) A language $L \subseteq \Sigma^*$ is regular iff the number of equivalence classes induced by the equivalence relation "$\equiv_L$" is finite. Furthermore, the number of equivalence classes equals to the state number of the minimal DFA accepting L.

Now we are ready to present our main result.

Theorem 11.6 If L is accepted by a 1QFAC $\mathcal{M}$ with bounded error, then L is regular and it holds that $kn = \Omega(log\ m)$ where k and n denote numbers of classical states and quantum basis states of $\mathcal{M}$, respectively, and m is the state number of the minimal DFA accepting L.

Proof. Let $\mathcal{V}' \subseteq \mathcal{V}$ (where $\mathcal{V}$ is given in Eq. (11.8)) satisfying for any two elements $|\phi_x\rangle, |\phi_y\rangle \in \mathcal{V}'$ it holds that $|\phi_x\rangle \neq |\phi_y\rangle \Leftrightarrow x \not\equiv_L y$. Then for two different elements $|\phi_x\rangle, |\phi_y\rangle \in \mathcal{V}'$ there exists $z \in \Sigma^*$ satisfying $xz \in L$ whereas $yz \notin L$ (or $xz \notin L$ whereas $yz \in L$). That is

$$\begin{aligned} \mathsf{Prob}_{\mathcal{A},a}(xz) &= \|P_a M(z)|\phi_x\rangle\|^2 \geq \lambda + \epsilon, \quad (11.18)\\ \mathsf{Prob}_{\mathcal{A},a}(yz) &= \|P_a M(z)|\phi_y\rangle\|^2 \leq \lambda - \epsilon \quad (11.19)\end{aligned}$$

for some $\lambda \in (0, 1]$ and $\epsilon > 0$. Therefore we have

$$\begin{aligned} \sqrt{2}\||\phi_x\rangle - |\phi_y\rangle\| &\geq \|M(z)|\phi_x\rangle - M(z)|\phi_y\rangle\| \quad (11.20)\\ &\geq |\mathsf{Prob}_{\mathcal{A},a}(xz) - \mathsf{Prob}_{\mathcal{A},a}(yz)| \quad (11.21)\\ &\geq 2\epsilon \quad (11.22)\end{aligned}$$

where the first inequality follows from Lemma 11.2 and the second follows from Lemma 11.4. In summary, we obtain that two different elements $|\phi_x\rangle$ and $|\phi_y\rangle$ from $\mathcal{V}'$ satisfy $\||\phi_x\rangle - |\phi_y\rangle\| \geq \sqrt{2}\epsilon$. Therefore, according to Lemma 11.3, we have that the number $|\mathcal{V}'|$ of elements in $\mathcal{V}'$ satisfies $|\mathcal{V}'| \leq (1+\frac{\sqrt{2}}{\epsilon})^{2kn}$, which means that the number of equivalence classes induced by the equivalence relation "$\equiv_L$" is upper bounded by $(1+\frac{\sqrt{2}}{\epsilon})^{2kn}$. Therefore, by Lemma 11.5 we have completed the proof. □

When the number of classical states equals one in a 1QFAC $\mathcal{M}$, $\mathcal{M}$ exactly reduces to an MO-1QFA. Therefore, as a corollary, we can obtain a precise relationship between the numbers of states for MO-1QFA and DFA that was also derived by Ablayev and Gainutdinova [2].

Corollary 11.7 If L is accepted by an MO-1QFA $\mathcal{M}$ with bounded error, then L is regular and it holds that $n = \Omega(\log m)$ where n denotes the number of quantum basis states of $\mathcal{M}$, and m is the state number of the minimal DFA accepting L.

11.2.3 The lower bound is tight

Although 1QFAC accept only regular languages as DFA, 1QFAC can accept some languages with essentially less number of states than DFA and these languages cannot be accepted by any MO-1QFA or MM-1QFA or multi-letter 1QFA. In this section, our purpose is to prove these claims, and we also obtain that the lower bound in Theorem 11.6 is tight.

First, we establish a technical result concerning the acceptability by 1QFAC of languages resulting from set operations on languages accepted by MO-1QFA and by DFA.

Lemma 11.8 Let Σ be a finite alphabet. Suppose that the language L_1 over Σ is accepted by a minimal DFA with n_1 states and the language L_2 over Σ is accepted by an MO-1QFA with n_2 quantum basis states with bounded error ϵ. Then the intersection $L_1 \cap L_2$, union $L_1 \cup L_2$, differences $L_1 \setminus L_2$ and $L_2 \setminus L_1$ can be accepted by some 1QFAC with n_1 classical states and n_2 quantum basis states with bounded error ϵ.

Proof. Let $A_1 = (S, \Sigma, \delta, s_0, F)$ be a minimal DFA accepting L_1, and let $A_2 = (Q, \Sigma, |\psi_0\rangle,$
$\{U(\sigma)\}_{\sigma\in\Sigma}, Q_{acc})$ be an MO-1QFA accepting L_2, where $s_0 \in S$ is the initial state, δ is the transition function, and $F \subseteq S$ is a finite subset denoting accepting states; the symbols in A_2 are the same as those in the definition of MO-1QFA as above.

Then by A_1 and A_2 we define a 1QFAC

$$\mathcal{A} = (S, Q, \Sigma, \Gamma, s_0, |\psi_0\rangle, \delta, \mathbb{U}, \mathcal{M})$$

accepting $L_1 \cap L_2$, where $S, Q, \Sigma, s_0, |\psi_0\rangle, \delta$ are the same as those in A_1 and A_2, $\Gamma = \{a, r\}$, $\mathbb{U} = \{U_{s\sigma} = U(\sigma) : s \in S, \sigma \in \Sigma\}$, and $\mathcal{M} = \{M_s : s \in S\}$ where $M_s = \{P_{s,a}, P_{s,r}\}$ and

$$P_{s,a} = \begin{cases} \sum_{p \in Q_{acc}} |p\rangle\langle p|, & s \in F; \\ O, & s \notin F, \end{cases}$$

where O denotes the zero operator, and $P_{s,r} = I - P_{s,a}$ with I being the identity operator.

According to the above definition of 1QFAC, we easily know that, for any string $x \in \Sigma^*$, if $x \in L_1$ then the accepting probability of 1QFAC $\mathcal{A}$ is equal to the accepting probability of MO-1QFA A_2; if $x \notin L_1$ then the accepting probability of 1QFAC $\mathcal{A}$ is zero. So, 1QFAC $\mathcal{A}$ accepts the intersection $L_1 \cap L_2$.

Similarly, we can construct the other three 1QFAC accepting the union $L_1 \cup L_2$, differences $L_1 \setminus L_2$, and $L_2 \setminus L_1$, respectively. Indeed, we only need define different measurements in these 1QFAC. If we construct 1QFAC accepting $L_1 \cup L_2$, then

$$P_{s,a} = \begin{cases} I, & s \in F; \\ \sum_{p \in Q_{acc}} |p\rangle\langle p|, & s \notin F. \end{cases}$$

If we construct 1QFAC accepting $L_1 \setminus L_2$, then

$$P_{s,a} = \begin{cases} \sum_{p \in Q \setminus Q_{acc}} |p\rangle\langle p|, & s \in F; \\ O, & s \notin F. \end{cases}$$

If we construct 1QFAC accepting $L_2 \setminus L_1$, then

$$P_{s,a} = \begin{cases} \sum_{p \in Q \setminus Q_{acc}} |p\rangle\langle p|, & s \notin F; \\ O, & s \in F. \end{cases}$$

□

Consider the regular language

$$L^0(m) = \{w0 : w \in \{0,1\}^*, |w0| = km, k = 1, 2, 3, \cdots\}.$$

Clearly, the minimal classical DFA accepting $L^0(m)$ has $m+1$ states, as depicted in Figure 11.1.

Indeed, neither MO-1QFA nor MM-1QFA can accept $L^0(m)$. We can easily verify this result by employing a lemma from [6, 12]. That is,

Lemma 11.9 ([6, 12]) Let L be a regular language, and let M be its minimal DFA containing the construction in Figure 3, where states p and q are distinguishable (i.e., there exists a string z such that either $\delta(p, z)$ or $\delta(q, z)$ is an accepting state). Then, L can not be accepted by MM-1QFA.

Proposition 11.10 Neither MO-1QFA nor MM-1QFA can accept $L^0(m)$.

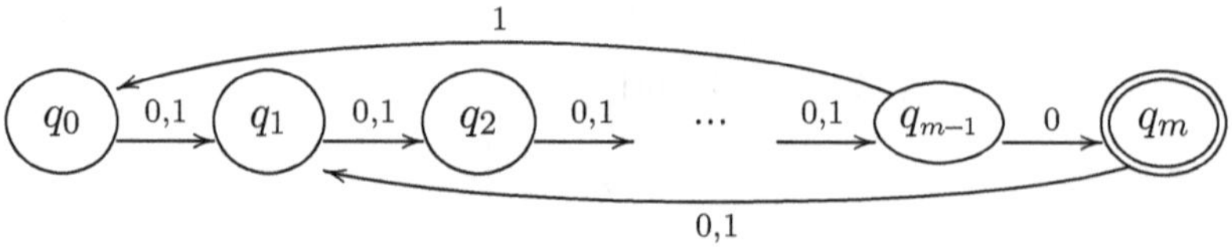

Figure 11.1: DFA accepting $L^0(m)$.

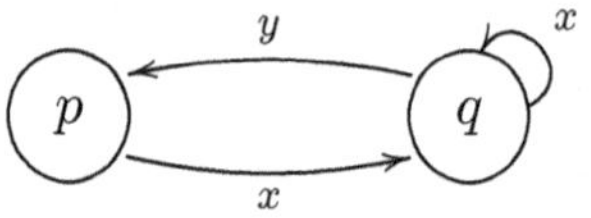

Figure 11.2: Construction not accepted by an MM-1QFA.

Proof. It suffices to show that no MM-1QFA can accept $L^0(m)$ since the languages accepted by MO-1QFA are also accepted by MM-1QFA [1, 6, 5]. By Lemma 11.9, we know that $L^0(m)$ can not be accepted by any MM-1QFA since its minimal DFA (see Figure 2) contains such a construction: For example, we can take $p = q_0, q = q_m, x = 0^m, y = 0^{m-1}1, z = \epsilon$. □

In the following we recall a relevant result.

Proposition 11.11 ([1]) Let the language $L_p = \{a^i : i \text{ is divisible by } p\}$ where p is a prime number. Then for any $\varepsilon > 0$, there exists an MM-1QFA with $O(\log(p))$ states such that for any $x \in L_p$, x is accepted with no error, and the probability for accepting $x \notin L_p$ is smaller than ε.

Indeed, from the proof of Proposition 11.11 by [1], also as Ambainis and Freivalds pointed out in [1] (before Section 2.2 in [1]), Proposition 11.11 holds for MO-1QFA as well.

Clearly, by the same technique used in the proof of Proposition 11.11 [1], one can obtain that, by replacing L_p with $L(m) = \{w : w \in \{0,1\}^*, |w| = km, k = 1, 2, 3, \cdots\}$ with m being a prime number, Proposition 11.11 still holds (by viewing all input symbols in $\{0,1\}$ as a). By combining Proposition 11.11 with Lemma 11.8, we have the following corollary.

Corollary 11.12 Suppose that m is a prime number. Then for any $\varepsilon > 0$, there exists a 1QFAC with 2 classical states and $O(\log(m))$ quantum basis states such that for any $x \in L^0(m)$, x is accepted with no error, and the probability for accepting $x \notin L^0(m)$ is smaller than ε.

Proof. Note that we have

$$L^0(m) = L^0 \cap L(m)$$

where $L^0 = \{w0 : w \in \{0,1\}^*\}$ is accepted by a DFA (depicted in Figure 11.3) with only two states and $L(m)$ can be accepted by an MO-1QFA with $O(\log(m))$ quantum basis states as shown in Proposition 11.11. Therefore, the result follows from Lemma 11.8. □

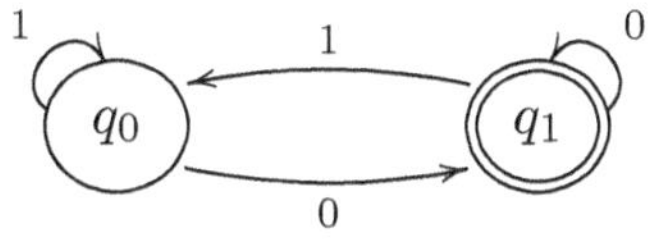

Figure 11.3: DFA accepting $\{0,1\}^*0$.

In summary, we have the following result.

Theorem 11.13 For any prime number $m \geq 2$, there exists a regular language $L^0(m)$ satisfying: (1) neither MO-1QFA nor MM-1QFA can accept $L^0(m)$; (2) the number of states in the minimal DFA accepting $L^0(m)$ is $m+1$; (3) for any $\varepsilon > 0$, there exists a 1QFAC with 2 classical states and $O(\log(m))$ quantum basis states such that for any $x \in L^0(m)$, x is accepted with no error, and the probability for accepting $x \notin L^0(m)$ is smaller than ε.

From the above result (see (2) and (3)) it follows that the lower bound given in Theorem 11.6 is tight, that is, attainable.

One should ask at this point whether similar results can be established for multi-letter 1QFA as proposed by Belovs et al. [4].

Recall that 1-letter 1QFA is exactly an MO-1QFA. Any given k-letter QFA can be simulated by some $k+1$-letter QFA. However, Qiu and Yu [33] proved that the contrary does not hold. Belovs et al. [4] have already showed that $(a+b)^*b$ can be accepted by a 2-letter QFA but, as proved in [17], it cannot be accepted by any MM-1QFA. On the other hand, a^*b^* can be accepted by MM-1QFA [1] but it can not be accepted by any multi-letter 1QFA [33], and furthermore, there exists a regular language that can not be accepted by any MM-1QFA or multi-letter 1QFA [33].

Let Σ be an alphabet. For string $z = z_1 \cdots z_n \in \Sigma^*$, consider the regular language

$$L_z = \Sigma^* z_1 \Sigma^* z_2 \Sigma^* \cdots \Sigma^* z_n \Sigma^*.$$

L_z belongs to piecewise testable set that was introduced by Simon [38] and studied in [30]. Brodsky and Pippenger [6] proved that L_z can be accepted by an MM-1QFA with $2n+3$ states.

Consider the following regular language $L(m) = \{w : w \in \Sigma^*, |w| = km, k = 1, 2, \cdots\}$. Then the minimal DFA accepting L_z needs $n+1$ states, and the minimal DFA accepting the intersection $L_z(m)$ of L_z and $L(m)$ needs $m(n+1)$ states. We will prove that no multi-letter 1QFA can accept $L_z(m)$. Indeed, the minimal DFA accepting $L_z(m)$ can be described by $A = (Q, \Sigma, \delta, q_0, F)$ where $Q = \{S_{ij} : i = 0, 1, \ldots, n; j = 1, 2, \ldots, m\}$, $\Sigma = \{z_1, z_2, \ldots, z_n\}$, $q_0 = S_{01}$, $F = \{S_{n1}\}$, and the transition function δ is defined as:

$$\delta(S_{ij}, \sigma) = \begin{cases} S_{n,(j \bmod m)+1}, & \text{if } i = n, \\ S_{i+1,(j \bmod m)+1}, & \text{if } i \neq n \text{ and } \sigma = z_{i+1}, \\ S_{i,(j \bmod m)+1}, & \text{if } i \neq n \text{ and } \sigma \neq z_{i+1}. \end{cases} \qquad (11.23)$$

The number of states of the minimal DFA accepting $L_z(m)$ is $m(n+1)$.

For the sake of simplicity, we consider a special case: $m = 2$, $n = 1$, and $\Sigma = \{0, 1\}$. Indeed, this case can also show the above problem as desired. So, we consider the following language:

$$L_0(2) = \{w : w \in \{0,1\}^*0\{0,1\}^*, |w| = 2k, k = 1, 2, \cdots\}.$$

The minimal DFA accepting $L_0(2)$ above needs 4 states and its transition figure is depicted by Figure 11.4 as follows.

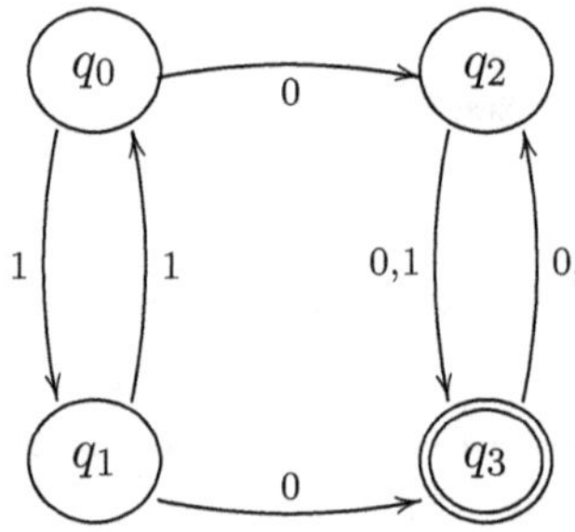

Figure 11.4: DFA accepting $w \in \{0,1\}^*0\{0,1\}^*$ with $|w|$ even.

We recall the definition of F-construction and a proposition from [4].

Definition 11.2 ([4]) A DFA with state transition function δ is said to *contain an F-construction* (see Figure 11.5) if there are non-empty words $t, z \in \Sigma^+$ and two distinct states $q_1, q_2 \in Q$ such that $\delta^*(q_1, z) = \delta^*(q_2, z) = q_2$, $\delta^*(q_1, t) = q_1$, $\delta^*(q_2, t) = q_2$, where $\Sigma^+ = \Sigma^* \backslash \{\epsilon\}$, ϵ denotes empty string.

We can depict F-construction by Figure 11.5.

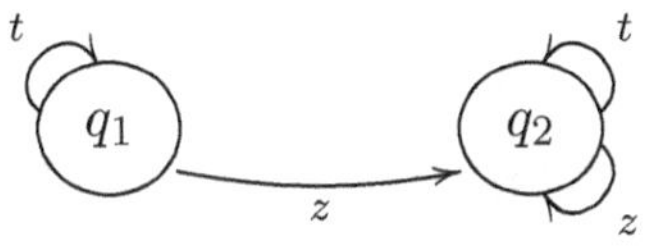

Figure 11.5: F-Construction.

Lemma 11.14 ([4]) *A language L can be accepted by a multi-letter 1QFA with bounded error if and only if the minimal DFA of L does not contain any F-construction.*

In Figure 11.4, there are an F-construction: For example, we consider q_0 and q_3, and strings 00 and 11, from the above proposition which shows that no multi-letter 1QFA can accept $L_0(2)$.

Therefore, similarly to Theorem 11.13, we have:

Theorem 11.15 If we have to restrict m to be a prime number, then for any string z with $|z| = n \geq 1$ there exists a regular language $L_z(m)$ that can not be accepted by any multi-letter 1QFA, but for every ε there exists a 1QFAC $\mathcal{A}_m$ with $n+1$ classical states (independent of m) and $O(\log(m))$ quantum basis states such that if $x \in L_z(m)$, x is accepted with no error, and the probability for accepting $x \notin L_z(m)$ is smaller than ε. In contrast, the minimal DFA accepting $L_z(m)$ has $m(n+1)$ states.

11.3 Quantum Turing machines with classical control

Quantum Turing machines were proposed originally by Deutsch [10]. One of the main problems with Deutsch proposal is that it is hard to adapt and extend classical computability results using his notion of quantum machine, namely because states of the Turing machine are quantum superpositions of classical states.

To address this problem, [29] proposed a notion of quantum Turing machine where termination is similar to a probabilistic Turing machine. However, it is also not easy to derive computability results when the function computed by a Turing machine is a random variable. To address this issue, [25] proposed a notion of quantum Turing machine with deterministic control, which we revise here.

A *deterministic-control quantum Turing machine* (in short, *dcq Turing machine*) is a variant of a binary Turing machine with two tapes, one classical and the other with quantum contents, which are infinite in both directions. Depending only on the state of the classical finite control automaton and the symbol being read by the classical head, the quantum head acts upon the quantum tape, a symbol can be written by the classical head, both heads can be moved independently of each other and the state of the control automaton can be changed.

A computation ends if and when the control automaton reaches the halting state ($\mathsf{q_h}$). Notice that the contents of the quantum tape do not affect the computation flow, hence the deterministic control and, so, the deterministic halting criterion. In particular, the contents of the quantum tape do not influence at all if and when the computation ends.

The quantum head can act upon one or two consecutive qubits in the quantum tape. In the former case, it can apply any of the following operators to the qubit under the head: identity (Id), Hadamard (H), phase (S) and π over 8 ($\pi/8$). In the latter case, the head acts on the qubit under it and the one immediately to the right by applying swap (Sw) or control-not ($\mathsf{c\text{-}Not}$) with the control qubit being the qubit under the head.

Initially, the control automaton is in the starting state ($\mathsf{q_s}$), the classical tape is filled with blanks (that is, with $\square$'s) outside the finite input sequence x of bits, the classical head is positioned over the rightmost blank before the input bits, the quantum tape contains three independent sequences of qubits – an infinite sequence of $|0\rangle$'s followed by the finite input sequence $|\psi\rangle$ of possibly entangled qubits followed by an infinite sequence of $|0\rangle$'s, and the quantum head is positioned over the rightmost $|0\rangle$ before the input qubits. In this situation, we say that the machine starts with input $(x, |\psi\rangle)$.

The control automaton is defined by the partial function

$$\delta : Q \times \mathbb{A} \rightharpoonup \mathbb{U} \times \mathbb{D} \times \mathbb{A} \times \mathbb{D} \times Q$$

where: Q is the finite set of control states containing at least the two distinct states $\mathsf{q_s}$ and $\mathsf{q_h}$ mentioned above; $\mathbb{A}$ is the alphabet composed of 0, 1 and $\square$; $\mathbb{U}$ is the set $\{\mathsf{Id}, \mathsf{H}, \mathsf{S}, \pi/8, \mathsf{Sw}, \mathsf{c\text{-}Not}\}$ of primitive unitary operators that can be applied to the quantum tape; and $\mathbb{D}$ is the set $\{\mathsf{L}, \mathsf{N}, \mathsf{R}\}$ of possible head displacements – one position to the left, none, and one position to the right.

For the sake of a simple halting criterion, we assume that $(\mathsf{q_h}, a) \notin \operatorname{dom}\delta$ for every $a \in \mathbb{A}$ and $(q, a) \in \operatorname{dom}\delta$ for every $a \in \mathbb{A}$ and $q \neq \mathsf{q_h}$. Thus, as envisaged, the computation carried out by the machine does not terminate if and only if the halting state $\mathsf{q_h}$ is not reached.

The machine evolves according to δ as expected:

$$\delta(q,a) = (U,d,a',d',q')$$

imposes that if the machine is at state q and reads a on the classical tape, then the machine applies the unitary operator U to the quantum tape, displaces the quantum head according to d, writes symbol a' on the classical tape, displaces the classical head according to d', and changes its control state to q'.

In short, by a dcq Turing machine we understand a pair (Q,δ) where Q and δ are as above.

Concerning computations, the following terminology is useful. The machine is said *to start from* $(x,|\psi\rangle)$ or to receive *input* $(x,|\psi\rangle)$ if: (i) the initial content of the classical tape is x surrounded by blanks and the classical head is positioned in the rightmost blank before the classical input x; (ii) the initial content of the quantum tape is $|\psi\rangle$ surrounded by $|0\rangle$'s and the quantum head is positioned in the rightmost $|0\rangle$ before the quantum input $|\psi\rangle$. Observe that the qubits containing the quantum input are not entangled with the other qubits of the quantum tape. When the quantum tape is completely filled with $|0\rangle$'s we say that the quantum input is $|\varepsilon\rangle$.

Furthermore, the machine is said *to halt at* $(y,|\varphi\rangle)$ or to produce *output* $(y,|\varphi\rangle)$ if the computation terminates and: (i) the final content of the classical tape is y surrounded by blanks and the classical head is positioned in the rightmost blank before the classical output y; (ii) the final content of the quantum tape is $|\varphi\rangle$ surrounded by $|0\rangle$'s and the quantum head is positioned in the rightmost $|0\rangle$ before the quantum output $|\varphi\rangle$. In this situation we may write

$$M(x,|\psi\rangle) = (y,|\varphi\rangle).$$

Clearly, the qubits containing the quantum output are not entangled with the other qubits of the quantum tape.

For each $n \in \mathbb{N}^+$, denote by H^n the Hilbert space of dimension 2^n. A unitary operator

$$U : \mathsf{H}^n \to \mathsf{H}^n$$

is said to be *dcq computable* if there is a dcq Turing machine (Q,δ) that, for every unit vector $|\psi\rangle \in \mathsf{H}^n$, when starting from $(\varepsilon,|\psi\rangle)$ produces the quantum output $U|\psi\rangle$. Note that the final content of the classical tape is immaterial.

A (classical) problem

$$X \subseteq \{0,1\}^*$$

is said to be *dcq decidable* if there is a dcq Turing machine (Q, δ) that, for every $x \in \{0,1\}^*$, when starting from $(x, |\varepsilon\rangle)$ produces a quantum output $|\varphi\rangle$ such that:

$$\begin{cases} \mathsf{Prob}\,(\mathsf{Proj}_1|\varphi\rangle = 1) > 2/3 & \text{if } x \in X \\ \mathsf{Prob}\,(\mathsf{Proj}_1|\varphi\rangle = 0) > 2/3 & \text{if } x \notin X \end{cases}$$

where Proj_1 is the projective measurement defined by the operator

$$\begin{pmatrix} 0 & 0 \\ 0 & 1 \end{pmatrix} \otimes \mathsf{ID}$$

using the adopted computational basis $\{|0\rangle, |1\rangle\}$, with the first factor acting on the first qubit of the quantum output (the qubit immediately to the right of the quantum head) and the identity acting on the remaining qubits of the output. Clearly, the possible outcomes of the measurement are the eigenvalues 0 and 1 of the defining operator.

Moreover, problem X is said to be (time) *dcq bounded error quantum polynomial*, in short in dcBQP, if there are polynomial $\xi \mapsto P(\xi)$ and a dcq Turing machine deciding X that, for each x, produces the output within $P(|x|)$ steps. In [25] it was established that the quantum computation concepts above coincide with those previously introduced in the literature using quantum circuits.

It is straightforward to see that dcq decidability coincides with the classical notion. It is enough to take into account that the dcq Turing machines can be emulated by classical Turing machines using a classical representation of the contents of the quantum tape that might be reached from $(x, |\varepsilon\rangle)$.

In the sequel we also need the following notion that capitalises on the fact that dcq Turing machines can work like classical machines by ignoring the quantum tape. A function

$$f : \{0,1\}^* \rightharpoonup \{0,1\}^*$$

is said to be *classically dcq computable* if there is a dcq Turing machine (Q, δ) that, for every $x \in \{0,1\}^*$, when starting from input $(x, |\psi\rangle)$ produces the classical output $f(x)$ if $x \in \mathrm{dom} f$ and fails to halt with a meaningful classical output if $x \notin \mathrm{dom} f$.

Theorem 11.16 (Polynomial translatability) There is a dcq Turing machine T such that, for any dcq Turing machine $M = (Q, \delta)$, there is a map

$$s : \{0,1\}^* \to \{0,1\}^*$$

which is classically dcq computable in linear time and fulfils the following conditions:

$$\begin{cases} \forall\, p, x \in \{0,1\}^*, |\psi\rangle \in \mathsf{H}^n, n \in \mathbb{N}^+ & M(p\Box x, |\psi\rangle) = T(s(p)\Box x, |\psi\rangle) \\ \exists\, c \in \mathbb{N}\ \ \forall\, p \in \{0,1\}^* & |s(p)| \leq |p| + c. \end{cases}$$

Moreover, there is a polynomial $(\xi_1, \xi_2, \xi_3) \mapsto P(\xi_1, \xi_2, \xi_3)$ such that if M starting from $(p\Box x, |\psi\rangle)$ produces the output in k steps then T produces the same output in at most

$$P(|p| + |x|, |Q|, k)$$

steps when starting from $(s(p)\Box x, |\psi\rangle)$.

Proof. Without any of loss of generality assume that

$$Q = \{q_0, q_1, \ldots, q_\nu, q_{\nu+1}\}$$

with $\mathsf{q_s} = q_0$ and $\mathsf{q_h} = q_{\nu+1}$. Hence, $|Q| = \nu + 2$. Consider the map

$$s = p \mapsto \underline{\delta}\, 111\, p : \{0,1\}^* \to \{0,1\}^*$$

where $\underline{\delta}$ encodes δ as follows:

$$\underline{\delta(q_0,0)}\,\underline{\delta(q_0,1)}\,\underline{\delta(q_0,\Box)} \ldots \underline{\delta(q_i,0)}\,\underline{\delta(q_i,1)}\,\underline{\delta(q_i,\Box)} \ldots \underline{\delta(q_\nu,0)}\,\underline{\delta(q_\nu,1)}\,\underline{\delta(q_\nu,\Box)}$$

with each

$$\underline{\delta(q,a)} = \underline{U}\,\underline{d}\,\underline{a'}\,\underline{d'}\,\underline{q'} \in \{0,1\}^*$$

where

$$\underline{U} = \begin{cases} 000 & \text{if} \quad U = \mathsf{Id} \\ 001 & \text{if} \quad U = \mathsf{H} \\ 010 & \text{if} \quad U = \mathsf{S} \\ 011 & \text{if} \quad U = \pi/8 \\ 100 & \text{if} \quad U = \mathsf{Sw} \\ 101 & \text{if} \quad U = \text{c-Not} \end{cases}$$

$$\underline{d} = \begin{cases} 00 & \text{if} \quad d = \mathsf{L} \\ 01 & \text{if} \quad d = \mathsf{N} \\ 11 & \text{if} \quad d = \mathsf{R} \end{cases}$$

$$\underline{a'} = \begin{cases} 00 & \text{if} \quad a' = 0 \\ 11 & \text{if} \quad a' = 1 \\ 10 & \text{if} \quad a' = \Box \end{cases}$$

$$\underline{d'} = \begin{cases} 00 & \text{if} \quad d' = \mathsf{L} \\ 01 & \text{if} \quad d' = \mathsf{N} \\ 11 & \text{if} \quad d' = \mathsf{R} \end{cases}$$

$$\underline{q'} = 1^{j+1}00$$

assuming that $\delta(q,a) = (U, d, a', d', q')$ and $q' = q_j$. Notice that one can identify in $s(p)$ the end of the encoding of $\underline{\delta}$ since each $\underline{\delta(q,a)}$ starts with $\underline{U}$ and the sequence 111 does not encode any gate. Clearly, as defined, s can be dcq computed in linear time and fulfils the conditions in the statement of the theorem by taking $c = |\underline{\delta}| + 3$.

It is necessary to encode in the classical tape of T the current classical configuration of M (composed of the current contents of the classical tape, the current position of the classical head and the current state of the control automaton). There is no need to encode the quantum configuration of M since in a dcq Turing machine it does not affect its transitions. In due course, when explaining how M computations are emulated by T computations, we shall see how quantum configurations of T are made to follow those of M. The following notation becomes handy for describing classical configurations of dcq Turing machines.

We write

$$w \overset{\overset{q}{\nabla}}{a} w'$$

for stating that the machine in hand is at state q, its classical head is over a tape cell containing symbol a, with the finite sequence w of symbols to the left of the head, with the finite sequence w' of symbols to the right of the head, and with the rest of the classical tape filled with blanks.

Before describing how a classical configuration of M is encoded in T we need to introduce some additional notation. Recall that a symbol $a \in \{0, 1, \Box\}$ is encoded as

$$\underline{a} = \begin{cases} 00 & \text{if} \quad a = 0 \\ 11 & \text{if} \quad a = 1 \\ 10 & \text{if} \quad a = \Box. \end{cases}$$

We denote by $\underline{a}_1$ and $\underline{a}_2$ the first and second bit of $\underline{a}$, respectively. The reverse encoding of a is $\underline{\underline{a}} = \underline{a}_2\underline{a}_1$. Given a string $w = w_1 \dots w_m \in \{0, 1, \Box\}^*$, we denote its encoding by $\underline{w} = \underline{w_1} \dots \underline{w_m}$ and its reverse encoding by $\underline{\underline{w}} = \underline{\underline{w_m}} \dots \underline{\underline{w_1}}$.

The classical configuration

$$w \overset{\overset{q_i}{\nabla}}{a} w'$$

of M should be encoded as the following classical configuration of T

$$\underline{w}\;\square\;\underbrace{\underbrace{1\ldots1}_{\nu+i-1}\;\square\;\underbrace{1\ldots1}_{i+1}}_{q_i}\;\overset{\overset{q'}{\nabla}}{\square}\;\underline{\delta}\;111\;\underline{\underline{w'}}\;\underline{a_2}\underline{a_1}$$

where q' is a state of T representing the stage where the machine is able to start emulating a transition of M. As we shall see later, whenever a transition of M has just been emulated by T and the resulting state is not the halting state of M, T is at state q'.

The initial classical configuration of M is

$$\overset{\overset{q_0}{\nabla}}{\square}\;p\;\square\;x$$

and, moreover, the initial classical configuration of T is

$$\overset{\overset{q'_0}{\nabla}}{\square}\;\underbrace{\underline{\delta}\;111\;p}_{s(p)}\;\square\;x,$$

where q'_0 is the initial state of T. The objective of this stage is to change the initial classical configuration of T to the encoding of the initial configuration of M, as described before, that is:

$$\underbrace{1\ldots1\;\square\;1}_{\text{encoding of } q_0}\;\overset{\overset{q'}{\nabla}}{\square}\;\underline{\delta}\;111\;\underline{\underline{\square p \square x}}.$$

Writing the encoding of q_0 can be done straightforwardly in $O(k)$ steps. It remains to describe how to encode $\square p \square x$ in reverse order within $O((|p|+|x|)^2)$ steps, keeping $\underline{\delta}\;111$ unchanged:

1. **Encoding of** x: Recall that $x = x_1 \ldots x_m$ has no blanks, and therefore $\underline{x} = x_1 x_1 \ldots x_m x_m$. The idea is to shift $x_2 \ldots x_m$ to the right, duplicate x_1 in the vacated cell and then iterate this process to $x_2 \ldots x_m$. First, the head moves on top of x_2 and copies the contents of $x_2 \ldots x_m$ one cell to the right, leaving the original cell of x_2 with a blank. Then the head moves back to x_1 and copies its contents to the cell on its right. The process is iterated for $x_2 \ldots x_m$ until the last symbol of x is reached. Since shifting to the right the contents of m cells, leaving the first one blank, can be done with a linear number of steps in m, this operation takes a quadratic number of steps on the size of x.

2. **Encoding the $\square$ in $p\square x$**: First, the encoding of x is shifted one cell to the right and then, the head is moved back to the top of the first two blanks separating p and $\underline{x}$. Finally, the head replaces the two blanks by 10 and it is parked in the 1. Note that this can be done with a linear number of steps on the size of x, and moreover, the encoding of $\square x$ has no blanks.

3. **Encoding of p**: Let l be the size of p. The encoding of p is similar to the encoding of x. First the encoding of $\square x$ is shifted three cells to the right. Then the head of the machine is moved to the beginning of p. Notice that the machine can identify it as the first cell on the right of $\underline{\delta}$ 111. Next, p is shifted one cell to the right (which leaves two blanks before $\square x$) and the head of the machine is moved to the cell containing p_l. The machine copies p_l to the two cell immediately on its right and writes $\square$ in the original cell. After these steps, the content of the classical tape is:

$$\underline{\delta}\ 111 \square\, p_1 \ldots p_{l-1} \overset{q'}{\overset{\nabla}{\square}} \underline{p_l \square x}.$$

Next $\underline{p_l \square x}$ is shifted one cell to the right and the process of writing the encoding of p_l in the tape is repeated for p_{l-1}, p_{l-2}, ... until p_1. The end of this construction is reached whenever the symbol $\square$ is placed after $\underline{\delta}$ 111 is read. Finally, $\underline{p \square x}$ is shifted two cells to the left.

4. **Reversing the encoding of $p\square x$**: Assume that $\underline{p\square x} = y_1 \ldots y_m$ (with $m = |\underline{p}| + |\underline{x}| + 2$) is the contents of the cells containing the encoding of $p\square x$. The objective is to replace $y_1 \ldots y_m$ by $y_m \ldots y_1$. First, the cell containing y_1 is replaced by a blank and y_1 is copied to the right cell of y_m. Second, the sequence $y_2 \ldots y_m$ is shifted one cell to the left. This process is repeated with $y_2 \ldots y_m$ in such a way that y_2 is copied to the left of y_1 and until the contents of the tape is $\square y_m \ldots y_1$. Finally, the blank symbol is removed when $y_m \ldots y_1$ is shifted one cell to the left. Observe that the operations leading to $\square y_m \ldots y_1$, take $O(m^2)$ steps. Moreover, the final shift is linear, and so the overall stage takes a quadratic number of steps.

5. **Placing the reverse encoding of a blank at the right end**: The head is moved to the right until the first blank is found. Then, the head writes a 0 and moves one cell to the right, where it writes a 1. Finally, the head is moved to the left until the first blank is found.

It is straightforward to check that the overall cost of these operations is quadratic on $|p| + |x|$ and that the five stages above require just a constant number of states in T (that is, the number of states does not depend on p and x).

Next, we describe the steps needed to emulate in T one step by M. Assume that the transition to be emulated is $\delta(q_i, a) = (U, d, a', d', q_j)$ and that T is at the following classical configuration:

$$\underline{w}\;\underbrace{\underbrace{1\cdots1}_{\nu-i-1}\;\square\;\underbrace{1\cdots1}_{i+1}}_{\text{encoding of } q_i}\;\overset{q'}{\overset{\nabla}{\square}}\;\underline{\delta}\;111\;\underline{\underline{w'}}\;\underline{a_2}\;\underline{a_1}.$$

The objective is to set T at the following classical configuration

$$\underline{w''}\;\underbrace{\underbrace{1\cdots1}_{\nu-j-1}\;\square\;\underbrace{1\cdots1}_{j+1}}_{\text{encoding of } q_j}\;\overset{q'}{\overset{\nabla}{\square}}\;\underline{\delta}\;111\;\underline{\underline{w'''}}$$

where, depending on the move of the classical head, three cases may occur:

- if $d' = N$ then $w = w''$ and $w''' = a'w$;
- if $d' = L$ then $w'' = w_1 \dots w_{|w|-1}$ and $w''' = w_{|w|}a'w'$;
- if $d' = R$ then $w'' = wa'$ and $w''' = w'_2 \dots w'_{|w'|}$.

Machine T performs the emulation of $\delta(q_i, a)$ as follows:

1. **Identifying the value** a: The head of the classical tape of T is moved to $\underline{a}_1$ which is the rightmost cell that is not blank. The head reads the contents of that cell and the contents of the cell on its left, which has $\underline{a}_2$, and goes to a different state of T depending on the value a. The cost of this operation is linear in the number of states of M and on the space used by M.

2. **Parking the head at the encoding of** $\delta(q_i, a)$ **in** $\underline{\delta}$: First the head is moved to the cell containing the rightmost 1 of the encoding of q_i. Notice that such encoding has at least one 1 to the right of the blank. Since the head starts from position $\underline{a}_2$, such 1 is on the left to the first blank that the head finds while reading the classical tape from right to the left. So, this operation is at most linear in

the size of the space used by M. Recall that the encoding $\underline{\delta}$ of δ is as follows:

$$\underline{\delta(q_0,0)\delta(q_0,1)\delta(q_0,\Box)}\ldots\underline{\delta(q_i,0)\delta(q_i,1)\delta(q_i,\Box)}\ldots\underline{\delta(q_\nu,0)\delta(q_\nu,1)\delta(q_\nu,\Box)}.$$

Moreover, each $\underline{\delta(q_i,a)}$ ends with 00 and starts with nine cells corresponding to $\underline{U}\cdot\underline{d}\cdot\underline{a'}\cdot\underline{d'}$ and a sequence of 1's, encoding the resulting state of that transition. This stage consists in a loop with progress variable, say r, starting from $r=1$ until $r=i+1$. The goal of the loop is to replace the r rightmost 1's of encoding of q_i by 0's while the 00, at the end of $\underline{\delta(q_{r-1},\Box)}$, are replaced by $\Box\Box$. The end of the loop $r=i+1$, is detected when a blank symbol is read in the encoding of q_i. For each value of r we keep only a pair of $\Box\Box$ in $\underline{\delta}$: those at the end of $\underline{\delta(q_{r-1},\Box)}$. When $r=i+1$, the encoding of $\delta(q_i,\Box)$ is marked in $\underline{\delta}$ with $\Box\Box$, and so, it remains to park the head in the first cell of the encoding of $\delta(q_i,a)$. This movement can be achieved taking into account the symbol a read in the previous stage. Observe that all the operations performed in this stage depend linearly on the space used by M (on the right of its classical head) and quadratically on the number of states of M.

3. **Identifying and applying U**: Using the first three cells of $\underline{\delta(q_i,a)}$, the machine T identifies the unitary transformation and applies it to its own quantum tape.

4. **Performing the d-move of the quantum head**: Using the fourth and fifth cells of $\underline{\delta(q_i,a)}$, T identifies the movement of the quantum head and operates accordingly on its own quantum head.

5. **Identifying and writing a'**: Using the sixth and seventh cells of $\underline{\delta(q_i,a)}$, T identifies the encoding of the symbol a' to be written under $\underline{a_2}\underline{a_1}$. The encoding of a' in $\underline{\delta(q_i,a)}$ is marked with two blanks and $\underline{a'}$ is copied in reversed order to $\underline{a_2}\underline{a_1}$, which are the two rightmost non-blank cells. After completing the last operation, the head returns to the original position and restores $\underline{a'}$ in $\underline{\delta(q_i,a)}$. Notice that the operations of this stage can be done in a linear number of steps on the space used by M the input and linearly in the number of states.

6. **Performing the d'-move of the classical head**: The ninth and tenth cells of $\underline{\delta(q_i,a)}$ store the movement of the classical head. If $d'=N$ nothing has to be done. W.l.o.g. assume $d'=R$. First,

the encoding of d' is marked with two blanks. Then the rightmost non-blank cells have to be copied (in reverse order) to the left of the leftmost non-blank cells. Clearly the rightmost non-blank cells have to be replaced by two blanks if $|w'| > 0$, and have to be replaced by 01 (the reverse encoding of a blank) if $|w'| = 0$. *Mutatis mutandis* if $d' = L$. This stage can be done in a linear number of steps on the space in the classical tape used by M.

7. **Updating the emulated state to** q_j: Assume that q_j is not the halting state and recall that q_j is encoded as $1^{j+1}00$ at the rightmost part of $\underline{\delta(q_i, a)}$. The idea is to update the emulated state q_i to q_j by replacing each 1 in $1^{j+1}00$ at $\underline{\delta(q_i, a)}$ by a $\square$ while updating the cells used to encode the current state of M. Given Stage 2, the cells used to encode q_i contain $1^{\nu-i-1}\square 0^{i+1}$. If $j \leq i$ then we replace $j + 1$ rightmost 0's by 1's and then place a $\square$ left to them. If $j > i$, then the $i + 1$ rightmost 0's are replaced by 1's and after, the blank has to be carried to the left while being replaced by a 1, until there are $(j + 1)$ 1's. This process ends when all 1's in $1^{j+1}00$ have been replaced by blanks. After the cells encoding the emulated state are updated, the encoding of q_j in $\underline{\delta(q_i, a)}$ is restored, by replacing the blanks by 1's. This stage does not depend on the input of M, but only quadratically in the number of states of M. If q_j is the halting state, we have to restore the contents of the classical tape to $wa'w'$, with the head positioned over a'. This corresponds to inverting the process used to prepare the initial configuration, erasing the encoding of q_i and δ. Such stage can be done in a number of steps quadratic to the space used by M and linearly in the number of states of M.

Finally, the overall emulation is polynomial (in fact quadratic) on $|p|+|x|$, ν and k since the space used by M is bounded by k. □

A machine T fulfilling the conditions of Theorem 11.16 is said to enjoy the *s-m-n* property. Any such machine is universal as shown in the next result.

Theorem 11.17 (Polynomial universality) Let T be a dcq Turing machine enjoying the *s-m-n* property. Then, for any dcq Turing machine $M = (Q, \delta)$, there is $p \in \{0, 1\}^*$ such that

$$M(x, |\psi\rangle) = T(p\square x, |\psi\rangle) \quad \forall\, x \in \{0,1\}^*, |\psi\rangle \in \mathsf{H}^n, n \in \mathbb{N}^+.$$

Moreover, there is a polynomial $(\xi_1, \xi_2, \xi_3) \mapsto P(\xi_1, \xi_2, \xi_3)$ such that if M, when starting from $(x, |\psi\rangle)$, produces the output in k steps then T

produces the same output in at most

$$P(|x|, |Q|, k)$$

steps when starting from $(p\square x, |\psi\rangle)$.

Proof. Consider M' such that $M'(\varepsilon\square x, |\psi\rangle) = M(x, |\psi\rangle)$. By applying Theorem 11.16 to M' and choosing $p = s(\varepsilon)$ the result follows. □

Acknowledgments

The authors acknowledge IT project QbigD funded by FCT PEst-OE/EEI/LA0008/2013 and UID/EEA/50008/2013. The authors also acknowledge the Confident project PTDC/EEI-CTP/4503/2014.

Bibliography

[1] A. Ambainis, R. Freivalds, One-way quantum finite automata: strengths, weaknesses and generalizations, in: Proceedings of the 39th Annual Symposium on Foundations of Computer Science, IEEE Computer Society Press, Palo Alfo, California, USA, 1998, pp. 332-341.

[2] F. Ablayev, A. Gainutdinova, On the Lower Bounds for One-Way Quantum Automata, in: Proceedings of the 25th International Symposium on Mathematical Foundations of Computer Science (MFCS'2000), Lecture Notes in Computer Science, Vol. 1893, Springer, Berlin, 2000, pp. 132-140.

[3] S. Bozapalidis, Extending stochasic and quantum functions, Theory Computing Systems 36 (2003) 183-197.

[4] A. Belovs, A. Rosmanis, and J. Smotrovs, Multi-letter Reversible and Quantum Finite Automata, in: Proceedings of the 13th International Conference on Developments in Language Theory (DLT'2007), Lecture Notes in Computer Science, Vol. 4588, Springer, Berlin, 2007, pp. 60-71.

[5] A. Bertoni, C. Mereghetti, B. Palano, Quantum Computing: 1-Way Quantum Automata, in: Proceedings of the 9th International Conference on Developments in Language Theory (DLT'2003), Lecture Notes in Computer Science, Vol. 2710, Springer, Berlin, 2003, pp. 1-20.

[6] A. Brodsky, N. Pippenger, Characterizations of 1-way quantum finite automata, SIAM Journal on Computing 31 (2002) 1456-1478.

[7] M. P. Ciamarra, Quantum Reversibility and a New Model of Quantum Automaton, in Proceeding of 13th International Symposium on Fundamentals of Computation Theory, Lecture Notes in Computer Science, Vol. 2138, Springer-Verlag, Berlin, 2001, pp. 376-379.

[8] R. Chadha, P. Mateus, and A. Sernadas. Reasoning about quantum imperative programs. *Electronic Notes in Theoretical Computer Science*, 158:19–40, 2006. Invited talk at the Twenty-second Conference on the Mathematical Foundations of Programming Semantics, May 24-27, 2006, Genova.

[9] R. Chadha, P. Mateus, A. Sernadas, and C. Sernadas. Extending classical logic for reasoning about quantum systems. In D. Gabbay K. Engesser and D. Lehmann, editors, *Handbook of Quantum Logic and Quantum Structures: Quantum Logic*, pages 325–372. Elsevier, 2009.

[10] D. Deutsh, Quantum theory, the Church-Turing principle and the universal quantum computer, Proceedings of the Royal Society of London Series A 400 (1985) 97-117.

[11] R.P. Feynman, Simulating physics with computers, International Journal of Theoretical Physics 21 (1982) 467-488.

[12] M. Golovkins, M. Kravtsev, Probabilistic reversible automata and quantum automata, in: Proc. 18th International Computing and Combinatorics Conference (COCOON'02), Lecture Notes in Computer Science, Vol. 2387, Springer, Berlin, 2002, pp. 574-583.

[13] C. Hermida and P. Mateus. Paracategories I: Internal paracategories and saturated partial algebras. *Theoretical Computer Science*, 309:125–156, 2003.

[14] C. Hermida and P. Mateus. Paracategories II: Adjunctions, fibrations and examples from probabilistic automata theory. *Theoretical Computer Science*, 311:71–103, 2004.

[15] J.E. Hopcroft, J.D. Ullman, Introduction to Automata Theory, Languages, and Computation, Addision-Wesley, New York, 1979.

[16] E. Jeandel, Topological Automata, Theory Computing Systems 40 (2007) 397-407.

[17] A. Kondacs, J. Watrous, On the power of finite state automata, in: Proceedings of the 38th IEEE Annual Symposium on Foundations of Computer Science, Miami Beach, Florida, USA, 1997, pp. 66-75.

[18] L.Z. Li, D.W. Qiu, Determining the equivalence for one-way quantum finite automata, Theoretical Computer Science 403 (2008) 42-51.

[19] L. Li, D. Qiu, X. Zou, L. Lvjun, L. Wu, and P. Mateus. Characterizations of one-way general quantum finite automata. *Theoretical Computer Science*, 419:73–91, 2012.

[20] P. Mateus, D.W. Qiu, L.Z. Li, On the complexity of minimizing probabilistic and quantum automata, Information and Computation 218 (2012) 36-53.

[21] P. Mateus and A. Sernadas. Exogenous quantum logic. In W. A. Carnielli, F. M. Dionísio, and P. Mateus, editors, *Proceedings of CombLog'04, Workshop on Combination of Logics: Theory and Applications*, pages 141–149, 1049-001 Lisboa, Portugal, 2004. Departamento de Matemática, Instituto Superior Técnico. Extended abstract.

[22] P. Mateus and A. Sernadas. Reasoning about quantum systems. In J. Alferes and J. Leite, editors, *Logics in Artificial Intelligence, Ninth European Conference, JELIA'04*, volume 3229 of *Lecture Notes in Artificial Intelligence*, pages 239–251. Springer, 2004.

[23] P. Mateus and A. Sernadas. Weakly complete axiomatization of exogenous quantum propositional logic. *Information and Computation*, 204(5):771–794, 2006. ArXiv math.LO/0503453.

[24] P. Mateus, A. Sernadas, and C. Sernadas. Precategories for combining probabilistic automata. *Electronic Notes in Theoretical Computer Science*, 29, 1999. Early version presented at FIREworks Meeting, Magdeburg, May 15-16, 1998. Presented at CTCS'99, Edinburgh, September 10-12, 1999.

[25] P. Mateus, A. Sernadas, and A. Souto. Universality of quantum Turing machines with deterministic control. *Journal of Logic and Computation*, 27(1):1–19, 2017.

[26] C. Moore, J.P. Crutchfield, Quantum automata and quantum grammars, Theoretical Computer Science 237 (2000) 275-306.

[27] C. Mereghetti, B. Palano, Quantum finite automata with control language, RAIRO-Inf. Theor. Appl. 40 (2006) 315-332.

[28] K. Paschen, Quantum finite automata using ancilla qubits, Technical report, University of Karlsruhe, 2000.

[29] S. Perdrix and P. Jorrand. Classically-controlled quantum computation. *Electronic Notes in Theoretical Computer Science*, 135(3):119–128, 2006.

[30] D. Perrin, Finite automata, In: J. van Leeuwen (Eds.), Handbook of Theoretical Computer Science, Elsevier Science, Holland, 1994, Chap. 1.

[31] D.W. Qiu, L.Z. Li, P. Mateus, J. Gruska, Quantum finite automata, in: Finite State Based Models and Applications (Edited by Jiacun Wang), CRC Handbook, 2012, pp. 113-144.

[32] D. Qiu, L. Li, P. Mateus, and A. Sernadas. Exponentially more concise quantum recognition of non-RMM regular languages. *Journal of Computer and System Sciences*, 81(2):359–375, 2015.

[33] D.W. Qiu, S. Yu, Hierarchy and equivalence of multi-letter quantum finite automata, Theoretical Computer Science 410 (2009) 3006-3017.

[34] D.W. Qiu, L.Z. Li, X. Zou, P. Mateus, J. Gruska, Multi-letter quantum finite automata: decidability of the equivalence and minimization of states, Acta Informatica 48 (2011) 271-290.

[35] M. O. Rabin, Probabilistic Automata, Information and Control, 6 (3) (1963) 230-245.

[36] L. Schröder and P. Mateus. Universal aspects of probabilistic automata. *Mathematical Structures in Computer Science*, 12(4):481–512, 2002.

[37] A. Sernadas, P. Mateus, and Y. Omar. Quantum computation and information. In M. S. Pereira, editor, *A Portrait of State-of-the-Art Research at the Technical University of Lisbon*, pages 46–65. Springer, 2007.

[38] I. Simon, Piecewise testable events, in: Proc. the 2nd GI conference, Lecture Notes in Computer Science, Vol. 33, Springer, New York, 1975.

[39] W.G. Tzeng, A Polynomial-time Algorithm for the Equivalence of Probabilistic Automata, SIAM Journal on Computing 21 (2) (1992) 216-227.

[40] S. Yu, Regular Languages, In: G. Rozenberg, A. Salomaa (Eds.), Handbook of Formal Languages, Springer-Verlag, Berlin, 1998, pp. 41-110.

[41] T. Yamakami, Analysis of quantum functions, Internat. J. Found. Comput. Sci. 14 (2003) 815-852,

[42] S.G. Zheng, D.W. Qiu, L.Z. Li, Jozef Gruska, One-way finite automata with quantum and classical states, In: H. Bordihn, M. Kutrib, and B. Truthe (Eds.), Dassow Festschrift 2012, Lecture Notes in Computer Science, Vol. 7300, Springer, Berlin, 2012, pp. 273–290.

[43] S. Zheng, D. Qiu, J. Gruska, L. Li, and P. Mateus. State succinctness of two-way finite automata with quantum and classical states. *Theoretical Computer Science*, 499:98–112, 2013.

12

Interacting electrons in one dimension: the two chain perylene-metal dithiolate series

Luís Alcácer[1], Rui T. Henriques[1] and Manuel Almeida[2]

[1]Instituto de Telecomunicações, Instituto Superior Técnico, Universidade de Lisboa, Portugal

[2]Centro de Ciências e Tecnologias Nucleares, Campus Tecnológico e Nuclear, Pólo de Loures do IST, Universidade de Lisboa, Portugal

Corresponding author contact: alcacer@lx.it.pt

Abstract

In this review, we attempt to present the most notable aspects of the story of the two chain, highly one-dimensional, organic conductors of the perylene metal-dithiolate series, which have been under investigation since the end of the 1960's until the present time, with the recent discovery of a superconducting phase. We will start with an overview of the developments of the research on these materials and then, after a brief introduction and a survey of the theoretical background, we will describe and discuss the structure, transport and magnetic properties, as well as the low temperature instabilities and phase transitions. Finally we will focus on the study of their behaviour under extreme conditions of low temperatures, high pressures and high magnetic fields up to 45 tesla.

Overview

"One-dimensional thought is systematically promoted by the makers of politics and their purveyors of mass information. Their universe of discourse is populated by self-validating hypotheses which, incessantly and monopolistically repeated, become hypnotic definitions of dictations."

Herbert Marcuse, One-Dimensional Man, Beacon Press, 1964.

In 1964, William Little, from Stanford University, proposed a model to synthesize a room temperature organic superconductor [1], which, a few years later, led to a new field of research entitled *one-dimensional conductors*. In those materials, the electrical conductivity is high along one direction and much lower in the other directions. This is possible in two cases: i) in solids in which flat organic molecules are stacked on top of each other, and through intermolecular interactions, electrons can travel along the staking axis, and ii) in organic polymers in which single carbon-carbon bonds alternate with double bonds allowing electrons to travel along the polymer chain. Nowadays, one can confine electrons in one or two dimensions by means of top-down approaches to nanotechnology, but that is not the subject of the present paper. The present paper deals with organic crystalline solids, with extreme anisotropy, in which the one-dimensional character of the interactions of the electrons with each other and with the lattice vibrations (the phonons) are the source of various types of competing instabilities and phase transitions. Take, for example, a linear chain of N sites, separated from each other by a distance a, the *unit cell* size, and with one electron per site, or unit cell. Since each site can accommodate two electrons, with opposite spins, the system would have $2N$ states available, but only N would be occupied. One would have, therefore, a *half-filled band* metal. However, the total energy of the electrons would decrease if the sites would arrange themselves in pairs, doubling the unit cell, which corresponds to opening a gap at the Fermi level, the last occupied energy level. In one-dimension, below a certain critical temperature, the energy needed to distort the lattice, doubling the unit cell, becomes lower than the *gain* in energy of the electrons, and a phase transition may occur. This is the Peierls transition, one of many that may occur.

A good example of a one-dimensional solid is the one, best known as $(TMTSF)_2PF_6$, which was the first organic superconductor to be discovered, in 1979, and has been a centre of attention since then. In the words of Paul Chaikin, a well known physicist, now at the New York University, this is "the most remarkable electronic material ever discovered. In one single crystal, one can observe all the usual competitions, all electron transport mechanisms known to man, plus somethings new".

In an instance of lucky serendipity, one of the present authors (LA), synhtesized the first members of a family of organic conductors based on the perylene molecule and metal-dithiolate complexes [2] (see Figure 12.1). At the time, it was not possible to prove that any of these materials was a metal and much less a superconductor, in spite of the many evenings spent cooling them down to liquid helium temperature. They always turned insulators! The first papers with a full description of their electrical and magnetic properties appeared only in 1974 and 1976 [3, 4] as the outcome of more detailed studies. This family of materials of general formula $(Per)_2[M(mnt)_2]$, with M = Pt, Au, etc., became of considerable interest later on for exhibiting metallic behaviour and very high one-dimensionality. The crystal structures are formed by chains of $[M(mnt)_2]$ units surrounded by perylene chains, which are 3/4 filled band metals and therefore highly conducting at room temperature. On the other hand, the $[M(mnt)_2]$ chains are insulating, with localized spins in the case of M=Ni and Pt, (S=1/2), for example, spins that *exchange fast*, with the itenerant electrons on the perylene chains. For more than 40 years, now, these materials have also been amazing us, with their remarkable properties, some of them typical of the extreme anysotropy, but mostly not expected as they came. The nature of the metal-insulator transition in the Pt compound, for example, is particularly compelling. At the temperature of 8 kelvin, the perylene chain tetramerizes, developing a *charge density wave* (CDW), with the opening of a gap. This lattice distortion drives a distortion of the $[M(mnt)_2]$ chains, which dimerize, leading to a spin-Peierls ground state (SP). The spin-Peierls state in the $[M(mnt)_2]$ chains is coupled to the charge density wave state in the perylene chain, up to magnetic fields of the order of 20 tesla, a situation not foreseen in the present theory, which predicts a decoupling at much lower fields. Under extreme conditions of low temperature, high pressure and high magnetic fields, these materials exhibit a rich variety of phenomena, including field induced charge density waves (FICDW), interference effects of electrons from "trajectories" on different Fermi surface sheets, and for our delight, superconductivy was finally observed, under pressure, in 2009, evolving in a rather unusual way, from the charge density wave state. This paper is about the wonderful materials of this family.

12.1 Introduction

On page 111, of his 1955 "Quantum Theory of Solids", [5], Rudolf Peierls states that "it is therefore likely that a one-dimensional model could never have metallic properties". This he concludes, after arguing that in a one-dimensional metal with a partly filled band the regular chain structure could not be stable, since one would always find a distortion corresponding to a suitable number of atoms in the unit cell for which a break would occur at or near the edge of the Fermi level [6].

In the early 1960's, a few years after the success of the BCS theory of superconductivity [7], there was the hope to learn how to produce materials which could be superconductors at room temperature. In 1964, William Little's seminal paper entitled "Possibility of Synthesizing an Organic Superconductor" [1] stimulated the search for new materials which could comply with his model system. Based on London's idea [8] that superconductivity might occur in organic macromolecules, he proposed a mechanism similar to the BCS theory of superconductivity and even suggested model molecules that could lead to room temperature superconductivity. His model structure was composed of a conjugated polymer chain dressed with highly polarizable molecules as side groups. The polymer chain would be a one-dimensional metal with a single mobile electron per C-H unit; electrons on separate units would be Cooper-*paired* by interacting with the exciton field on the polarizable side groups. However, in spite of much effort, nobody succeeded in making Little's molecule, nor room temperature organic superconductors!

The first organic conductor, a perylene-bromine complex, had been reported in 1954 by H. Akamatu, H. Inokuchi and Y. Matsunaga [9]. As many important discoveries, this was a case for serendipity. At the time, nobody would believe that an organic substance could ever be a conductor. Linus Pauling, who was visiting Japan at the time, was truly surprised. In the late 1960's and early 1970's, several organic semiconductors were reported and the subject became of much interest, especially due to William Little's paper.

Before going into more detail on the $(Per)_2[M(mnt)_2]$ family, (see Figure 12.1), it should be mentioned that, in the meanwhile, several organic conductors were synthesized, in particular TTF-TCNQ, a charge-transfer salt of tetrathiafulvalene (TTF) and tetracyanoquinodimethane (TCNQ) reported to exhibit *superconducting fluctuations and Peierls instability* [10]. Superconductivity was not confirmed in this compound, but the discussion it triggered led to considerable experimental and theoretical developments, and the search for organic superconductivy was on its way.

At last, in 1979, the first organic superconductor was synthesized

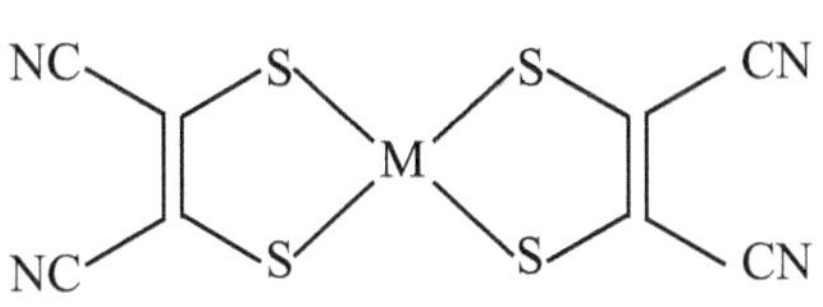

$M(mnt)_2$

(mnt=maleonitriledithiolate)

M=Ni, Pt, Au, Co, Fe, Cu, Pd

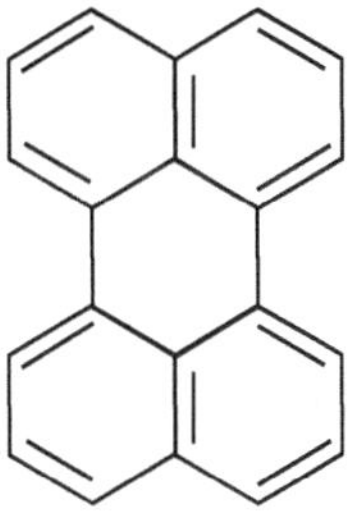

Perylene

Figure 12.1: The constituent molecules of the $(Per)_2[M(mnt)_2]$ charge transfer salts (M = Ni, Pt, Au, Co, Fe, Cu, Pd).

by Klaus Bechgaard and found to have a transition temperature of T_c = 1.1 K, under an external pressure of 6.5 kbar [11]. Such finding gave rise to the dawn of a new era in the field of organic conductors and superconductors. Several series of organic superconductors were synthesized and the needed theory was developed.

Little's idea was based on conjugated polymers, for which single carbon-carbon bonds alternate with double bonds, and, in principle, they should be metals or semiconductors, depending on the unit cell. Polyacetylene was the paradigm compound and had been under investigation for some time [12], but it was shown to become highly conducting only through heavy doping [13]. That accidental discovery led to the thorough study of conducting polymers, which were the subject of the 2000 Nobel Prise in Chemistry awarded to Alan Heeger, Alan MacDiarmid and Hideki Shirakawa. A new technology emerged and is now the important field of *Organic Electronics*.

12.2 The Interesting Physics: 1 D Instabilities

In a one-dimensional crystal, fluctuations destroy all long-range order, and since there are always fluctuations at any finite temperature, such a system cannot be ordered except at zero temperature. This is what makes the 1D real systems interesting—they are not strictly 1D and they exhibit many interesting kinds of fluctuations, instabilities and phase transitions. This is what happens in solids with extreme anisotropy, as in most organic conductors and superconductors, where the electron-electron interactions mediated by phonons originate structural instabilities and phase transitions, namely of the metal-insulator and superconducting type. The needed theoretical framework is identical to that of superconductivity and all these *critical phenomena* benefit of a common formalism. The anisotropy in the properties of these solids, such as the electrical conductivity along the chain, which can attain values as high as 10^5 times the value in the perpendicular directions, is due to the highly directional $\pi - \pi$ interactions along the stacks of flat molecules or along the chains of carbon atoms in conjugated polymers.

One can consider several types of instabilities in a 1 D crystal, namely:

1. In the presence of electron-phonon interactions, the ground state is unstable relative to the development of *charge density waves* (CDW) with wave vector q twice the Fermi wave vector ($q = 2k_F$). This is the *Peierls instability* and competes, in general, with the BCS superconducting instability.

2. In the presence of magnetic interactions, the ground state is unstable relative to the development of *spin density waves*, (SDW), without lattice modulation, and *spin-Peierls instabilities*, when there is a lattice distortion.

3. In a 1 D system with short range interactions, the thermal fluctuations destroy the long range order at any temperature $T > 0$.

4. In a 1 D system, an arbitrarily small disorder induces electron localization and the consequent transition from a metal into an insulator.

Figure 12.2 illustrates how, due to the electron-phonon interaction, a regular linear chain of molecules with one electron per molecule, in a metallic state, (half filled band system) undergoes a Peierls distortion to a *charge density wave* (CDW) state, by doubling the size of the unit cell, and becomes an insulator.

In Figure 12.2 a), a chain of molecules along the x-axis, equally spaced by a lattice parameter a, with one electron per unit cell, is a metal at high temperature with constant electron density, n. If there is one electron per unit cell in the high temperature state, the energy band is half-filled and $k_F = \pi/2a$, as shown in Figure 12.2 b). At low temperature, the lattice parameter is modulated and tends to double due to the electron-phonon interaction, and a *charge density wave* (CDW) is generated (Figure 12.2 c)). Whenever the number of electrons per unit cell is less than two, the band is partially filled and an identical situation can occur. In general, if the degree of band filling is $1/n'$, $k_F = \frac{1}{n'}\frac{\pi}{a}$ and the stable charge density wave will have a wave length $n'a$. However, we should take into account that, in a band more than half filled,

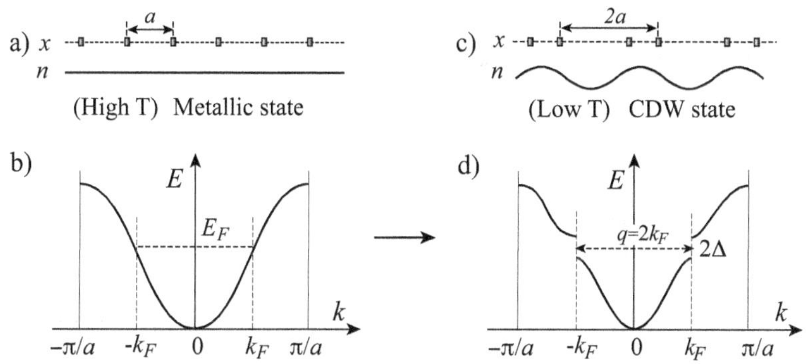

Figure 12.2: a) 1 D lattice, electron density n, corresponding to one electron per site, and b) electron dispersion on the regular chain with a half filled band. c) and d) Peierls modulated in the electron-phonon system. States with $|k| \lesssim k_F$ gain energy by the $q = 2k_F$ distortion (lattice doubling), leading to the opening of an energy gap at k_F.

holes should be considered instead of electrons, and the degree of filling would be $1 - n'$. A 3/4 filled band would correspond to a 1/4 *hole* filled band, and the charge density wave (of positive charge) would have a wave length $4a$, giving rise to a tetramerization (see below). Since in one dimension the elastic energy needed to modulate the crystal lattice is less than the energy gain in the conduction electrons, when the temperature is lowered, the opening of the gap, 2Δ, at k_F, drives a structural instability and eventually a structural phase transition.

12.2.1 The Peierls Instability

As mentioned before, Peierls [5] and Fröhlich [6] predicted that a 1 D electron system in a deformable lattice is unstable relative to a modulation of the lattice with wave vector $q = 2k_F$, in which k_F is the Fermi wave vector. For an overview of the theoretical background we will follow the excellent theoretical introduction of D. Jérôme and H. J. Schulz [14] to the instabilities of quasi-one-dimensional electron systems originally published in 1982 in *Advances in Physics* and which should be referred to for more detail. A one-dimensional electron system can be described by the following Hamiltonian, in second quantization,

$$H = \sum_{ks} \varepsilon_k \, a^+_{ks} a_{ks} + \sum_q \omega_q \, b^+_q b_q + \frac{g}{\sqrt{L}} \sum_{kqs} a^+_{k+q,s} a_{ks} \left(b_q + b^+_{-q} \right) \quad (12.1)$$

where a^+_{ks} and a_{ks} are the creation and annihilation operators for an electron in state k, with spin s, and energy ε_k; b^+_q and b_q are the creation and annihilation operators for a phonon with wave vector q and energy ω_q; L is the length of the system. For simplicity we make $\varepsilon_{k_F} = 0$ and $\hbar = 1$. The first two terms describe the non-interacting electron and phonon systems, respectively, and the third term is the electron-phonon interaction, with coupling constant g, usually a function of k and q. In the present context, electrons with $|k| \approx k_F$ and phonons with $|q| \approx 2k_F$ are the most important and, therefore, the wave vector dependence on g can be ignored. The sums are limited to the Brillouin zone.

The lattice oscillations u can be written in terms of the phonon operators, as

$$u(x) = \sum_q \frac{1}{\sqrt{2L\omega_q}} \left(b_q + b_{-q}^+\right) e^{iqx} \tag{12.2}$$

A modulation of the lattice with wave vector $q = 2k_F$ is described by $\langle b_{2k_F}\rangle = \langle b_{-2k_F}^+\rangle \propto \sqrt{L}$, *i.e.*, the phonon modes with wave vector $\pm k_F$ are occupied.

In the mean field approximation, one can approximately describe the system by the Hamiltonian

$$\begin{aligned} H = &\sum_{ks} \varepsilon_k a_{ks}^+ a_{ks} + 2\omega_{2k_F} \left|\langle b_{2k_F}\rangle\right|^2 + \\ &+ \frac{2g}{\sqrt{L}} \sum_{ks} \left(a_{k+2k_F,s}^+ a_{ks} \langle b_{2k_F}\rangle + a_{k-2k_F,s}^+ a_{ks} \langle b_{2k_F}^+\rangle \right) \end{aligned} \tag{12.3}$$

this meaning that the lattice modulation mixes the k states with the $k \pm 2k_F$ states. More important is the mixing of the quasi-degenerate states with $|k| \approx k_F$ that can be approximately treated in the free electron model and by linearizing ε_k in the vicinity of the Fermi points, *i.e.*, $\varepsilon_k = v_F(|k| - k_F)$, v_F being the Fermi velocity, giving

$$H = \sum_{ks} E_k\, c_{ks}^+ c_{ks} + L \frac{\omega_{2k_F}}{2g^2} |\Delta|^2 \tag{12.4}$$

with

$$E_k = \text{sign}\left(|k| - k_F\right) \left[v_F^2 \left(|k| - k_F\right)^2 + |\Delta|^2 \right]^{1/2} \tag{12.5}$$

c_{ks}^+ and c_{ks} are now the creation and annihilation operators for electrons in states which are linear combinations of the old states k e $k \pm 2k_F$, and in which we define the *order parameter*

$$\Delta = \frac{2g\,\langle b_{2k_F}\rangle}{\sqrt{L}} \tag{12.6}$$

As it can be seen in Figure 12.2 d), the energy of the electrons is lowered and the lattice modulation entails the opening of a gap of width $2|\Delta|$ at the Fermi level, as shown from equation (12.5). At the same time, the elastic energy increases due to the modulation of the lattice described by the second term of equation (12.4).

The overall effect on the total energy of the electronic system can be calculated to give

$$\begin{aligned} E_{el}(\Delta) &= 2 \sum_{k=-k_F}^{k_F} E_k \\ &= -\frac{L\,n\,E_F}{2}\left[1 + \frac{|\Delta|^2}{E_F^2}\ln\left(\frac{2\,E_F}{|\Delta|}\right) + \text{higher order terms}\right] \end{aligned} \tag{12.7}$$

In this equation, E_F is the Fermi energy, measured relative to the bottom of the band, and n is the electron density. From equation (12.7), one can see that the energy gained by the electrons is, for small $|\Delta|$, proportional to $-|\Delta|^2 \ln|\Delta|$, which is always bigger than the loss in elastic energy which is only proportional to $|\Delta|^2$. From this, one can conclude that, in the mean field approximation, *the system is unstable relative to a modulation of the lattice, for an arbitrarily small coupling.* To get the value for the gap, one can minimize the energy and obtain

$$|\Delta| = 2\,E_F\,e^{-1/\lambda}, \qquad \lambda = \frac{2\,g^2}{\pi\,v_F\,\omega_{2k_F}} \tag{12.8}$$

where λ is the wave length of the lattice modulation. It should be noted that the logarithmic term in (12.7) only appears if the gap opens exactly at $\pm k_F$. In other words, the wave length of the lattice modulation is determined by the electronic band filling and is equal to π/k_F. The expectation value of the modulation in the ground state is

$$\langle u(x)\rangle = \sqrt{\left(\frac{2}{\omega_{2k_F}}\right)}\frac{|\Delta|}{g}\cos(2k_F\,x + \phi) \tag{12.9}$$

where ϕ is the phase of Δ (*i.e.*, $|\Delta|e^{i\phi}$): there is a modulation of the electron density as a consequence of the lattice modulation, usually known as a *charge density wave*, (CDW).

The excited states depend only on the amplitude of Δ and not on the phase. Due to the translational symmetry, the CDW can move along the crystal, transporting a d.c. current, which ideally would be infinite, but which is suppressed in real crystals due to several mechanisms such as impurities and lattice locking.

The description of the Peierls ordered state for $T = 0$ is qualitatively valid for temperatures near zero, while the long range order is not destroyed. At sufficiently high temperatures, the complete destruction of the order is expected, and the thermally excited electrons above the gap have energies $\approx |\Delta|$ and will gain energy if the gap decreases. A decrease of $|\Delta|$ enables the additional excitation through the gap, and the mechanism predicts the disappearance of the gap and of the long range order above a certain temperature.

In these 1 D systems, a sharp dip in the phonon spectrum, at the wave vector $q = 2k_F$, known as the Kohn anomaly [15], is particularly noticeable (Figure 12.3). It is due to the possibility of exciting all the electrons from one side of the Fermi distribution into the other side with the single wave vector and very little energy.

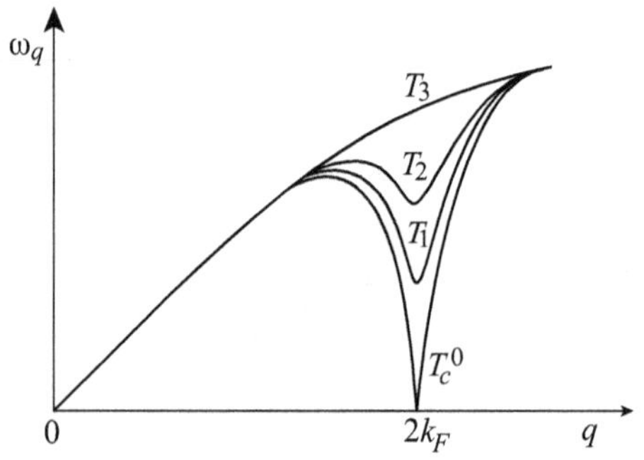

Figure 12.3: Evolution of the Kohn anomaly, in the phonon spectrum, with the temperature ($T_3 > T_2 > T_1 > T_c^0$).

The phonons with wave vector near $q = 2k_F$ involve displacements of the lattice and are collective oscillations strongly coupled to the lattice, as well as to the charge density. It is this coupling that is responsible for the dip in the phonon spectrum. As the temperature is lowered, the dip increases and, at the critical temperature T_c^0, one finally has $\omega_{2k_F} = 0$, indicating the instability of the system relative to a modulation with wave vector $q = 2k_F$—the complete softening of the lattice and a static distortion appears. This is one explanation for charge density waves in solids. The wave vectors at which a Kohn anomaly is possible are the nesting vectors of the Fermi surface, that is, vectors that connect most of the points of the Fermi surface. For a one-dimensional chain of atoms, this vector would be $2k_F$, connecting entirely the two flat Fermi Surfaces.

The description of the Peierls transition is very similar to the BCS theory of superconductivity, namely:

- In both cases, the gap opens on the whole Fermi surface inducing a rapid decrease in the energy of the electrons.

- In both cases, the instability is expressed by a correlation function which diverges logarithmically.

- The transition temperature and the gap at $T = 0$ only differ by a proportionality factor, following the equation $2|\Delta_{(T=0)}|/T_c = 3.5$.
- The effect of impurities in the Peierls transition is analogous to the effect of magnetic impurities in a superconductor.

12.2.2 Competing Instabilities in a 1 D Electron Gas

In the framework of mean-field theory, the electron-electron interaction near the Fermi level can be characterized by four coupling constants, g_1, g_2, g_3 and g_4, their values determining the most stable phases. g_1 corresponds to the electron-electron interaction with momentum transfer $\pm 2k_F$ (backward scattering), g_2 and g_4 have small momentum transfer (forward scattering) involving electrons at both k_F and $-k_F$, (g_2), or only on one side, (g_4). For g_3, the total electron momentum is changed by $4k_F$, being only allowed if $4k_F$ is a reciprocal lattice vector (umklapp scattering), *i.e.* for a half-filled band.

Just to have a qualitative notion of the problem, one can, in the mean-field approximation, look for the possibility of opening a gap at the Fermi level, by evaluating the expectation values of some operators which may originate stable ordered states. Let us then consider the following operators

$$O_{CDW}(q) = \sum_{ks} b^{+}_{-k_F+k-q,s}\, a_{k_F+k,s} \tag{12.10}$$

$$O_{SDW\alpha}(q) = \sum_{kss'} b^{+}_{-k_F+k-q,s}\, \sigma_\alpha^{ss'}\, a_{k_F+k,s'} \quad (\alpha = x, y, z) \tag{12.11}$$

$$O_{SS}(q) = \sum_{ks} b^{+}_{-k_F-k,-s}\, a_{k_F+k+q,s} \tag{12.12}$$

$$O_{TS\alpha}(q) = \sum_{kss'} b^{+}_{-k_F-k,-s}\, \sigma_\alpha^{ss'}\, a_{k_F+k+q,s'} \quad (\alpha = x, y, z) \tag{12.13}$$

where a and b are operators which describe electrons moving to the right and to the left, respectively, and $\sigma_\alpha^{ss'}$ are elements of the spin Pauli matrices. The operators $O_{CDW}(q)$ and $O_{SDW_\alpha}(q)$ are the Fourier components of the charge and spin densities, with wave vector $2k_F + q$, respectively. The $O_{SS}(q)$ and the three possible $O_{TS_\alpha}(q)$ are the *singlet* and *triplet* operators of the Cooper pairs of the superconducting state. To look for ordered states, one would have to compute the expectation values of these operators. If, for example, one considers the *singlet* superconducting state, the Hamiltonian leads to a solution which is similar to the

Peierls transition with a gap $|\Delta_S|$ in the Fermi surface which opens at $q = 0$.

$$|\Delta_{SS}| = 2\,E_F\,e^{-\frac{1}{\lambda_{SS}}}$$
$$\lambda_{SS} = \frac{g_1 + g_2}{2\pi\,v_F} \tag{12.14}$$

for $g_1 + g_2 < 0$. For $g_1 + g_2 > 0$, there is no self-consistent solution with finite $|\Delta|$.

The transition temperature is given by

$$T_c = \frac{2C}{\pi} E_F\,e^{\frac{1}{\lambda_{SS}}} \;, \qquad (C = 1.781) \tag{12.15}$$

The same argument can be applied to the other phase transitions, replacing λ_{SS} by the respective λ parameters:

$$\lambda_{CDW} = -\frac{2g_1 - g_2}{2\pi\,v_F} \tag{12.16}$$

$$\lambda_{SDW\alpha} = \frac{g_2}{2\pi\,v_F} \tag{12.17}$$

$$\lambda_{TS\alpha} = -\frac{g_1 - g_2}{2\pi\,v_F} \tag{12.18}$$

Figure 12.4 schematically represents the domains in the $g_1 - g_2$ plane for which the various O operators give stable phases.

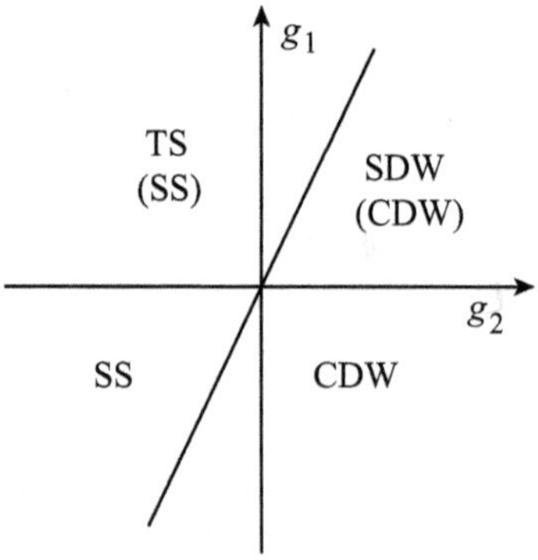

Figure 12.4: Phase diagram in the $g_1 - g_2$ plane showing the domains for which the various operators give stable phases.

The more stable phase is the one which will have the highest transition temperature. Besides the CDW of Peierls type all other phases are possible depending on the g_1, g_2, g_3, g_4 coupling constants. The superconducting *singlet* (SS, CDW) and *triplet* (TS, SDW) occur for $g_1 < 0$ and $g_1 > 0$ respectively. The density wave and superconducting phases are separated by the line $g_1 = 2g_2$ (Fig.12.4).

12.3 The Perylene-Metal Dithiolate Series

12.3.1 Crystal Structure, Energy Bands and Fermi Surface

This family of charge-transfer salts appears in several phases with different crystal structures. There are also perylene-metal dithiolate salts with other stoichiometries. In this paper, we will focus on the so called α-phases which are the most studied and, so far, the most interesting.

The α-$(Per)_2[M(mnt)_2]$, with M=Ni, Au, Cu Pt or Pd, form an isostructural series, crystallizing in the monoclinic system (space group $P2_1/c$) [16, 17, 18]. The Fe and Co compounds are dimerized, the b axis being doubled. The projection of the crystal structure of the platinum compound along the b axis is illustrated in Fig.12.5. It shows that

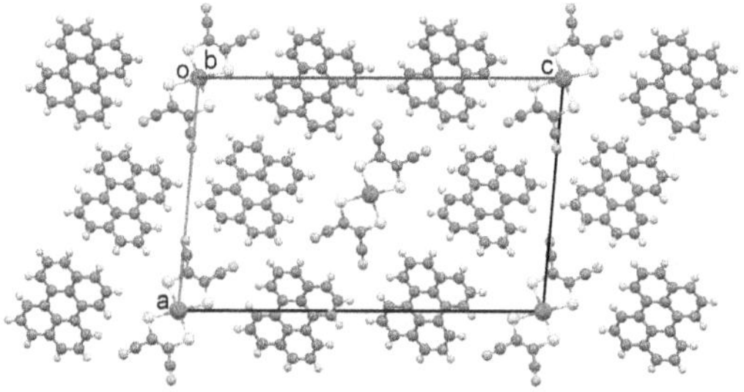

Figure 12.5: Crystal Structure of the α-phase of the $(Per)_2[Pt(mnt)_2]$. Projection along the b axis.

there are segregated regular stacks of $[Pt(mnt)_2]$ and perylene parallel to b. Each column of $[Pt(mnt)_2]$ is surrounded by six columns of perylene and each perylene stack has three stacks of perylene and three stacks of anions as nearest neighbours. The dihedral angle between the two mean planes of the molecules is 67°. Interchain coupling is weak, with intermolecular distances only slightly smaller than the sum of the van der Waals radii of the nitrogen and carbon atoms of the two molecules. In the stack, adjacent overlapping $[Pt(mnt)_2]$ anions are separated by 3.65 Å, the distance between consecutive Pt atoms is 4.19 Å and the tilt of the plane with respect to b is 29°. The perylene molecules present a very good graphite-like mode of overlap at the distance of 3.32 Å, shorter than in graphite (3.35 Å), the tilt of the plane with respect to b being 37°. This mode of stacking allows very short distances between carbon atoms ($\approx$ 3.3 Å) and therefore there is consistent evidence for strong interactions. In table 12.1, the cell parameters for the different $(Per)_2[M(mnt)_2]$ compounds are compaired [18, 19].

Table 12.1: Cell parameters for $(Per)_2[M(mnt)_2]$ [18, 19]. In the Co and Fe salts, the b axis is doubled, when compared with the others [19].

M	Au	Pt	Pd	Ni	Fe	Co
a (Å)	16.602	16.612	16.469	17.44	17.600	17.75
b (Å)	4.191	4.194	4.189	4.176	8.136	8.22
c (Å)	30.164	30.211	30.057	25.18	30.088	30.88
β (°)	118.69	118.70	118.01	91.57	123.72	123.0

For the α-phases of $(Per)_2[M(mnt)_2]$, the energy band in the perylene chain is quarter filled and the band in the $[M(mnt)_2]$ chain is half filled. The perylene chain is metallic at room temperature and undergoes a metal-insulator transition at low temperature. The $[M(mnt)_2]$ chain, for M=Pt, Ni, Pd (S=1/2), behaves as a Mott-insulator at room temperature [16] and undergoes a spin-Peierls transition, at low temperature.

The band structure of the perylene chain was first calculated, by Veiros *et al* [20] and more recently by Canadell *et al* [21] (using the extended Huckel approach with both single-ζ and double-ζ type atomic basis sets). These calculations estimate the intrachain transfer integral (t_b) as 148.6 meV (or 353.8 meV for the double-ζ basis set), with the other transfer integrals in the range of 0 - 2.4 meV (or 0 - 8 meV for the double-ζ). This confirms that these materials are nearly perfect one-dimensional conductors, and that the conduction is mainly along the stacking direction of the perylene molecules, which are partially oxidized, $(Per)^{+1/2}$. These values of the transfer integrals are in agreement with the experimental values of the bandwidth, W, which are in the range 0.4 - 0.6 eV ($W = 4t$) depending on the metal M, as estimated from both thermopower and magnetic susceptibility measurements [22, 23, 24]. They also agree with the experimental measurements of the anisotropy of the electrical conductivity estimated as of $\sigma_b/\sigma_a \approx 10^3$ [25].

Since there are four perylene stacks per unit cell, there must be four HOMO bands crossing the Fermi level. If we would neglect the transverse interactions, the Fermi surface would be the superposition of four planes at $\pm k_F = \pm 0.375\ b^*$, (by convention, $(b \cdot b^*) = 2\pi$, $b = 4.1891$ Å; $k_F = \frac{1}{4}b^* = \frac{\pi}{2b} = 0.375$ Å^{-1} along b^*), but because of these transverse interactions, the Fermi surface splits into four sheets with some warping as shown in Figure 12.6. The warping of the sheets is very small, of the order of 2 meV along a^* and is even smaller along c^*.

With these relatively weak interchain interactions, the four Fermi surface sheets may touch each other, and by hybridization between them

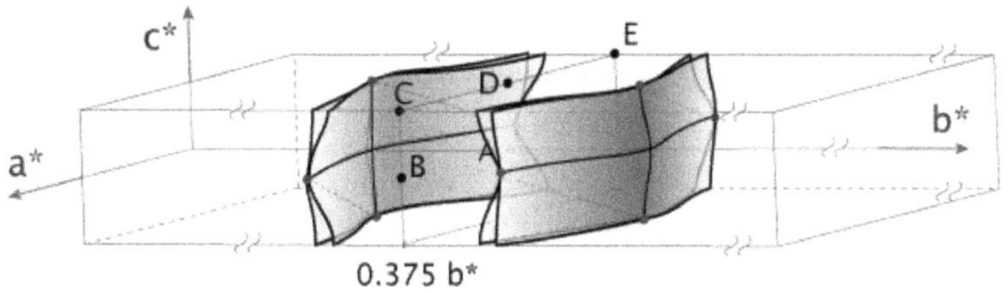

Figure 12.6: Estimated Fermi surface of the perylene chain in $(Per)_2[Pt(mnt)_2]$. Note that this is a largely enlarged picture around $k_F = 0.375\ b^*$. (After reference [21]).

there could be regions with closed pockets. Although these effects usually would be ignored, they may become important at very low temperatures, as suggested by experimental results described below.

The geometry of the Fermi surface was also tested with detailed measurements of angular effects in the magnetoresistance, but the results are difficult to interpret because there are simultaneously angular effects and orbital effects with quantum interference [26] (see below).

12.3.2 Transport and Magnetic Properties

The α-$(Per)_2[M(mnt)_2]$ (M=Pt, Au, Ni, Cu) compounds show room temperature conductivities of the order of 700 S/cm along the high conductivity b axis and the anisotropy in the a, b plane is estimated to be of the order of 900 - 1000 (see Figure 12.7).

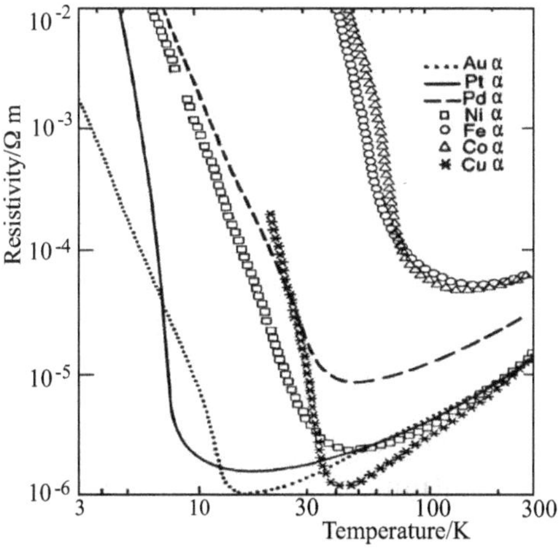

Figure 12.7: Resistivities as a function of temperature for the α-phases of the $(Per)_2[M(mnt)_2]$ family (after references [17, 25, 27, 23]).

The room temperature conductivity of the Pd compound is of the order of 300 S/cm and the Co and Fe compounds (with a crystal structure where the b axis is doubled) present a relatively lower conductivity. In all members of the α series, the resistivity, $\rho(T)$, decreases with the temperature, with the conductivities following a metallic regime as $\sigma \sim T^{-\alpha}$, with $\alpha = 1.5$, reaches a minimum at a temperature T_ρ and goes through a metal-insulator transition at T_c, corresponding to the maximum of $d \ln \rho / d(1/T)$.

These results are compiled in Table 12.2.

Table 12.2: Metal-insulator transition temperatures of the α-$(Per)_2[M(mnt)_2]$.

M	Au	Pt	Pd	Ni	Cu	Fe	Co
T_c/ K	12	8	28	25	33	58	73

The thermopower, S, in the metallic regime, is very similar for all the α-phases and in the range of 32 - 42 μV K^{-1} at room temperature, and it decreases upon cooling, as expected for metals, and in agreement with the resistivity data. At the metal-insulator transition temperature, T_c, $dS/d(1/T)$ also shows an anomaly and, at lower temperature, the thermopower varies as $1/T$, as expected for a semiconductor. The positive sign of the thermopower in the metallic state is indicative of *hole* type conduction, which is consistent with a 3/4 filled band of the perylene chains. From the linear regime at high temperatures, it is possible to estimate a bandwidth, $W = 4t$, of the order of 0.6 eV for the series, with the exception of the Co and Fe compounds where it is of the order of 0.5 eV.

The magnetic properties, namely the magnetic susceptibility, χ, and electron spin resonance, ESR, spectra have been studied in great detail in all the members of the series [3, 4, 16, 25, 28]. It is important to mention that, through the combination of the susceptibility data with the ESR spectra, is was possible to separate the contributions to the susceptibility from both the itinerant electrons in the perylene chain and the localised electrons in the $[M(mnt)_2]$ chain, for the paramagnetic species where M=Ni, Pd, Pt. The $[Au(mnt)_2]$, being diamagnetic, was used as a term of comparison. Figure 12.8 schematically represents the general features of the magnetic properties for the salts with paramagnetic anions ($S = 1/2$).

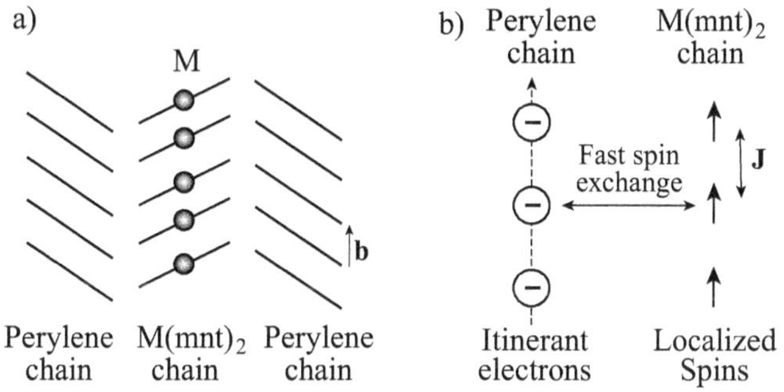

Figure 12.8: a) In $(Per)_2[M(mnt)_2]$ (M=Ni, Pt, Pd), the crystal structures are formed by chains of $[M(mnt)_2]$ units surrounded by perylene chains. b) Interactions along and between the chains: the perylene chain is a 3/4 filled band metal, while the $[M(mnt)_2]$ chain is an insulating array of localized spins, with fast spin exchange with the itinerant electrons.

As we will discuss below in some detail, it is the particular combination of the structural and the electronic and magnetic properties and, in particular, the features of the Fermi surface, that make this series of quasi one-dimensional organic conductors so full of surprises.

12.3.3 Charge Density Wave Formation: Precursor Effects

As discussed in section 2 and illustrated in Figure 12.2, it would be expected to observe charge density wave formation and Peierls instabilities in the α-phases of the $(Per)_2[M(mnt)_2]$ family, as a result of the strong one-dimensionality evidenced by the crystal structure and high anisotropy in the conductivity. A schematic representation of the possible charge density wave formation and lattice modulation is shown in Figure 12.9 for the 3/4 $(Per)^{0.5+}$ filled band, which is expected to be unstable to tetramerization of the perylene chains (CDW).

In fact, metal-insulator transitions are clearly observed in the plots of Figure 12.7 for all members of the family, and the transition temperatures are given in Table 12.2, where T_c corresponds to the CDW formation (M=Au: T_{CDW} = 12 K; M=Pt: T_{CDW} = 8 K); $2\Delta_0 = \zeta k_B T$, with $5 < \zeta < 10$. In the Pt compound, (and for the Ni and Pd), as the tetramerization of the perylene chain occurs, there is a simultaneous dimerization of the $[Pt(mnt)_2]$ chains, corresponding to a spin-Peierls instability.

In order to study the expected structural instabilities of these 1D conductors, (see section 2.1 and Figure 12.3), Henriques *et al* [17] performed

X-ray diffuse scattering measurements in $(Per)_2[M(mnt)_2]$ (M=Pt, Pd, Au), at several temperatures until reaching below the observed metal-insulator transitions, with the exception of the Pt compound.

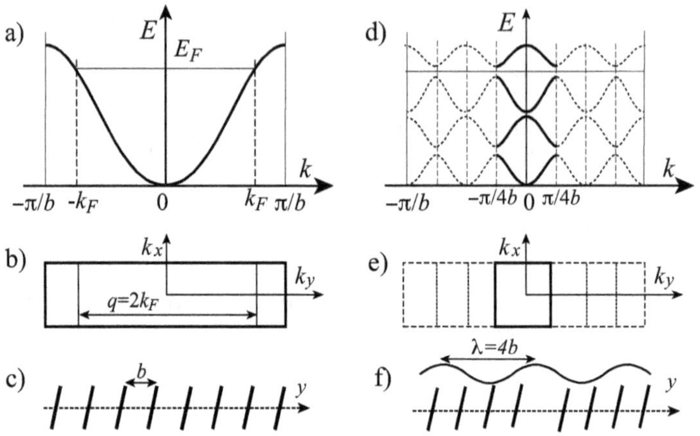

Figure 12.9: CDW formation for a 3/4 filled band (1/4 filled hole band) in a regular $(Per)^{0.5+}$ chain. a,b,c) Electron dispersion of the regular chain, Brillouin zone showing the $q = 2k_F$ phonon wave vector, and the regular lattice with cell parameter b, respectively. d,e,f) Opening of the gap at $k_F = \frac{\pi}{4b}$, CDW formation with $\lambda = 4b$, and tetramerization of the perylene chain.

These structural fluctuations are observed in the metallic state of $(Per)_2[Pt(mnt)_2]$, as diffuse lines, at the wave vector $\frac{1}{2}b^*$, as shown, at 10 K, in Figure 12.10, but start to appear at 25 K, well above the critical temperature $T_c = T_{CDW} = 8$ K, which unfortunately could not be reached with the available experimental set-up. These diffuse scattering lines, attributed to the $[M(mnt)_2]$ stacks, are also observed in $(Per)_2[Pd(mnt)_2]$, and, in this case, they condense in superstructure reflections below 28 K, characteristic of a spin-Peierls transition.

A puzzling feature shown in $(Per)_2[Pt(mnt)_2]$ (Figure 12.10), and not shown in $(Per)_2[Pd(mnt)_2]$, is the presence of sharp diffuse lines passing through the main Bragg reflections, belonging to diffuse sheets with the reduced wave vector $q_b = 0b^*$. Their origin could not be understood at the time. Another unexpected result of this investigation is that no diffuse lines were detected in $(Per)_2[Au(mnt)_2]$ down to 12 K, contrary to $(Per)_2[Pt(mnt)_2]$ and $(Per)_2[Pd(mnt)_2]$ which show 1D structural fluctuations below 25 K and 100 K respectively.

The instabilities in one-dimensional electronic systems may occur at the $q_b = 2k_F$ or the $q_b = 4k_F$ wave vectors, where k_F is the Fermi

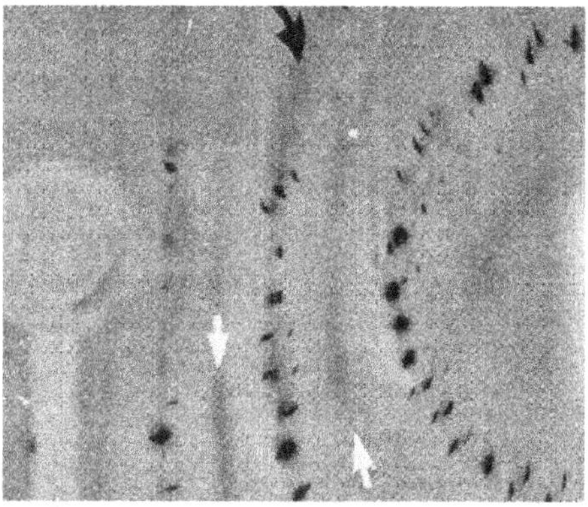

Figure 12.10: X-ray pattern of $(Per)_2[Pt(mnt)_2]$ showing at 10 K (these lines appear below 25 K), the broad $\frac{1}{2}b^*$ diffuse scattering (white arrows) and the sharp $0b^*$ diffuse scattering (black arrow). On this pattern, the chain axis, b, is horizontal. This pattern is free from $\lambda/2$ contamination. After reference [17].

wave vector. Through coupling with the phonons, these (CDW or SDW) instabilities may induce structural instabilities at the same wave vector.

Considering that the conduction band (from the perylene chain) is 3/4 filled, corresponding to the formula $(Per)_2^+[M(mnt)_2]^-$ with one electron per dithiolate, we will be in the presence of a $q_b = 2k_F$ instability. This corresponds to $S = 1/2$ for each $[Pt(mnt)_2]^-$ and $[Pd(mnt)_2]^-$ anion, and a closed-shell ion, with spin pairing ($S = 0$) for the $[Au(mnt)_2]^-$ anion. This is in agreement with the magnetic properties of these charge transfer salts, showing a Curie-Weiss law (at high temperature) for the Pt [29, 16] and Pd [4] species, and no such behaviour for the Au species, which show only a small temperature independent susceptibility typical of metals due to the perylene stacks [30].

Similar diffuse scattering studies were performed in $(Per)_2[M(mnt)_2]$ (M=Cu, Ni, Co and Fe) [31]. The metal-insulator transitions observed at 33, 25, 73 and 58 K for the Cu, Ni, Co and Fe salts, respectively, were found to correspond to the tetramerization on the perylene chains in the case of the Ni and Cu derivatives, and to the dimerization for the Fe and Co derivatives (where the b axis was doubled). In all cases, the periodic lattice distortion can be associated to a $2k_F$ Peierls instability of the perylene chains. The Ni compound shows, in addition, a $4k_F$ distortion, corresponding to the dimerization of the $[Ni(mnt)_2]^-$ chain of localized 1/2 spin, leading to a non magnetic spin-Peierls like ground state.

A remarkable phenomenon, which was observed, was the non-linear electrical transport, both in broad brand and narrow band noise in the

CDW state of $(Per)_2[M(mnt)_2]$, (M=Au, Pt, Co), that provided, for the first time in an organic conductor, clear evidence of coherent CDW motion as the origin of non-linear transport. These non-linear effects are denoted by a large increase of current above a critical threshold field, associated with an increase of broad and narrow band noise. In the Pt compound, the presence of narrow band noise with frequencies proportional to the excess current was in excellent agreement with the coherent motion of a $2k_F$ CDW ($\lambda = 4b = 16.8$ Å) [32, 33, 34, 35, 36].

12.3.4 Nature of the Phase Transition in $(Per)_2[Pt(mnt)_2]$

The most remarkable and unique feature of the $(Per)_2[M(mnt)_2]$ 1D solids is the fact that the perylene chain is a quarter (hole) filled band metal, at room temperature, while the band in the $[M(mnt)_2]$ chain is half filled, with localized spins for M=Ni, Pt and Pd (S=1/2), exhibiting fast spin exchange with the itinerant electrons on the perylene chain (see Figure 12.8). For M=Au, the chain is diamagnetic, making a good reference to study the peculiar magnetic properties of the series [16]. In this section, we will focus on the nature of the metal-insulator transition in the platinum compound, which has been the most studied (along more than 40 years) and is still a matter of controversy. Figure 12.11 illustrates the cascade of phenomena observed in $(Per)_2[Pt(mnt)_2]$, when going from room temperature to low temperatures.

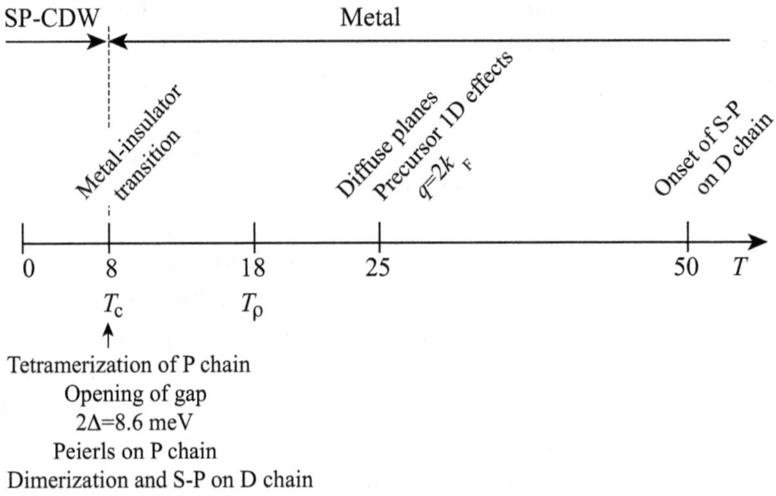

Figure 12.11: Cascade of phenomena (precursor effects and transition) for the $(Per)_2[Pt(mnt)_2]$ compound as the temperature is lowered. The P chain is the perylene chain, while the D chain is the $[Pt(mnt)_2]$ chain.

The perylene chain is metallic at room temperature and, at $T \sim 8$ K, undergoes a Peierls (CDW) transition to an insulating state, where it tetramerizes with wave vector $q_{Per} = \pi/2b$ and the opening of a gap of 8.6 meV. On the other hand, the [Pt(mnt)$_2$], (S = 1/2), chain is a Mott insulator that undergoes a spin-Peierls transition, dimerizing with wave vector $q_{Pt} = \pi/b$ and forming a spin singlet. An interesting aspect is that even though $q_{Pt} = 2q_{Per}$, diffuse X-ray scattering, specific heat, and electrical transport indicate that both the CDW and SP transitions occur at the same, or at very close temperatures, ($T_c = T_{SP\text{-}CDW}$) [17, 31, 37]. This is an indication that the two chains are strongly coupled, even with the mismatch in the q vectors.

The transition is associated to precursor effects when coming from high to low temperatures: at $T \sim 50$ K, the magnetic susceptibility, χ, is affected by the onset of the spin-Peierls instability on the dithiolate chain, while at $T \approx 25$ K, diffuse planes appearing on the X-ray diffraction pattern are a symptom of the displacive distortion at the phonon wave vector $q = 2k_F$, k_F being the Fermi wave vector for the electron dispersion (on the perylene chain).

The nature of the transition, and its precursor effects, has been studied again and again, using many experimental techniques, including electron spin resonance, magnetic susceptibility, transport properties (resistivity and thermopower) and X-ray diffuse scattering, as previously discribed, as well as specific heat [37] and nuclear magnetic resonance (NMR), both ^{1}H [38, 39] and ^{195}Pt NMR [39].

To make the story short, the present understanding based on, particularly, the comparison of the NMR data with all other results, one concludes that the tetramerization on the perylene chain, corresponding to the Peierls transition, 'forces' the dimerization on the [Pt(mnt)$_2$] chain, corresponding to a spin-Peierls transition (see Figure 12.12).

The details of the coupling between the Peierls transition in the perylene chain and the spin-Peierls transition in the [Pt(mnt)$_2$] is not yet clear, but it appears that a unique combination of SP and CDW order parameters arises. Based on NMR results, in particular, on the linear scaling of $1/T_1$ *vs.* $T\ \chi(T)$ [38], which show that the spin degrees of freedom remain diffusive and weak until $T_{SP\text{-}CDW}$, it is not likely that the SP dimerization of [Pt(mnt)$_2$] would be a result of quantum spin dynamics alone. However, electron paramagnetic resonance (EPR) studies indicate that an exchange interaction between the itinerant and the localized spins, associated with the onset of the $T_{SP\text{-}CDW}$ transition, might be involved in the cooperative SP-CDW transition. One [40] and two-chain [38] 1D Kondo lattice models have been considered, but at present

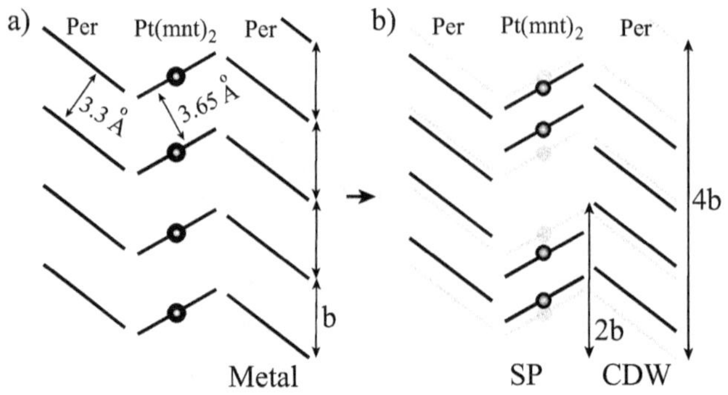

Figure 12.12: a) Configurations of the perylene chains (outer planes) and $[Pt(mnt)_2]$ chain (inner plane) along b, in the metallic state ($T > T_c$). b) Configurations at $T < T_c$: Spin-Peierls (SP) in the $[Pt(mnt)_2]$ chain (dimerized) and CDW in the perylene chain (tetramerized).

there is no complete theory that treats the two-chain $(Per)_2[Pt(mnt)_2]$ case. It has been shown [41] that, even at high magnetic fields, this coupling persists, and that the initial tetramerization in the perylene chains may promote the stabilization of the SP dimerization and the resulting spin-singlet ground state, in spite of the predictions of mean-field theory [42] that place the spin-Peierls decoupling well below the CDW phase boundary, as shown in Figure 12.13, which represents the temperature *vs.* magnetic field phase diagram.

Three regions of the $(Per)_2[Pt(mnt)_2]$ phase diagram are shown: the metallic phase (M) is generally above 8 K; the SP-CDW phase is below 8 K and below 20 - 25 T, and the high-field (HF) phase is below 8 K and above 20 - 25 T. The SP and CDW order parameters remain coupled at high fields. All NMR spectra and spin relaxation data indicate that the full spin-Peierls state forms below the CDW transition seen in transport measurements, at least at low magnetic fields. The CDW ground state formation appears to be a necessary condition to drive the formation of the spin-Peierls ground state, and the commensurability (2:1) of the two order parameter periodicities is likely to favour this effect.

12.3.5 $(Per)_2[M(mnt)_2]$ under Extreme Conditions

The behaviour of matter under *extreme conditions*, meaning, in particular, very low temperatures, high pressures and high magnetic fields is of the greatest importance, for both theoretical and technological development. The history of superconductivity is a good example.

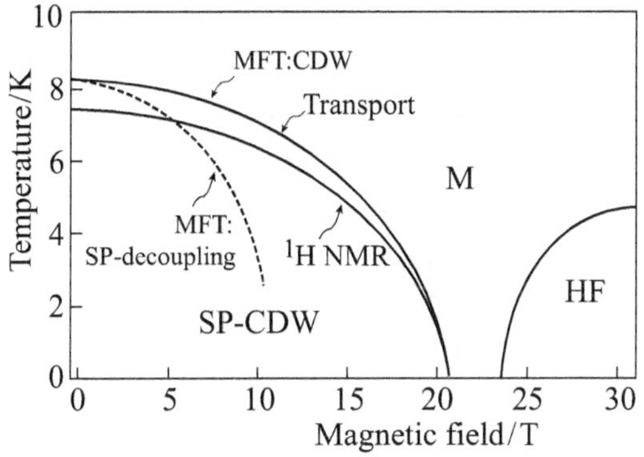

Figure 12.13: Schematic temperature-magnetic field phase diagram of $(Per)_2[Pt(mnt)_2]$ showing the spin-Peierls (SP), charge density wave (CDW), M (paramagnetic-metallic), and HF (high field) phase boundaries, derived from the onset of splitting of NMR spectra, and metal-insulator transitions from electrical transport measurements, for a field applied along the a-axis. 'MFT:CDW' and 'MFT:SP-decoupling' are the mean-field theoretical predictions. (After references [41, 39]).

The search for superconductivity in organic solids led to the study of the transport properties under high pressure, based on the belief that forcing the molecules in a chain to come closer would lead to higher transfer integrals and eventually to a superconducting phase. This was actually achieved with the observation of superconductivity, for the first time in an organic solid, namely $(TMTSF)_2PF_6$, [11], the first of the series of the so called Bechgaard superconducting salts.

Pressure effects on the transport properties of $(Per)_2[M(mnt)_2]$, for M=Pt and Au, were first studied by Henriques *et al*, [43], at pressures up to 27 kbar. It was observed that the room temperature conductivity measured along the needle axis, (b-axis), increases with pressure and tends to saturate above approximately 9 kbar. The results show that the evolution of the electronic correlations taking place in the neighbouring conducting chains was quite unusual, when compared with other 1D organic conductors, and further studies were needed.

The low T_c values of the Peierls transitions in these compounds (8 K in Pt and 12 K in Au) elect them as ideal systems to test the effect of high magnetic fields on a CDW. Initial measurements up to 8 T [44]) detected, for the first time in a CDW system, a sensible decrease of the transition temperature with magnetic field. Larger effects were latter detected in measurements up to 18 T which, although not large enough for a complete suppression of the CDW, put into evidence a larger and anisotropic dependence in the Pt compound, which was ascribed to the coupling to the magnetic chains undergoing a spin-Peierls transition [45]. The total suppression of the CDW was finally observed in 2004 [46] when studying the high field behaviour of the CDW ground state in several of the elements of the series. At very high magnetic fields (up to 45 T) a cascade of transitions in the platinum compound, tentatively ascribed to field induced charge density wave (FICDW) phases, was discovered [47]. Figure 12.14 is a schematic representation of the phase diagram of the platinum compound, where field induced charge density wave (FICDW) state phases are shown at high magnetic fields (up to 45 T).

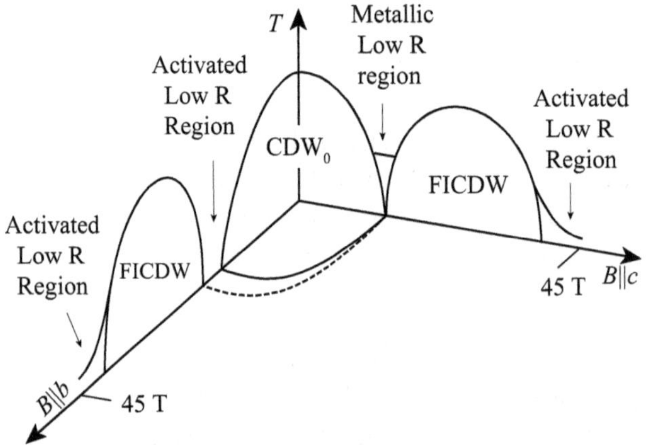

Figure 12.14: Schematic representation of the phase diagram of $(Per)_2[Pt(mnt)_2]$, where field induced charge density wave (FICDW) phases are shown at magnetic fields up to 45 T. (After reference [47])

The theory of CDW under high magnetic fields is still incomplete, but considerable achievements have been reached, particularly by Lebed *et al* [48, 49, 50, 51], namely for the $(Per)_2[Pt(mnt)_2]$.

12.3.6 Quantum Interference Effects

The interference of electrons was first observed in 1927 by Davisson and Germer [52] in the diffraction pattern of electrons incident on nickel, and was the confirmation of the de Broglie hypothesis that material particles, such as electrons, behaved as waves with wave length $\lambda = h/p$ ($p = mv$). A double slit Young type experiment with electrons was first reported by Jönsson in 1961 [53]. Quantum interference of electron waves in a metal has been observed by Stark and Friedberg [54] and later by Sandesara and Stark [55], and is known as the Stark interference effect. The Stark interference effect is observed as large amplitude oscillations in the transverse magnetoresistance signal, periodic in inverse field, which result from the direct interference of normal-state electron waves, corresponding to two electron "trajectories" on different Fermi-surface (FS) sheets, with a phase difference of $\Psi = 2\pi\phi/\phi_0$, where ϕ/ϕ_0 is the ratio of the magnetic flux enclosed between the electron paths to the magnetic flux quantum $\phi_0 = h/e$.

The Stark interference effect has been observed through magneto-transport measurements in $(Per)_2[Au(mnt)_2]$, a one dimensional organic conductor, by suppressing the charge density wave (CDW) state with pressure (above 3 kbar), below the transition temperature $T_{CDW} = 12$ K [56], see Figure 12.15.

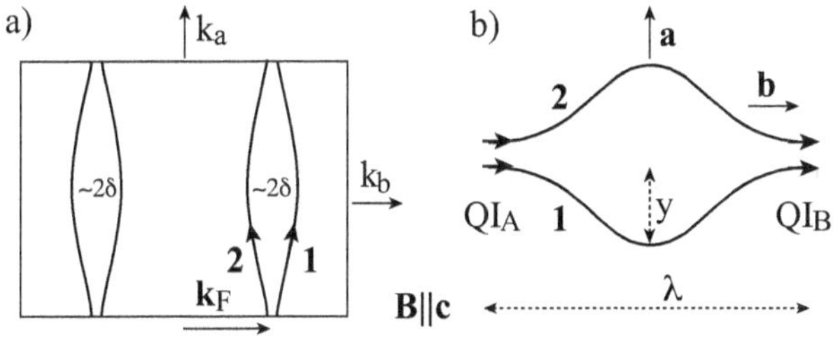

Figure 12.15: a) Simplified Fermi surface for $(Per)_2[Au(mnt)_2]$ after Ref. [21]. The lattice parameters are $(a, b, c) = (16.6, 4.19, 26.6)$ Å and the bandwidths, $(t_a, t_b, t_c) \sim (2, 150, < 0.1)$ meV; $k_F = 0.56$ Å$^{-1} \pm \delta$, $\delta \sim 0.002\ k_b$, and $v_F \sim 1.6 \times 10^5$ m/s. b) Real-space motion of the carriers with wavelength $\lambda = 2\pi\hbar/eBa \sim 2.5/B$ (μm T) and amplitude $y = 4t_a/ev_FB \sim 50/B$ (nm T). Transmission or reflection of the carriers can occur at the interference nodes QI_A and QI_B. (After reference [56]).

The quantum interference (QI) oscillation amplitude exhibits a temperature dependent scattering rate, indicative of an inhomogeneous metal CDW ground state. The QI oscillation frequency reveals a Fermi-surface topology of the unnested Fermi surface, in agreement with that recently computed by Canadell *et al* [21] and schematically shown in Figure 12.6.

12.3.7 Superconductivity at Last

Finally, there is the discovery of superconductivity in the Au compound, with the peculiarity that it occurs in the vicinity of the CDW state [57]. As shown in Figure 12.16a), the resistance of $(Per)_2[Au(mnt)_2]$, under a pressure of 5.3 kbar, drops sharply below 0.35 K, indicative of the onset of superconductivity, although a residual resistance remains in the sample. In the inset of Figure 12.16 a), the magnetoresistance at 25 mK is displayed showing, at low fields, the critical-field behaviour characteristic of superconductivity, and at higher fields the oscillations characteristic of the Stark interference.

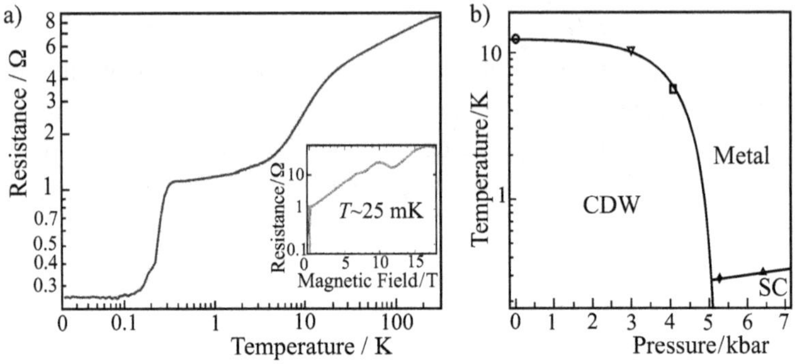

Figure 12.16: a) Resistance vs. temperature for $(Per)_2[Au(mnt)_2]$ at 5.3 kbar. Inset: magnetoresistance, showing the SC critical field and Stark effect. b) CDW-SC phase diagram (after reference [57]).

$(Per)_2[Au(mnt)_2]$ has a non-magnetic anion, and exhibits a Peierls type transition to a CDW ground state below 12 K [22]. The resulting insulating CDW state is suppressed in high magnetic fields [58, 47]. The CDW state can also be suppressed under high pressure, but, under some conditions, a residual "quantum melted" CDW character remains [59]. Under pressures above 3 kbar, the magnetoresistance shows a metallic state with long range order, exhibiting a quantum interference Stark effect. The onset of superconductivity occurs at $\approx$ 300 mK, and the upper critical field (perpendicular to the perylene stacking axis) is 50 mT. Although to fully demonstrate superconductivity will require measurements, such as heat capacity, or magnetic susceptibility, which are difficult to do under high pressure, the critical field phase diagram suggests BCS-like superconducting behaviour, (see figure 3 of reference [57]) with the parameters $B_0 = 49.4$ mT and $T_c = 0.313$ K. Figure 12.16 b) schematically shows the proposed CDW-SC phase diagram. The metal-CDW transitions were determined from the standard $d\ln(R)/d(1/T)$ peaks on the resistance, and the metal-superconductor transition from the onset of superconductivity. An interesting aspect in this compound is that superconductivity emerges from a charge density wave state, at variance with most other superconductors, where superconductivity emerges from antiferromagnetic states.

12.4 Concluding Remarks

The enthusiastic expectations of the early nineteen seventies did not bring about room temperature superconductors, but led to new theoretical developments in the science of condensed matter and to promising emergent technologies. On the scientific side, the results have shown how impressive is the physics dealing with the motion of electrons in one dimension. In the particular case of the two-chain highly one-dimensional conductors of the $(Per)_2[M(mnt)_2]$ family, the work done, along these more than 40 years, has shown how rich in physical phenomena are these amazing materials. The story told in this paper is far from exhaustive and it is certainly unfinished—there will be more surprises as theoretical and experimental methods progress. There are still no theoretical answers to many questions, such as, for example, the nature of the mechanism of coupling of the charge density wave (CDW) and the spin-Peierls (SP) state in the perylene and in the $[M(mnt)_2]$ chains, respectively, and the origin of the field induced spin density wave (FISDW) state under high magnetic fields.

There were many actors in this story, which includes several PhD thesis. The development of the research in organic conductors, in Portugal, benefited from several close collaborations, mainly with the groups of Denis Jérôme and Jean-Paul Poujet at the Université Paris-Sud at Orsay and the group of the late James Brooks, from the National High Magnetic Field Laboratory (NHMFL) in Tallahasse (USA). As for the best part of the theoretical work, it is due to the precious contribution of Claude Bourbonnais from the Université de Sherbrooke in Canada and also to Enric Canadell from Institut de Ciència de Materials de Barcelona (ICMAB-CSIC). A word of recognition is also due to the many co-authors who contributed for the progress of research in this family of compounds.

Acknowledgements

This work was supported by FCT-Portugal under contracts UID/EEA/50008/2013 and UID/Multi/04349/2013.

Bibliography

[1] W. A. Little. Possibility of synthesizing an organic superconductor. *Physical Review*, 134:A1416–A1424, 1964.

[2] L. Alcácer. *Highly conductive molecular complexes*. PhD thesis, University of California, Riverside, 1970.

[3] L. Alcácer and A. H. Maki. Electrically conducting metal dithiolate-perylene complexes. *Journal of Physical Chemistry*, 78:215, 1974.

[4] L. Alcácer and A. H. Maki. Magnetic properties of some electrically conducting perylene-metal dithiolate complexes. *J. Phys. Chem.*, 80(17):1912–1916, 1976.

[5] R. E. Peierls. *Quantum Theory of Solids.* Oxford University Press, 1955.

[6] H. Fröhlich. On the theory of superconductivity: The one-dimensional case. *Proc. R. Soc. A*, 223:296, 1954.

[7] J. Bardeen, L. N. Cooper, and J. R. Schrieffer. Theory of superconductivity. *Physical Review*, 108:1175–1204, 1957.

[8] F. London. *Superfluids*, volume 1. John Wiley & Sons, Inc., New York, 1950.

[9] H. Akamatu, H. Inokuchi, and Y. Matsunaga. Electrical conductivity of the perylene-bromine complex. *Nature*, 173:168–169, 1954.

[10] L. B. Coleman, M. J. Cohen, D. J. Sandman, F. J. Yamagishi, A. F. Garito, and A. J. Heeger. Superconducting fluctuations and Peierls instability in an organic solid. *Solid State Communications*, 12:1125, 1973.

[11] D. Jérôme, A. Mazaud, M. Ribault, and K. Bechgaard. Superconductivity in a synthetic organic conductor $(TMTSF)_2PF_6$. *Journal de Physique Lettres*, 41(4):L95–98, 1980.

[12] H. Shirakawa and S. Ikedo. Infrared spectra of poly(acetylene). *Polym. J.*, 2:231–244, 1971.

[13] Hideki Shirakawa, Edwin J. Louis, Alan G. MacDiarmid, Chwan K. Chiang, and Alan J. Heeger. Synthesis of electrically conducting organic polymers: halogen derivatives of polyacetylene, $(CH)_x$. *J. Chem. Soc., Chem. Commun.*, 16:578–580, 1977.

[14] D. Jérôme and H. J. Schulz. Organic conductors and superconductors. *Advances in Physics*, 51(1):293–479, 2002.

[15] W. Kohn. Image of the Fermi surface in the vibration spectrum of a metal. *Phys. Rev. Lett.*, 2:393, 1959.

[16] L. Alcácer, H. Novais, S. Flandrois, C. Coulon, D. Chasseau, , and J. Gautier. Synthesis, structure and preliminary results on electrical and magnetic properties of (Perylene)$_2$[Pt(mnt)$_2$]. *Solid State Commun.*, 35:945, 1980.

[17] R. T. Henriques, L. Alcácer, J. P. Pouget, and D. Jérôme. Electrical conductivity and x-ray diffuse scattering study of the family of organic conductors (perylene)$_2$M(mnt)$_2$, (M=Pt, Pd, Au). *Journal of Physics C: Solid State Physics*, 17(29):5197, 1984.

[18] A. Domingos, R. T. Henriques, V. Gama, M. Almeida, A. Lopes Vieira, and L. Alcácer. Crystalline structure/transport properties realtionship in the (Perylene)$_2$M(mnt)$_2$ family (M=Au, Pd, Pt, Ni). *Synthetic Metals*, pages B411–B416, 1988.

[19] V. Gama, R. T. Henriques, M. Almeida, and L. Alcácer. Transport and magnetic properties of the new members from the series of organic conductors (Per)$_2$M(mnt)$_2$, M=Cu, Ni, Co, Fe, Pd, Au, Pt. *Synthetic Metals*, 42(3):2553–2556, 1991.

[20] L. F. Veiros, M. J. Calhorda, and E. Canadell. Electronic structure of the 1/1 mixed molecular and polymeric conductor (Perylene)$_2$Co(mnt)$_2$(CH$_2$Cl$_2$)$_{0.5}$ and comparison with the 2/1 α-(Perylene)$_2$M(mnt)$_2$ phases. *Inorg. Chem.*, 33(19):4290–4294, 1994.

[21] E. Canadell, M. Almeida, and J. S. Brooks. Electronic band structure of α-(Per)$_2$M(mnt)$_2$ compounds. *Eur. Phys. J. B*, 42:R453, 2004.

[22] M. Almeida and R.T. Henriques. *Organic Conductive Molecules and Polymers*, volume 1, chapter 2, pages 87–149. John Wiley and Sons, Chichester, 1997.

[23] V. Gama, M. Almeida, R. T. Henriques, I. C. Santos, A. Domingos, S. Ravy, and J. P. Pouget. Low dimensional molecular conductors (Per)$_2$M(mnt)$_2$, M=Cu and Ni: Low and high conductivity phases. *J. Phys. Chem.*, 95:4263–4267, 1991.

[24] R. T. Henriques, M. Almeida, M. J. Matos, L. Alcácer, and C. Bourbonnais. Thermoelectric power of the (perylene)$_2$M(mnt)$_2$ family (M=Pt, Au, Pd). *Synth. Met.*, 19:379, 1987.

[25] L. Alcácer. Metal-insulator transitions in the perylene-dithiolate family. *Molecular Crystals and Liquid Crystals*, 120(1):221–228, 1985. A7.

[26] D. Graf, J. S. Brooks, E. S. Choi, M. Almeida, R. T. Henriques, J. C. Dias, and S. Uji. Geometrical and orbital effects in a quasi-one dimensional conductor. *Physical Review B*, 80:155104–155108, 2009.

[27] V. Gama, R. T. Henriques, and M. Almeida. New members of the series of quasi-one-dimensional metals (perylene)$_2$[M(mnt)$_2$], M=Ni and Cu. *Fizika*, 21:19–22, 1989.

[28] R. T. Henriques, L. Alcácer, M. Almeida, and S. Tomic. Transport and magnetic properties on the family of perylene-dithiolate conductors. *Molecular Crystals and Liquid Crystals*, 120:237–241, 1985.

[29] L. Alcácer, H. Novais, and F. Pedroso. *Molecular Metals, W. E. Hatfield ed.*, chapter Perylene-Dithiolate Complexes; The platinum salt, pages 415–418. Plenum, New York, 1979.

[30] S. Flandrois. Unpublished results.

[31] V. Gama, R. T. Henriques, M. Almeida, and J. P. Pouget. Diffuse X-ray scatterinmg evidence for Peierls and 'spin-Peierls' like transitions in the organic conductors (Per)$_2$[M(mnt)$_2$] (M= Cu, Ni, Co and Fe). *Synthetic Metals*, 55-57:1677–1682, 1993.

[32] E. B. Lopes, M. J. Matos, R. T. Henriques, M. Almeida, and J. Dumas. Charge density wave nonlinear transport in the molecular conductor (Perylene)$_2$Au(mnt)$_2$, (mnt=maleonitriledithiolate). *Europhysics Letters*, 27:241–246, 1994.

[33] E. B. Lopes, M. Matos, R. T. Henriques, M. Almeida, and J. Dumas. CDW nonlinear transport in the organic systems (Per)$_2$M(mnt)$_2$. *Synthetic Metals*, 70:1267–1270, 1995.

[34] E. B. Lopes, M. J. Matos, R. T. Henriques, M. Almeida, and J. Dumas. Charge density wave dynamics in the molecular conductor (Perylene)$_2$Pt(mnt)$_2$, (mnt = maleonitriledithiolate). *Physical Review B*, 52(R2237-R2240), 1995.

[35] J. Dumas, N. Thirion, E. B. Lopes, M. J. Matos, R. T. Henriques, and M. Almeida. Wavelet analysis of the charge density wave dynamics in the molecular conductor (Perylene)$_2$Pt(mnt)$_2$, (mnt = maleonitriledithiolate). *Journal de Physique I, France*, 5:539–545, 1995.

[36] E. B. Lopes, M. J. Matos, R. T. Henriques, M: Almeida, and J. Dumas. CDW dynamics in quasi-one-dimensional conductors $(\mathrm{Per})_2\mathrm{M}(\mathrm{mnt})_2$, M=Au and Pt. *Synthetic Metals*, 86:2163–2164, 1997.

[37] G. Bonfait, M. J. Matos, R. T. Henriques, and M. Almeida. Spin-Peierls instability in $(\mathrm{Per})_2[\mathrm{M}(\mathrm{mnt})_2]$ compounds probed by specific heath. *Journal de Physique IV, Colloque C2, 3*, pages 251–254, 1993.

[38] C. Bourbonnais, R. T. Henriques, P. Wzieteck, D. Kongle, J. Voiron, and D. Jérôme. Nuclear and electronic resonance approaches to magnetic and lattice fluctuations in the two-chain family of organic compounds $(\mathrm{perylene})_2[\mathrm{Pt}(\mathrm{S}_2\mathrm{C}_2(\mathrm{CN})_2)_2]$ (M=Pt, Au). *Physical Review B*, 44:641–651, 1991.

[39] E. L. Green, L. Lumata, J. S. Brooks, P. Kuhns, A. Reyes, S. E. Brown, and M. Almeida. ^{1}H and ^{195}Pt NMR study of the parallel two-chain compound $\mathrm{Perylene}_2[\mathrm{Pt}(\mathrm{mnt})_2]$. *Crystals*, 2:1116–1135, 2012.

[40] J. C. Xavier, R. G. Pereira, E. Miranda, and I. Affleck. Dimerization induced by the RKKY interaction. *Physical Review Letters*, 90(24):247204, 2003.

[41] E. L. Green, L. L. Lumata, J. S. Brooks S, P. L. Kuhns, A. P. Reyes, M. Almeida, M. J. Matos, R. T. Henriques, J. A. Wright, and S. E. Brown. Interaction of magnetic field-dependent Peierls and spin-Peierls ground states in $(\mathrm{Perylene})_2[\mathrm{Pt}(\mathrm{mnt})_2]$. *Physical Review B, Rapid Communications*, 84:121101 (R), 2011.

[42] J. W. Bray, L. V. Interrante, I. S. Jacobs, and J. C. Bonner. *Extended Linear Chain Compounds, edited by T. S. Miller*, chapter The Spin-Peierls Transition, pages 353–415. Plenum Press, New York and London, 1983.

[43] R. T. Henriques, L. Alcácer, D. Jérôme, C. Bourbonnais, and C. Weyl. Electrical conductivity of $(\mathrm{perylene})_2\mathrm{M}(\mathrm{mnt})_2$, (M=Pt, Au) under pressure. *Journal of Physics C: Solid State Physics*, 19(24):4663, 1986.

[44] G. Bonfait, E. B. Lopes, M. J. Matos, R. T. Henriques, and M. Almeida. Magnetic field dependence of the metal-insulator transition in $(\mathrm{Per})_2\mathrm{Pt}(\mathrm{mnt})_2$ and $(\mathrm{Per})_2\mathrm{Au}(\mathrm{mnt})_2$. *Solid State Communications*, 80:391–394, 1991.

[45] M. Matos, G. Bonfait, R. T. Henriques, and M. Almeida. Modification of the magnetic field dependence of the Peierls transition by a magnetic chain. *Physical Review B*, 54:15307–15313, 1996.

[46] D. Graf, E. S. Choi, J. S. Brooks, M. Matos, R. T. Henriques, and M. Almeida. High magnetic field induced charge density wave state in a quasi-one-dimensional organic conductor. *Phys. Rev. Lett.*, 93(7):076406, 2004.

[47] J. S. Brooks, D. Graf, E. S. Choi, M. Almeida, J. C. Dias, R. T. Henriques, and M. Matos. Magnetic field dependence of CDW phases in $Per_2M(mnt)_2$, (M = Pt, Au). *Journal of Low Temperature Physics*, 142:787–803, 2006.

[48] A. G. Lebed. Theory of magnetic field-induced charge-density wave phases. *JETP Letters*, 78:138, 2003.

[49] A. G. Lebed and S. Wu. Soliton wall superlattice charge-density-wave phase in a magnetic field. *JETP Letters*, 86:135–138, 2007.

[50] A. G. Lebed and S. Wu. Soliton wall superlattice in quasi-one-dimensional conductor $(Per)_2Pt(mnt)_2$. *Physical Review Letters*, 99:026402, 2007. Also: arXiv:cond-mat/0702500v2.

[51] A. G. Lebed. Universal theory of magnetic field-induced charge-density wave phases: Theory versus experiment. *Physical Review Letters*, 103:046401, 2009.

[52] C. Davisson and L. H. Germer. Diffraction of electrons by a crystal of nickel. *Phys. Rev.*, 30:705, 1927.

[53] C. Jönsson. Elektroneninterferenzen an mehreren künstlich hergestellten feinspalten. *Z. Phys.*, 161:454–474, 1961. English translation: Am. J. Phys. 42, 4-11 (1974).

[54] R. W. Stark and C. B. Friedberg. Quantum interference of electron waves in a normal metal. *Phys. Rev. Lett.*, 26(10):556–559, 1971.

[55] N. B. Sandesara and R. W. Stark. Macroscopic quantum coherence and localization for normal-state electrons in Mg. *Phys. Rev. Lett.*, 53(17):1681–1683, 1984.

[56] D. Graf, J. S. Brooks, E. S. Choi, M. Almeida, R. Henriques, J. C. Dias, and S. Uji. Quantum interference in a quasi-one-dimensional metal. *Physical Review B*, 75:245101–245108, 2007.

[57] D. Graf, J. S. Brooks, M. Almeida, J. C. Dias, S. Uji, T. Terashima, and M. Kimata. Evolution of superconductivity from a charge density wave ground state in pressurized $(Per)_2[Au(mnt)_2]$. *Europhysics Letters*, 85:27009, 2009.

[58] D. Graf, J. S. Brooks, E. S. Choi, S. Uji, J. C. Dias, M. Matos, and M. Almeida. Suppression of a charge density wave ground state in high magnetic fields: spin and orbital mechanisms. *Physical Review B*, 69:125113, 2004.

[59] N. Mitsu, K. Yamaya, M. Almeida, and R. T. Henriques. Pressure effect on charge density wave in $Per_2M(mnt)_2$, (M=Pt, Au). *Journal of Physics and Chemistry of Solids*, 66:1567–1570, 2005.

13

LEIBNIZIAN GROUNDS FOR THE IDEA OF UNIVERSAL MACHINE

Olga Pombo

Centro de Filosofia das Ciências da Universidade de Lisboa
Faculdade de Ciências da Universidade de Lisboa, Portugal

opombo@fc.ul.pt

Abstract

After short considerations on the concept of machine and on the paradoxes of its articulation with men's life, I will do a brief presentation of the main thesis of extended mind theory and I will try to show how the leibnizian theory of symbolism offers a fundamental basis for the contemporary idea of the continuity man-machine.

13.1 Machines

We know that the word machine comes from the Latin *machina* which, by its turn, comes from the Greek $\mu\alpha\chi\alpha\nu\acute{\alpha}$ and $\mu\eta\chi\alpha\nu\acute{\eta}$, a derivation of $\mu\tilde{\eta}\chi o\zeta$, meaning "means, expedient, remed". However, above this large, open etymological sense, Physics – as an inclusive, all-encompassing discipline as it has been up until the end of the XX century – was able to impose a much more restricted and technical conception of machine as any device capable of changing the direction or the intensity of a force by means of some work.

However strict, this conception of machine gave rise to a much optimistic view of the value of machines in men's life. From all the classical machines – the Archimedean lever, pulley and screw (III century BC), and the Ieron of Alexandria's wind wheel and wedge (10 – 75 BC) – to Galileu's inclined plane and Watt's steam engine (1736 – 1819), machines were positively praised because they substitute human labor, they liberate mankind of hard activities, they improve the production of merchandises and they deeply increase the profit of *bourgeoisie*.

Diderot (1713 – 1784) is an eloquent example of this optimistic view. In his monumental *Encyclopédie des Sciences, des Arts et des Métiers* (1751), mechanical arts are highly valued by its utility to mankind. Because the *Encyclopédie* is at the service of people, it must be open not only to the sciences but also to the mechanical arts and labor activities of ignored artists and artisans who contribute for progress as much as the men of science or the poets. As D'Alembert writes in the *Discours Préliminaire* (1751), "the discovery of the compass was not less relevant to humanity than the explanation of that needle's properties by Physics" (D'Alembert, 1965: 56).

That valorization of mechanical arts and labor activities is the deep reason for Diderot's detailed description of all kinds of machines, from the simplest to the most complicated ones.[1]

In fact, in addition to discursive descriptions, Diderot provided remarkable impressed pictures showing in a much ostensive, didactic and theatrical way each represented machine.[2] First he presented a general

[1]This is the case of the celebrated entrance "Bas" in which Diderot follows the expertise of M. Barrat who taught him the functioning procedures of the fantastic machine of making socks. For further developments, cf. Pombo (2006: 194-251).

[2]We could say that the *Encyclopédie* is, in itself, a machine, a vast procedure aiming to grasp, to penetrate, to decipher, to represent and to systematize all the secrets of Nature and Arts. All may be seen, open, shown, exposed to light of reason, both the interior of factories, ateliers and laboratories to the most antique agriculture and manufacture devices, the geological deepness of earth, of mines, of bodies, of machines (Cf. *ibid*).

view of the machine, usually together with horizontal or vertical cuts; later, the diverse elements of the whole mechanism; lastly, several layers of its internal organization and of its productive design.[3]

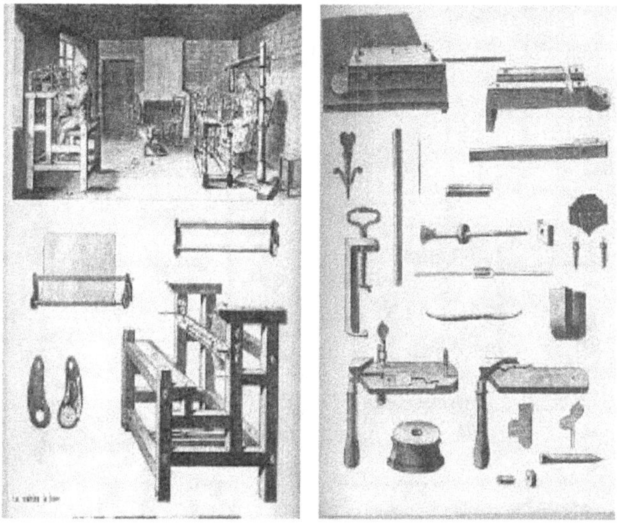

Figure 13.1: Public domain images from Diderot's and d'Alembert's *Encyclopédie*, volume 2 (from Wikimedia Commons).

Nonetheless, one hundred years later, views on machines begin to change. Optimism begins to give the place to a critical perspective towards machines. Stuart Mill, in his influential *Principles of Political Economy* (1849) did not hesitate to question the real value of machines for men's life. As he states: "It is questionable if all the mechanical inventions yet made have lightened the day's toil of any human being" (Mill, 1849: IV, 6.9). And Karl Marx (1818-1883), in the XV chapter of his outstanding *Das Kapital* (1867), precisely untitled *Machinery and large scale industry*, had no doubts to denounce machines as means for the production of the surplus-value: "The objective of machines is to make cheaper the merchandise, to diminish the part of labor which the worker needs for himself, and to enlarge the part of work which he gives for free to the capitalist" (Marx, 1867: IV, 15, 3).

[3]Corresponding to a period of developed manufacture economy, anterior to the introduction of steam engine, the *Encyclopédie* conceives technical labor, not anymore in its theological meaning, as divine punishment, not yet in its romantic dimension, as getaway of the agriculture tasks by which men may only be in harmony towards Nature, but as a form of progressive humanization of world, as exteriorization of knowledge, practical extension allowing to take off from science all its technical utility (Cf. *ibid*).

The movement of English industrial workers who, in the beginning of the nineteen century, used to break at night the boss's machines in which they have work during the day – the so called Luddism – was not just the first worker's movement fighting for better work conditions. It was a protest against the substitution of human labor by machines, a sign (and a symbol) of the each day larger and insidious role played by machines in men's life.[4]

Figure 13.2: Public domain image of Luddites smashing a loom (from Wikimedia Commons, original unknown).

On the contrary, the last decades of the XIX century were happy times marked by the emergence of many beneficial machines able to extraordinarily enhance the living conditions of mankind and to facilitate the day life of millions of human beings.[5]

That development became more and more exponential and prodigious during the XX century. But, at the same time, during that same XX century, we were confronted with the coming out of the most dreadful machines[6] of which resulted unexpected dangerous results.

[4]The name of that movement comes from the worker Ned Ludd, leader of the pressure group who used to break at night the machines in which they have worked during the day. After the night assault to the William Cartwright's manufactory, in April 1812, a major process against luddites was put forward. Seventy four workers were accused of having attempt against the factory, thirteen were condemned to dead and two were deported to colonies. For further developments and actual impact, cf. Sale (1995).

[5]Namely, the telephone (1878), the automobile (1886), the photographic machine (1888), the cinematograph (1895) or the radiography (1895).

[6]Such as the fighting cars of the first world war, the ballistic missiles of 1938, the atomic bomb of 1945, the nuclear reactors of 1956, the drones first used in Balkan, Afghanistan and Iraq wars.

Chaplin, in his masterpiece, *Modern Times* (1936), expressed this fear towards machines in strong poetic images, hard to forget. Forced to be seated at a modern eating machine, he is still able to smile. But he cannot avoid us to weep by seeing him tragically lost and victim of the blind, metallic moving parts of that powerful machinery.

That is to say, the machines which Physics allowed us to construct have produced an enormous ambiguity. The XX century hesitates between the euphoric, apologetic delight about the constantly new technological progresses of more and more sophisticated and highly helpful machines, and the fear, the regret, the disappointment face to the inhuman, alienating, polluting nature of some harmful machines.

13.2 Universal Machines

It is in this very context that appears the Universal Machine. Precisely in the same moment Chaplin produces *Modern Times* (1936), the universal machine is designed in a concise article – *On Computable Numbers* – published by Turing (1912 – 1954) at the *Proceedings of the London Mathematical Society* (1936).

ON COMPUTABLE NUMBERS, WITH AN APPLICATION TO THE ENTSCHEIDUNGSPROBLEM

By A. M. TURING.

[Received 28 May, 1936.—Read 12 November, 1936.]

The "computable" numbers may be described briefly as the real numbers whose expressions as a decimal are calculable by finite means. Although the subject of this paper is ostensibly the computable *numbers*, it is almost equally easy to define and investigate computable functions of an integral variable or a real or computable variable, computable predicates, and so forth. The fundamental problems involved are, however, the same in each case, and I have chosen the computable numbers for explicit treatment as involving the least cumbrous technique. I hope shortly to give an account of the relations of the computable numbers, functions, and so forth to one another. This will include a development of the theory of functions of a real variable expressed in terms of computable numbers. According to my definition, a number is computable if its decimal can be written down by a machine.

Apparently, it was just a brilliant article by a brilliant young mathematician of 24 years old. But the fact is that such short article provided the central concept and the mathematical theory necessary for the construction of the computer, a machine able to change the face of the world and to radically transform what men think about men.

It is true that universal machine does not make it disappear the above mentioned ambiguity. On a certain sense, it makes it even more obvious. In fact, when, in 1996, exactly sixty years after Turing's seminal article, *Deep Blue* triumphed over Kasparov, we all felt that this was not an innocent chess game. Something of high relevance had just happened[7].

What is necessary to realize is that universal machine is a new type of machine. It cannot be contained inside the definition we inherited from Physics. Universal machine is not just a device able to change the direction or intensity of a force by means of work. Universal machine is an intelligent, conceptual, cognitive machine. It is less an instrument, a tool, a resource, an artificial apparatus able to substitute human work and more a device which prolongs, complements, enlarges human activities and capacities.

In fact, with the computer men's relation to machines has decisively changed. A new theorization of the idea of machine begins to be imposed. Now, I believe, we need to think out the concept of machine without plunging neither in catastrophist pessimism nor in technological triumphalism. We require tranquility and backwards capacity for questioning machines in their origins, in their grounds, in their novelties and continuities, in their monstrous proliferation. And we have to enlarge our conception of machine, to overcome the strict sense we inherited from Physics and to come across a much more distended idea of machine. We have to understand that among all the cultural artifacts produced by mankind, machines are extensions, expansion devices which do not only substitute but prolong, complement and extend our capacities.

13.3 Extended Mind

One of the most meaningful references is the celebrated article *The Extended Mind* published by Andy Clark and David J. Chalmers in 1998.

The more important points are precisely the claim for the enlargement of the concept of machine and its independence towards the concept of technology (aspirin is a technological product but the "coup de point" is a machine as well as the plow, the cart, the pencil, the notebook or the computer).

> "*Consider the use of* **pen** *and* **paper** *to perform long multiplication, the use of physical re-arrangements of letter tiles to prompt word recall in Scrabble, the general paraphernalia of language,* **books, diagrams,** *and* **culture**.

[7]Kasparov's words at the end of the game are filled by such awareness: he had been the last human to win chess championship.

> *If the resources of my calculator or my **Filofax** are always there when I need them, then they are coupled with me as reliably as we need. They are part of the basic package of my cognitive resources.*
> *A **notebook**, for example, is a central part of my identity as a cognitive agent"*
> *In all these cases the individual brain performs some operations, while others are delegated to manipulations of **external media**"* (Clark and Chalmers, 1998: 2, 8, 20, 2 respectively) (our emphasis).

Two main theses are here present: 1) machines are not simply means to reach ends, mere tools or instruments which substitute human labor. They are extensions, expansions of our capacities of perception, memory and calculus. They are extrinsic devices; mundane, tangible procedures which prolong, amplify, enlarge, complement and extend our mental capacities; 2) cognitive machines (such as the pencil, the typewriter, the *filofax* or the computer) operate on basis of a language and of a writing.

> *"Language appears to be a central means by which cognitive processes are **extended** into the world.*
> *Without language, we might be much more akin to discrete Cartesian 'inner' minds, in which high-level cognition relies largely on internal resources. But the advent of **language** has allowed us to spread this burden into the world.*
> *Language, thus construed, is not a mirror of our inner states but a **complement** to them.*
> *It serves as a tool [machine] whose role is to **extend cognition** in ways that on-board devices cannot."* (Clark and Chalmers, 1998: 9 and 19–20) (our emphasis).

Now, it is precisely in this moment that I would like to make a step back to Leibniz (1646 – 1716). I believe that our going-back capacity for questioning the primordial groundings of the idea of universal machine may be is worthwhile.

13.4 Leibniz

Leibniz was a theoretical thinker always committed with practical action. Further the design of extremely ambitious projects such as the *characteristica universalis*[8], the *encyclopedia universalis*, the *scientia generalis*

[8]The *Characteristica Universalis* was in fact thought out as a machine similar to a microscope or a telescope. As Leibniz writes: "Humanum Organi genus novum,

– too ambitious and innovative to be achieved – he has promoted great realizations such as academies, scientific journals or irenic institutions[9] and dedicated much of his time to invent functional solutions for technical problems such as the wind-driven propellers for the extraction of silver and ore in the mines of Harz, water pumps and other hydraulic machines, lamps, submarines, portable watches of which he left detailed design.[10]

We also know that Leibniz was one of the first who had the idea of a logical machine able to enhance human intellectual capabilities and make more easy, quick and rigorous the realization of calculus and invention. Since his *De Arte Combinatoria* of 1660, he conceives and develops a set of combinatorial, synthetic and inventive procedures on basis of what he calls since then the *alphabet of human thoughts.*[11] The idea was that the establishment of a combinatory apparatus which, further the demonstrative logics of Aristotle, would not be limited to the analysis of the truths already known but would make possible the discovery of new truths.

However, Leibniz is mentioned as a pioneer of the history of computer mostly by his invention of the *Machina Arithmetica.* Leibniz thought it out since 1671 but only in 1672 he advanced its construction after having been informed, during his stay in Paris, about a calculating machine previously invented by Pascal. Leibniz decided immediately to ameliorate Pascal machine, only able to add and subtract. In 1673, Leibniz presented in the French Academy and in the Royal Society a wooden prototype of his arithmetic machine which had the ability, not only of adding and subtracting as the *Pascalina* did, but also of multiplying and dividing. Two years later, in 1675, he presented a metal prototype[12] to

---

plus multo mentis potentiam aucturum, quam vitra optica oculos juverunt, tantoque superius Microscopiis aut Telescopiis quanto praestantior est ratio visu" (Leibniz, GP 7: 187).

[9]Leibniz aimed to create national scientific societies in Dresden, Saint Petersburg, Vienna, and Berlin. Of those such only the Berlin Academy of Sciences, was indeed created in 1700 by Leibniz who designed its first statutes and served as its first President up until his dead in 1716. On Leibniz's projects for scientific societies, see Couturat (1901: 501–528, IV Appendices untitled "Leibniz fondateur d'Académies"). On Leibniz irenic project and many attempts of unifying Christianity, cf. Baruzzi (1907).

[10]Many scholars have underlined the practical activity which runs parallel to Leibniz theoretical thinking. See the case of Elster (1975) or Manuel Sanchez Rodriguez and Sergio Rodero Cilleros (eds)(2010).

[11]Cf. Leibniz (GP 4: 72–73). For further developments on this Leibnizian project, cf. Pombo (1987: 86–91 and 171–174).

[12]Leibniz ordered manufacturing probably ten other prototypes of his arithmetic machine of which two are still conserved (one at the *Landesbibliothek*, Hannover, and other at the *Deutsches* Museum, *München*).

Arnauld e Huygens in the French Academy and, in 1676, made a full demonstration of his arithmetic machine again in the Royal Society.

Leibniz was committed to this project all along his life. In his *De Progressione Dyadica* (1679) he provides a full description of his arithmetic machine operating via binary arithmetic he had just discovered.[13]

The machine was based on punctured devices whose holes would be open when corresponding to 1 and close when corresponding to 0, a surprisingly modern mechanism which was to be continuously developed up to the XX century. Other detailed descriptions may be found in his *Machina arithmetica in qua non aditio tantum et subtractio sed et multiplicatio nullo, divisio vero paene nullo animi labore peragantur* (1685) and, much later, in his *Brevis Descriptio Machinae Arithmeticae, cum Figura*(1710).

Figure 13.3: Public domain image of Leibnitz's stepped reckoner mechanism with the housing removed (from Wikimedia Commons).

It is true that, as many other of Leibniz projects, the arithmetical machine remained unfinished. However, Leibniz fundamental importance for the development of universal machines comes from his deep comprehension of their symbolic groundings, that is, from his theory of symbolism, close to the extended mind contemporary claims.

[13] In a memory written late, in 1703, untitled *Explication de l'arithmétique binaire qui se sert des seules charactères 0 et 1, avec des remarques sur son utilité, et sur ce qu'elle le sens das anciennes figures chinoises de FoHi* (Leibniz, GM 7: 223–7), Leibniz explains in detail his discovery of binary system and its analogy with the FoHi hexagrams which the Jesuit Bouvet have sent him from China.

13.5 Llull, Hobbes and Leibniz

Leibniz work in this domain has two main roots.[14] The first root is **Ramon Llull** (1232–1315) whose *Ars Magna* constitutes the remotest proposal of mechanization of logical procedures, a proposal which Leibniz knew well, stoutly criticizes and quotes since his *De Arte Combinatoria* (1660). Llull's central idea is that it would be possible, by the combination of a set of simple terms, to establish all possible propositions and thus to discover all possible statements and demonstrate all possible truths to which human knowledge can aspire. For the accomplishment of this project, Llull proposes a set of categories, a system of notations, a finite number of syntactic rules and points to a complex system of combinatorial mechanical procedures of automatic application (a set of material circles, rotating in concentric movement of superposition in order to allow the combination of the symbols marked in their limits).

Leibniz criticizes the incompleteness and imprecise nature of Llull's categories, the arbitrariness of the system of signs he has elected and the methodological solutions proposed by Llull. Instead, Leibniz proposes a much deeper analysis of the primitive terms, claims for a non-arbitrary system of signs and, by taking mathematics as the model, looks for submitting all human intellectual activity to calculatory processes.

The second main root of Leibniz theory of symbolism is **Thomas Hobbes**. For Hobbes (1588 – 1679), language is not a mere communication tool but above all a cognitive device. As Hobbes states in his treatise of *Human Nature* (1650): "it is by the very names that we are able to stabilize one representation" (V, §4).

We need words to fix our thoughts. We need words to think. We could not think without words. To think is to work out (to calculate) through words or, as Hobbes says in his masterpiece *Leviathan* (1651), "Reason is nothing but Reckoning (that is Adding and Subtracting) of Consequences of general names agreed upon, for the marking and signifying of our thoughts" (*Leviathan*: 11).

That is, only language provides the symbolic elements upon which the activity of calculus may be realised. Language is the sensible support for thought. It provides the material, signifying conditions required for the development of calculation.

Hobbes is here giving a significant contribution to Leibniz who will fully adopt Hobbes' cognitive conception of language. And indeed, Leibniz recognizes his heritage from Hobbes precisely in this point. As he states:

[14]In another paper, I claimed, not of two but of three roots of Leibniz's computational conception of reason, the third being the XVII century projects of philosophical language. Cf. Pombo (2010).

> "*Names are not only signs of my present thoughts for the others but notes of my previous thoughts to myself, as* ***Thomas Hobbes has demonstrated***".[15]

But **Leibniz** will give an important step further. He will work out the cognitive conception of language formulated by Thomas Hobbes however building a new theory of symbolism which makes possible to explore a set of epistemic and heuristic consequences of which Hobbes never suspected. As Leibniz states:

> "*When I think on one thousand or on a chiligone, I do it without contemplating those ideas, without putting me in the need of thinking what it is 10 and 100, because I suppose I know it and I do not have the need of conceiving it at this moment".*[16] *(Leibniz, GP 4: 450–451).*

For Hobbes, we need words to think what we are able to think. For Leibniz, we need words to think what we are not able to think (the chiligone, great numbers). For Hobbes, only with language we are able to think. For Leibniz, with language we are able to think what we will never be able to think otherwise.

13.6 Leibniz's theory of blind thought

That is the main point of Leibniz's celebrated theory of *cogitation caecae*, one of the greatest discoveries of Leibniz's philosophy of language.[17]

> "*In general, and above all if the analysis is too long, we do not simultaneously see all the nature of the thing but we use signs instead of the things (...)* ***I call this knowledge as blind or symbolic****; we make use of it in algebra and arithmetic's and in almost all domains*".[18](our emphasis)

[15]"Verba enim non tantum signa sunt cogitationis meae praensentis ad alios, sed et notae cogitationis meae praeteritae ad me ipsum, ut demonstravit Thomas Hobbes" (Leibniz, Ak. VI, 1. 278).

[16]"Lors que je pense à mille ou à un chiligone, je le fais sans en contempler l'idée, sans me mettre en peine de penser ce que c'est que 10 et 100, parce que je suppose de le savoir et ne dois pas d'avoir besoin à present de m'arrester à le concevoir.

[17]Which we have studied in Pombo (1998).

[18]"Plerumque autem, praesertim in Analysi longiore, non totam simul naturam rei intuemur, sed rerum loco signis utimur, quorum solemus praetermittere, scientes aut credentes nos eam habere in potestate (...) qualem cogitationem caecam vel symbolicam appellare soleo, qua et in Algebra et in Arithmetica utimur, imo fere ubique" (Leibniz, GP 4: 423).

Men cannot think simultaneous and constantly the greatest part of his ideas. However men has the possibility of thinking those ideas through the symbols which represent them, that is, men has the possibility of investing the symbols with a much larger meaning than the one he has in the moment. As Leibniz said early in the *De Arte* (1660):

> "*Nobody may calculate, especially with great numbers, without names or numerical signs since it would be necessary to distinctively imagine, instead of the number, all the unities in it contained. Who could distinctively imagine all the unities included in 1.000.000.000.000 unless having the age of Mathusalem?*"[19]

Against Descartes who claimed the need of seeing it all with the yeas of soul, who grounded mathematics in the evidence of its propositions, Leibniz accepts to go on progressing through a though that is blind, that is, he aims to progress without seeing nothing. Just with the external, material, sensible support of symbolism.

> "*The true method must provide us with a* Filum Ariadnes, *that is to say a kind of sensitive and rude means that guides mind in the same way as lines drawn in geometry and as the form of operations that are prescribed to apprentices in arithmetic*".[20](our emphasis)

The true method does not entail, as in Descartes, the confidence in the intuitive rightness of natural light. The true method requires the construction of an artificial symbolic device able to prolong, expand, extend natural reason. It is precisely in the systematic recovery to symbolism that, according to Leibniz, rests the secret of mathematics. They are more than a chain of intuitive reasons, as Descartes wanted. They are a machine operating with symbols, they bring with them their own procedures of control and confirmation.The following text of the *Preface à la Science Générale* (1677) is eloquent:

[19]"Quemad modum enim nemo computare posset, praesatim numeros ingentes, sine nominibus vel signis numeralibus, loco numeri enim deberet sibi distincte imaginari omnes in eo comprehensas unitates. Quis autem nisi tempore aetatis Methusalae imaginabitur sibi distincte unitates quae sunt in 1.000.000.000.000 et si posset tamen progrediendum priorum obliviscertur" (Ak 6.2: 481).

[20]"La veritable méthode nous doit fournir un *Filum Ariadnes*, c'est à dire, un certain moyen sensible et grossier, qui conduise l'esprit, comme sont les lignes tracés en geométrie et les formes des opérations qu'on prescrit aux apprentifs en Arithmetique" (Leibniz, GP 7: 22). (our emphasis)

> "*The reason why the art of demonstrating has been until now found only in mathematics (...) is this: Mathematics carries its own test with it. For when I am presented with a false theorem, I do not need to examine or even to know the demonstration, since I shall discover its falsity a posteriori by means of an easy experiment that is, by a calculation, costing no more than* ***paper and ink***". [21] (our emphasis)

Thought operations may – and must – be realized directly on the symbols without being necessary to go back to the ideas they are supposed to substitute. That is the secret: to affix reasoning to the manipulation of the symbols, "to oblige reasoning to leave visible traces on the paper". [22]

> "*(...) so, we may make sensible the analysis of thought and we may guide it, as by a* ***mechanical filum***" [23] (our emphasis)

This is exactly what Turing did. He fully realized that computing is based on external linguistic encodings of human mental states, that is, well defined mathematical signs connected by precise operational rules. As he stressed:

> "*Computing is normally done by writing certain symbols on paper (...) The behavior of the computer at any moment is determined by the symbols which he is observing*" (Turing, 1936: 249–250)

13.7 Final Remarks

Here the four points I would like to reach with these quick questioning on the concept of machine:

1. Machines are extensions, expansion devices which do not only substitute but prolong, complement and extend our capacities

[21] "Or la raison pour quoy l'art de démonstrer ne se trouve jusqu'ici que dans les mathématiques (...) est que les mathématiques portent leur épreuve avec elles: car quand on me présente un théorème faux, je n'ay pas besoin d'en examiner ny même d'en sçavoir la démonstration, puisque j'en découvriray la fausseté à posteriori par une expérience aisée, qui ne coîte rien que de *l'encre et du papier*" (Leibniz, C: 154). (our emphasis)

[22] As Leibniz says: "le secret est de fixer le raisonnement, et de l'obliger à laisser comme des traces visibles sur le papier, pour estre examiné à loisir" (Leibniz, C : 9).

[23] (...)hinc analysin cogitationum possumus sensibilem reddere, et velut quodam filo mechanico dirigere" (Leibniz, C: 351). Leibniz also uses *filum cogitandi* (Leibniz, C: 420) and *filum meditandi* (Leibniz, GP 7: 14). (our emphasis)

2. There is continuity between the most elementary gestures of cultural artifact production and the most sophisticated machines which surround us.

3. Neither catastrophist pessimism nor a technological triumphalism; neither unlearned rage nor erudite nostalgia.

4. We may became amazed, overwhelmed, but we do not need to became afraid, scared.

Bibliography

BARUZI, Jean (1907), *Leibniz et l' Organization Religieuse de la Terre*, Paris: Felix Alcan.

CLARK, Andy and CHALMERS, David J. (1998), "The Extended Mind", *Analysis*, 58: 10–23.

COUTURAT, Louis (1961), *La Logique de Leibniz d'après des Documents Inédits.* Hildesheim: Georg Olms Verlag.

D'ALEMBERT, Jean le Rond (1751), *Discours Préliminaire de l'Encyclopédie*, Paris: Gonthier (1965).

ELSTER, Jon (1975), *Leibniz et la Formation de l'Esprit Capitaliste*, Paris: Aubier Montaigne.

HOBBES, Thomas (1651), *Leviathan*, edited by C. B. Macpherson, London: Penguin Books, Pelican Classics (1968).

HOBBES, Thomas (1650), *Human Nature or the Fundamental Elements of Policy*, V, §1, in *The Elements of Law Natural and Politic*, ed. F. Tönies, London: Frankcass (1969).

SALE, Kirkpatrick (1995), *Rebels against the Future: the Luddites and their War on the Industrial Revolution: Lessons for the Computer Age*, London: Basic Books.

LEIBNIZ, *Gottfried Wilhelm Leibniz Samtliche Schrifften und Briefe, Akademie der Wissenschaften zu Berlin*, Reihe I–VI, Darmstadt: Reichl (1923 segs). **[Ak]**

LEIBNIZ, *Opuscules et Fragments Inédits de Leibniz. Extraits des Manuscrits de la Bibliothèque Royale de Hannover par Louis Couturat.* Paris: Alcan, 1903. **[C]**

LEIBNIZ, *Gottfried Wilhelm Leibniz. Mathematische Schriften, Hrsg. v. Carl Immanuel Gerhardt.* 1–7. Hildesheim: Olms, 1962. **[GM]**

LEIBNIZ, *Die Philosophischen Schriften von Gottfried Wilhelm Leibniz. Hrsg v. Carl Immanuel Gerhardt.* 1–7. Hildesheim: Olms, 1960. **[GP]**

LLULL, Ramón (1308), *Ars Brevis* (translation, introduction and notes by Armand Llinarès), Paris: Cerf (1991).

MANUEL SANCHEZ RODRIGUEZ and SERGIO RODERO CILLEROS (2010), *Leibniz en la Filosofia Y la Ciencia Modernas*, Granada: Comares.

MARX, Karl (1867), *Capital* (English translation by Bem Fowkes), London: Penguim Classics (1976).

MILL, Stuart (1849), *Principles of Political Economy with some of their Applications to Social Philosophy*, London: Longmans (1909).

POMBO, Olga (1987), *Leibniz and the Problem of a Universal Language*, Münster: Nodus PubliKationen.

POMBO, Olga (1985), "Linguagem e Verdade em Hobbes", *Filosofia n° 1*: 45–61.

POMBO, Olga (1998), "La Théorie Leibnizienne de la Pensée Aveugle en tant que Perspective sur quelques-unes des Apories Linguistiques de la Modernité", *Cahiers Ferdinand Saussure*, 51: 63–75.

POMBO, Olga, (2006), "Para uma História da Ideia de Enciclopédia. Alguns Exemplos", in O. Pombo; A. Guerreiro e A. Franco Alexandre (Eds.), *Enciclopédia e Hipertexto*, Lisboa: Editora Duarte Reis, pp. 194–251.

POMBO, Olga (2010), "Three Roots for Leibniz's Contribution to the Computational Conception of Reason", in F. Ferreira; B. Löwe; E. Mayordomo; L. M. Gomes (Eds.), *Programs, Proofs, Processes. 6th Conference on Computability in Europe*, CiE 2010, Berlin: Springer, pp. 352–361.

TURING, Allan (1936), "On Computable Numbers", *Proceedings of the London Mathematical Society*, 12: 230–265.